Ken Erickson

S_l	scattering matrix element
t	time
T	oscillation period
T	transmission coefficient
T	kinetic energy (Section 3.4)
T_T, T_R	transition matrix elements
U	voltage
v_0	group velocity
v_p	phase velocity
V	potential (energy)
V_{eff}	effective potential (energy)
w	average energy density
W	energy
W_l, W_{lm}	coefficients in the angular decomposition of a wave packet
x	position
$\langle x \rangle$	position expectation value
Y_{lm}	spherical harmonic
Z	atomic number
α	fine-structure constant
δ_l	scattering phase shift
Δk	wave number uncertainty
Δp	momentum uncertainty
Δx	position uncertainty
ε_0	vacuum permittivity
$\eta_{\mathbf{k}}(\mathbf{r})$	scattered wave
$\eta_l(\mathbf{r})$	scattered partial wave
ϑ	polar angle
ϑ	scattering angle
λ	wavelength
μ	reduced mass
ν	frequency
ρ	probability density
σ_0	width of ground state of harmonic oscillator (Chapter 6)
σ_k	width in wave number
σ_l	partial cross section
σ_p	width in momentum
σ_x	width in position
σ_{tot}	total cross section
$\varphi(x)$	stationary wave function
$\varphi_{\mathbf{p}}(\mathbf{r})$	stationary harmonic wave function
ϕ	azimuthal angle
$\psi(x,t)$	time-dependent wave function
$\psi_{\mathbf{p}}(\mathbf{r},t)$	harmonic wave function
ω	angular frequency
Ω	solid angle
∇	nabla (or del) operator
∇^2	Laplace operator

THE PICTURE BOOK OF QUANTUM MECHANICS

Siegmund Brandt

Hans Dieter Dahmen

Department of Physics
Siegen University • Siegen, Germany

JOHN WILEY & SONS

New York • Chichester • Brisbane • Toronto • Singapore

Library of Congress Cataloging in Publication Data:

Brandt, Siegmund.
The picture book of quantum mechanics.

Includes index.
1. Quantum theory. 2. Wave mechanics.
I. Dahmen, Hans Dieter, 1936– II. Title.
QC174.12.B73 1985 530.1′2 85-6388
ISBN 0-471-81776-7

Printed in the United States of America

10 9 8 7 6 5 4 3 2 1

To Renate Brandt
and Ute Dahmen

Preface

Students of classical mechanics can rely on a wealth of experience from everyday life to help them understand and apply mechanical concepts. Even though a stone is not a mass point, the experience of throwing stones certainly helps them to understand and analyze the trajectory of a mass point in a gravitational field. Moreover, students can solve many mechanical problems on the basis of Newton's laws and, in doing so, gain additional experience. When studying wave optics, they find that their knowledge of water waves, as well as experiments in a ripple tank, are very helpful in forming an intuition about the typical wave phenomena of interference and diffraction.

In quantum mechanics, however, beginners are without any intuition. Because quantum-mechanical phenomena happen on an atomic or a subatomic scale, we have no experience of them in daily life. The experiments in atomic physics involve more or less complicated apparatus and are by no means simple to interpret. Even if students are able to take Schrödinger's equation for granted, as many students do Newton's laws, it is not easy for them to acquire experience in quantum mechanics through the solution of problems. Only very few problems can be treated without a computer. Moreover, when solutions in closed form are known, their complicated structure and the special mathematical functions, which students are usually encountering for the first time, constitute severe obstacles to developing a heuristic comprehension. The most difficult hurdle, however, is the formulation of a problem in quantum-mechanical language, for the concepts are completely different from those of classical mechanics. In fact, the concepts and equations of quantum mechanics in Schrödinger's formulation are much closer to those of optics than to those of mechanics. Moreover, the quantities that we are interested in—such as transition probabilities, cross sections, and so on—usually have nothing to do with mechanical concepts such as the position, momentum, or trajectory of a particle. Nevertheless, actual insight into a process is a prerequisite for understanding its quantum-mechanical description and interpreting basic properties in quantum mechanics like position, linear and angular momentum, as well as cross sections, lifetimes, and so on.

Actually, students must develop an intuition of how the concepts of classical mechanics are altered and supplemented by the arguments of optics in order to acquire a roughly correct picture of quantum mechanics. In particular, the time evolution of microscopic physical systems has to be studied to establish how it corresponds to classical mechanics. Here computers and computer graphics offer incredible help, for they produce a large number of examples which are very detailed and which can be looked at in any phase of their time development. For instance, the study of wave packets in motion, which is practically impossible without the help of a computer, reveals the limited validity of intuition drawn from classical mechanics and gives us insight into phenomena like the tunnel effect and resonances, which, because of the importance of interference, can be understood only through optical analogies. A variety of systems in different situations can be simulated on the computer and made accessible by different types of computer graphics.

Some of the topics covered are

- scattering of wave packets and stationary waves,
- the tunnel effect,
- decay of metastable states,
- bound states in various potentials,
- energy bands,
- distinguishable and indistinguishable particles,
- angular momentum,
- three-dimensional scattering,
- cross sections and scattering amplitudes,
- eigenstates in three-dimensional potentials, for example, in the hydrogen atom,
- partial waves and resonances.

The graphical aids range from

- time evolutions of wave functions for one-dimensional problems,
- parameter dependences for studying, for example, the scattering over a range of energies,
- three-dimensional surface plots for presenting two-particle wave functions,

to

- ripple tank pictures to illustrate three-dimensional scattering.

Whenever possible, how particles of a system would behave according to classical mechanics has been indicated by their positions or trajectories. In passing, the special functions typical for quantum mechanics, such as Legendre, Hermite, and Laguerre polynomials, spherical harmonics, and spherical Bessel functions, are also shown in sets of pictures.

The text presents the principal ideas of wave mechanics. The introductory Chapter 1 lays the groundwork by discussing the particle aspect of light, using the fundamental experimental findings of the photoelectric and Compton effects and the wave aspect of particles as it is demonstrated by the diffraction of electrons. The theoretical ideas abstracted from these experiments are introduced in Chapter 2 by studying the behavior of wave packets of light as they propagate through space and as they are reflected or refracted by glass plates. The photon is introduced as a wave packet of light containing a quantum of energy.

To indicate how material particles are analogous to the photon, Chapter 3 introduces them as wave packets of de Broglie waves. The ability of de Broglie waves to describe the mechanics of a particle is explained through a detailed discussion of group velocity, Heisenberg's uncertainty principle, and Born's probability interpretation. The Schrödinger equation is found to be the equation of motion.

Chapters 4 through 8 are devoted to the one-dimensional quantum-mechanical systems. Study of the scattering of a particle by a potential helps us understand how it moves under the influence of a force and how the probability interpretation operates to explain the simultaneous effects of transmission and reflection. We study the tunnel effect of a particle and the excitation and decay of a metastable state. A careful transition to a stationary bound state is carried out. Quasi-classical motion of wave packets confined to the potential range is also examined.

Chapters 7 and 8 cover two-particle systems. Coupled harmonic oscillators are used to illustrate the concept of indistinguishable particles. The striking differences between systems composed of different particles, systems of identical bosons, and systems of identical fermions obeying the Pauli principle are demonstrated.

Three-dimensional quantum mechanics is the subject of Chapters 9 through 13. We begin with a detailed study of angular momentum and discuss methods of solving the Schrödinger equation. The scattering of plane waves is investigated by introducing partial-wave decomposition and the concepts of differential cross sections, scattering amplitudes, and phase shifts. Resonance scattering, which is the subject of many fields of physics research, is studied in detail in Chapter 13. Bound states in three dimensions are dealt with in Chapter 12. The hydrogen atom and the motion of wave packets on elliptical orbits under a harmonic force are among the topics covered.

The last chapter is devoted to results obtained through experiments in atomic, molecular, solid-state, nuclear, and particle physics. They can be qualitatively understood with the help of the pictures and the discussion in the body of the book. Thus examples for

- typical scattering phenomena,
- spectra of bound states and their classifications with the help of models,
- resonance phenomena in total cross sections,
- phase shift analyses of scattering and Regge classification of resonances,
- radioactivity as decay of metastable states,

taken from the fields of atomic and subatomic physics, are presented. Comparing these experimental results with the computer-drawn pictures of the book and their interpretation gives the reader a glimpse of the vast fields of science that can be understood only on the basis of quantum mechanics.

There are more than a hundred problems at the ends of the chapters. Many are designed to help students extract the physics from the pictures. Others will give them practice in handling the theoretical concepts. On the endpapers of the book are a list of frequently used symbols, a short list of physical constants, and a brief table converting SI units to particle physics units. The constants and units will make numerical calculations easier.

All computer-drawn figures were produced with an interactive computer program developed especially for this book. Figure 9.5, the one exception, was made by Dr. Peter Janzen. The hand-drawn figures and the lettering of the others were done by Manfred Euteneuer. Rüdiger Schütz helped with some technical points of the computer graphs. Gertrud Kreuz carefully typed the manuscript. Professor Diethard H. Schiller, Professor Fritz W. Bopp, and Dr. Hans-Jürgen Meyer read the manuscript and offered helpful criticism. We are grateful to all of them for their kind cooperation.

We are particularly grateful to Professor Eugen Merzbacher for his kind interest in our project and for many valuable suggestions which helped to improve the book.

Siegmund Brandt
Hans Dieter Dahmen

Siegen, Germany

Contents

1. Introduction

The basic fields of classical physics are mechanics and heat on the one hand and electromagnetism and optics on the other. Mechanical and heat phenomena involve the motion of particles as governed by Newton's equations. Electromagnetism and optics deal with fields and waves, which are described by Maxwell's equations. In the classical description of particle motion, the position of the particle is exactly determined at any given moment. Wave phenomena, in contrast, are characterized by interference patterns which extend over a certain region in space. The strict separation of particle and wave physics loses its meaning in atomic and subatomic processes.

Quantum mechanics goes back to Max Planck's discovery in 1900 that the energy of an oscillator of *frequency* ν is quantized. That is, the energy emitted or absorbed by an oscillator can take only the values $0, h\nu, 2h\nu, \ldots$. Only multiples of *Planck's quantum of energy*

$$E = h\nu$$

are possible. *Planck's constant*,

$$h = 6.262 \cdot 10^{-34}\ \mathrm{W\,s}$$

is a fundamental constant of nature, the central one of quantum physics. Often it is preferable to use the *angular frequency* $\omega = 2\pi\nu$ of the oscillator and to write Planck's quantum of energy in the form

$$E = \hbar\omega$$

Here

$$\hbar = \frac{h}{2\pi}$$

is simply Planck's constant divided by 2π. Planck's constant is a very small quantity. Therefore the quantization is not apparent in macroscopic systems. But in atomic and subatomic physics Planck's constant is of fundamental importance. In order to make this statement more precise, we shall look at experiments showing the following fundamental phenomena:

- the photoelectric effect,
- the Compton effect,
- the diffraction of electrons.

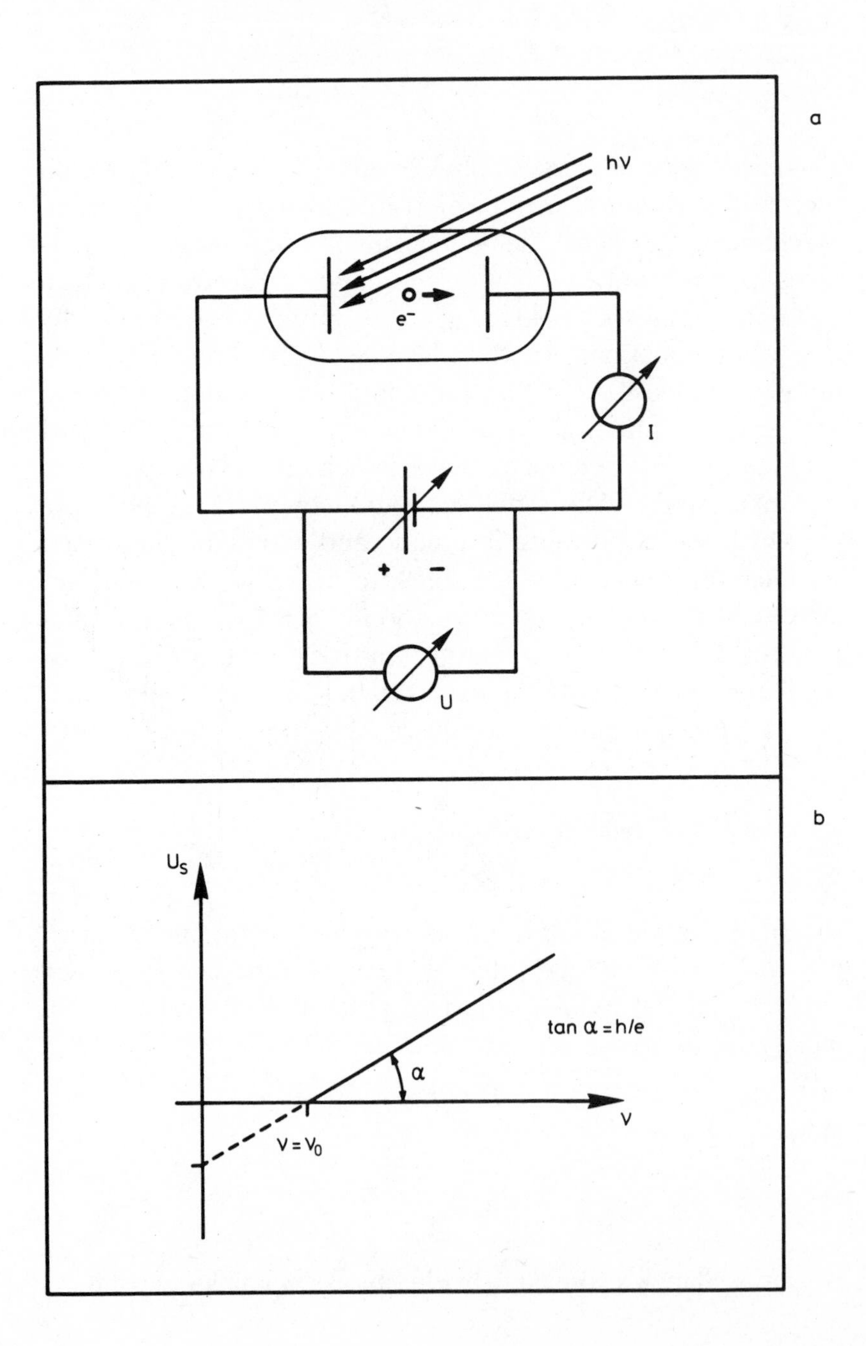

Figure 1.1 **Photoelectric effect. (a) The apparatus to measure the effect consists of a vacuum tube containing two electrodes. Monochromatic light of frequency ν shines on the cathode and liberates electrons which may reach the anode and create a current I in the external circuit. The flow of electrons in the vacuum tube is hindered by the external voltage U. It stops once the voltage exceeds the value U_s. (b) There is a linear dependence between the frequency ν and the voltage U_s.**

The photoelectric effect was discovered by Heinrich Hertz in 1887. It was studied in more detail by Wilhelm Hallwachs in 1888 and Philipp Lenard in 1902. We discuss here the quantitative experiment, which was first carried out in 1916 by R. A. Millikan. His apparatus is shown schematically in Figure 1.1a. Monochromatic light of variable frequency falls onto a photocathode in a vacuum tube. Opposite the photocathode there is an anode—we assume cathode and anode to consist of the same metal—which is at a negative voltage U with respect to the cathode. Thus the electric field exerts a repelling force on the electrons of charge $-e$ that leave the cathode. Here $e = 1.609 \cdot 10^{-19}$ coulomb is the elementary charge. If the electrons reach the anode, they flow back to the cathode through the external circuit, yielding a measurable current I. The kinetic energy of the electrons can therefore be determined by varying the voltage between anode and cathode. The experiment yields the following findings.

1. The electron current sets in, independent of the voltage U, at a frequency ν_0 that is characteristic for the material of the cathode. There is a current only for $\nu > \nu_0$.
2. The voltage U_s at which the current stops flowing depends linearly on the frequency of the light (Figure 1.1b). The kinetic energy E_{kin} of the electrons leaving the cathode then is equal to the potential energy eU_s of the electric field between cathode and anode:

$$E_{\text{kin}} = eU_s$$

If we call h/e the tangent of the straight line representing the relation between the frequency of the light and the voltage,

$$U_s = \frac{h}{e}(\nu - \nu_0)$$

we find that light of frequency ν transfers the kinetic energy eU_s to the electrons kicked out of the material of the cathode. When light has a frequency less than ν_0, no electrons leave the material. If we call

$$h\nu_0 = eU_k$$

the ionization energy of the material that is needed to free the electrons, we must conclude that light of frequency ν has energy

$$E = h\nu = \hbar\omega$$

with

$$\omega = 2\pi\nu, \qquad \hbar = \frac{h}{2\pi}$$

3. The number of electrons set free is proportional to the intensity of the light incident on the photocathode.

In 1905 Albert Einstein explained the photoelectric effect by assuming that light consists of quanta of energy $h\nu$ which act in single elementary processes. The *light quanta* are also called *photons* or γ-quanta. The number of quanta in the light wave is proportional to its intensity.

If the light quanta of energy $E = h\nu = \hbar\omega$ are particles, they should also have momentum. The relativistic relation between the energy E and momentum p of a particle of rest mass m is

$$p = \frac{1}{c}\sqrt{E^2 - m^2c^4}$$

where c is the speed of light in vacuum. Quanta moving with the speed of light must have rest mass zero, so that we have

$$p = \frac{1}{c}\sqrt{\hbar^2\omega^2} = \hbar\frac{\omega}{c} = \hbar k$$

where $k = \omega/c$ is the wave number of the light. If the direction of the light is $\mathbf{k}/k$, we find the vectorial relation $\mathbf{p} = \hbar\mathbf{k}$. To check this idea one has to perform an experiment in which light is scattered on free electrons. The conservation of energy and momentum in the scattering process requires that the following relations be fulfilled,

$$E_\gamma + E_e = E'_\gamma + E'_e$$

$$\mathbf{p}_\gamma + \mathbf{p}_e = \mathbf{p}'_\gamma + \mathbf{p}'_e$$

where E_γ, $\mathbf{p}_\gamma$ and E'_γ, $\mathbf{p}'_\gamma$ are the energies and the momenta of the incident and the scattered photon, respectively. E_e, $\mathbf{p}_e$, E'_e, and $\mathbf{p}'_e$ are the corresponding quantities of the electron. The relation between electron energy E_e and momentum p_e is

$$E_e = c\sqrt{p_e^2 + m_e^2c^2}$$

where m_e is the rest mass of the electron. If the electron is initially at rest, we have $\mathbf{p}_e = 0$, $E_e = m_ec^2$. Altogether, making use of these relations, we obtain

$$c\hbar k + m_ec^2 = c\hbar k' + c\sqrt{\mathbf{p}_e'^2 + m_e^2c^2}$$

$$\hbar\mathbf{k} = \hbar\mathbf{k}' + \mathbf{p}'_e$$

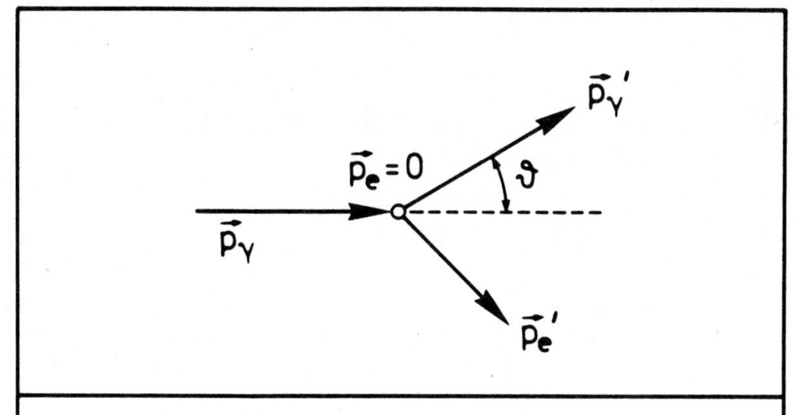

a

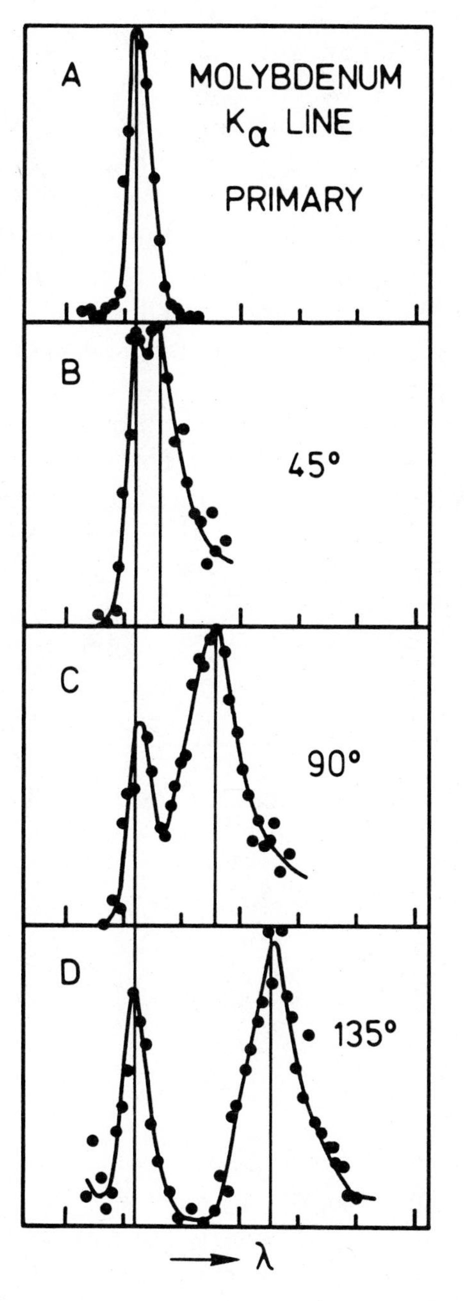

b

Figure 1.2 **The Compton effect. (a) Kinematics of the process. A photon of momentum p_γ is scattered by a free electron at rest, one with momentum $p_e = 0$. After the scattering process the two particles have the momenta p'_γ and p'_e, respectively. The direction of the scattered photon forms an angle ϑ with its original direction. From energy and momentum conservation in the collision, the absolute value p'_γ of the momentum of the scattered photon and the corresponding wavelength $\lambda' = h / p'_\gamma$ can be computed.**

(b) Compton's results. Compton used monochromatic X-rays from the K_α-line of molybdenum to bombard a graphite target. The wavelength spectrum of the incident photons shows the rather sharp K_α-line at the top. Observations of the photons scattered at three different angles ϑ (45°, 90°, 135°) yielded spectra showing that most of them had drifted to the longer wavelength λ'. There are also many photons at the original wavelength λ, photons which were not scattered by single electrons in the graphite. From A. H. Compton, *The Physical Review* **22** (1923) 409, copyright © 1923 by the American Physical Society, reprinted by permission.

as the set of equations determining the wavelength $\lambda' = 2\pi/k'$ of the scattered photon as a function of the wavelength $\lambda = 2\pi/k$ of the initial photon and the scattering angle ϑ (Figure 1.2a). Solving for the difference $\lambda' - \lambda$ of the two wavelengths, we find

$$\lambda' - \lambda = \frac{h}{m_e c}(1 - \cos\vartheta)$$

This means that the angular frequency $\omega' = ck' = 2\pi c/\lambda'$ of the light scattered at an angle $\vartheta > 0$ is smaller than the angular frequency $\omega = ck = 2\pi c/\lambda$ of the incident light.

Arthur Compton carried out an experiment in which light was scattered on electrons; he reported in 1923 that the scattered light had shifted to lower frequencies ω' (Figure 1.2b). The photoelectric effect and the Compton scattering experiment prove that light must be considered to consist of particles which have rest mass zero, move at the speed of light, and have energy $E = \hbar\omega$ and momentum $\mathbf{p} = \hbar\mathbf{k}$. They behave according to the relativistic laws of particle collisions. The propagation of photons is governed by the wave equation following from Maxwell's equations. The intensity of the light wave at a given location is a measure of the photon density at this point.

Once we have arrived at this conclusion, we wonder whether classical particles such as electrons behave in the same way. In particular, we might conjecture that the motion of electrons should be determined by waves. If the relation $E = \hbar\omega$ between energy and angular frequency also holds for the kinetic energy $E_{\mathrm{kin}} = \mathbf{p}^2/2m$ of a particle moving at nonrelativistic velocity, that is, at a speed small compared to that of light, its angular frequency is given by

$$\omega = \frac{1}{\hbar}\frac{p^2}{2m} = \frac{\hbar k^2}{2m}$$

provided that its wave number k and wavelength λ are related to the momentum p by

$$k = \frac{p}{\hbar}, \qquad \lambda = \frac{h}{p}$$

Thus the motion of a particle of momentum p is then characterized by a wave with the *de Broglie wavelength* $\lambda = h/p$ and an angular frequency $\omega = p^2/(2m\hbar)$. The concept of matter waves was put forward in 1923 by Louis de Broglie.

If the motion of a particle is indeed characterized by waves, the propagation of electrons should show interference patterns when an electron beam suffers diffraction. This was first demonstrated by Clinton Davisson and Lester Germer in 1927. They observed interference patterns in an experiment in which a crystal was exposed to an electron beam. In their experiment the regular lattice of atoms in a crystal acts like an optical grating. Even simpler conceptually is diffraction from a sharp edge. Such an experiment was performed by Hans Boersch in 1943. He mounted a platinum foil with a sharp edge in the beam of an electron microscope and used the magnification of the microscope to enlarge the interference pattern. Figure 1.3b shows his result. For comparison it is juxtaposed to Figure 1.3a indicating the pattern produced by

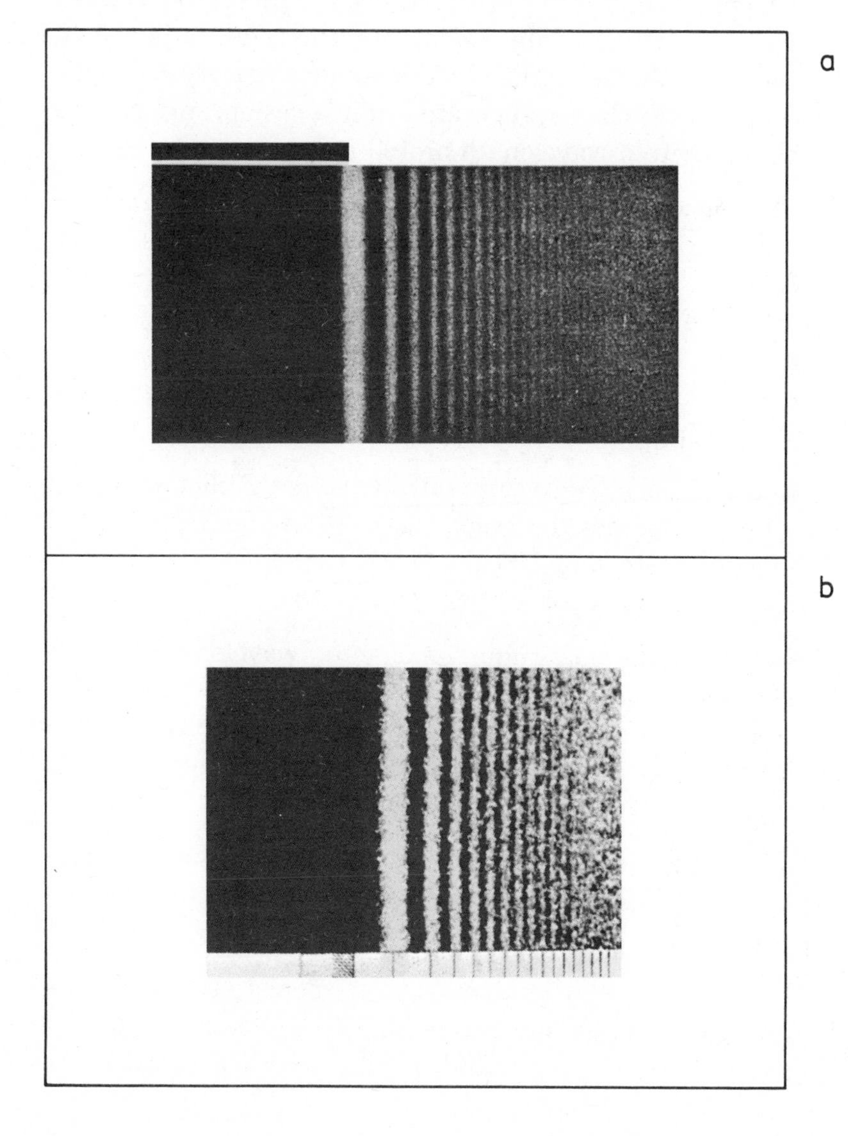

Figure 1.3 **(a) Interference pattern caused by the scattering of red light on a sharp edge. The edge is the border line of an absorbing half-plane, the position of which is indicated at the top of the figure. (b) Interference pattern caused by the scattering of electrons on a sharp edge.** *Sources*: (a) From R. W. Pohl, *Optik and Atomphysik*, ninth edition, copyright © 1954 by Springer-Verlag Berlin, Göttingen, Heidelberg, reprinted by permission. (b) From H. Boersch, *Physikalische Zeitschrift*, **44** (1943) 202, copyright © 1943 by S. -Hirzel-Verlag, Leipzig, reprinted by permission.

visible light diffracted from a sharp edge. The wavelength determined in electron diffraction experiments is in agreement with the formula of de Broglie.

Problems

1.1 Thirty percent of the 100-W power consumption of a sodium lamp goes into the emission of photons with the wavelength $\lambda = 589$ nm. How many photons are emitted per second? How many hit the eye of an observer—the diameter of the pupil is 5 mm—stationed 10 km from the lamp?

1.2 The minimum energy $E_0 = h\nu_0$ needed to set electrons free is called the work function of the material. For cesium it is $3.2 \cdot 10^{-19}$ J. What is the minimum frequency and the corresponding maximum wavelength of light that make the photoelectric effect possible? What is the kinetic energy of an electron liberated from a cesium surface by a photon with a wavelength of 400 nm?

1.3 The energy $E = h\nu$ of a light quantum of frequency ν can also be interpreted in terms of Einstein's formula $E = Mc^2$, where c is the velocity of light in a vacuum. (See also the introduction to Chapter 14.) What energy does a blue quantum ($\lambda = 400$ nm) lose by moving 10 m upward in the earth's gravitational field? How large is the shift in frequency and wavelength?

1.4 Many radioactive nuclei emit high-energy photons called γ-rays. Compute the recoil momentum and velocity of a nucleus possessing 100 times the proton mass and emitting a photon of 1-MeV energy.

1.5 Calculate the maximum change in wavelength experienced by a photon in a Compton collision with an electron initially at rest. The initial wavelength of the photon is $\lambda = 2 \cdot 10^{-12}$ m. What is the kinetic energy of the recoil electron?

1.6 Write the equations for energy and momentum conservation in the Compton scattering process when the electron is not at rest before the collision.

1.7 Use the answer to problem 1.6 to calculate the maximum change of energy and wavelength of a photon of red light ($\lambda = 8 \cdot 10^{-7}$ m) colliding head on with an electron of energy $E_e = 20$ GeV. (Collisions of photons from a laser with electrons from the Stanford linear accelerator are in fact used to prepare monochromatic high-energy photon beams.)

1.8 Electron microscopes are chosen for very fine resolution because the de Broglie wavelength $\lambda = h/p$ can be made much shorter than the wavelength of visible light. The resolution is roughly λ. Use the relativistic relation $E^2 = p^2c^2 + m^2c^4$ to determine the energy of electrons needed to resolve objects of the size 10^{-6} m (a virus), 10^{-8} m (a DNA molecule), 10^{-15} m (a proton). Determine the voltage U needed to accelerate the electrons to the necessary kinetic energy $E - mc^2$.

1.9 What are the de Broglie frequency and wavelength of an electron moving with a kinetic energy of 20 keV, which is typical for electrons in the cathode-ray tube of a color television set?

2.

Light Waves, Photons

2.1 Harmonic Plane Waves, Phase Velocity

Many important aspects and phenomena of quantum mechanics can be visualized by means of *wave mechanics*, which was set up in close analogy to wave optics. Here the simplest building block is the harmonic plane wave of light in a vacuum describing a particularly simple configuration in space and time of the electric field **E** and the magnetic field **B**. If the x-axis of a rectangular coordinate system has been oriented parallel to the direction of the wave propagation, the y-axis can always be chosen to be parallel to the *electric field strength* so that the z-axis is parallel to the magnetic field strength. With this choice the field strengths can be written as

$$E_y = E_0 \cos(\omega t - kx), \qquad B_z = B_0 \cos(\omega t - kx)$$

$$E_x = E_z = 0, \qquad B_x = B_y = 0$$

They are shown in Figures 2.1 and 2.2. The quantities E_0 and B_0 are the maximum values reached by the electric and magnetic fields, respectively. They are called *amplitudes*. The *angular frequency* ω is connected to the *wave number* k by the simple relation

$$\omega = c|k|$$

The points where the field strength is maximum, that is, has the value E_0, are given by the *phase* of the cosine function,

$$\delta = \omega t - kx = 2l\pi$$

where l takes the integer values $l = 0, 1, 2, \ldots$. Therefore

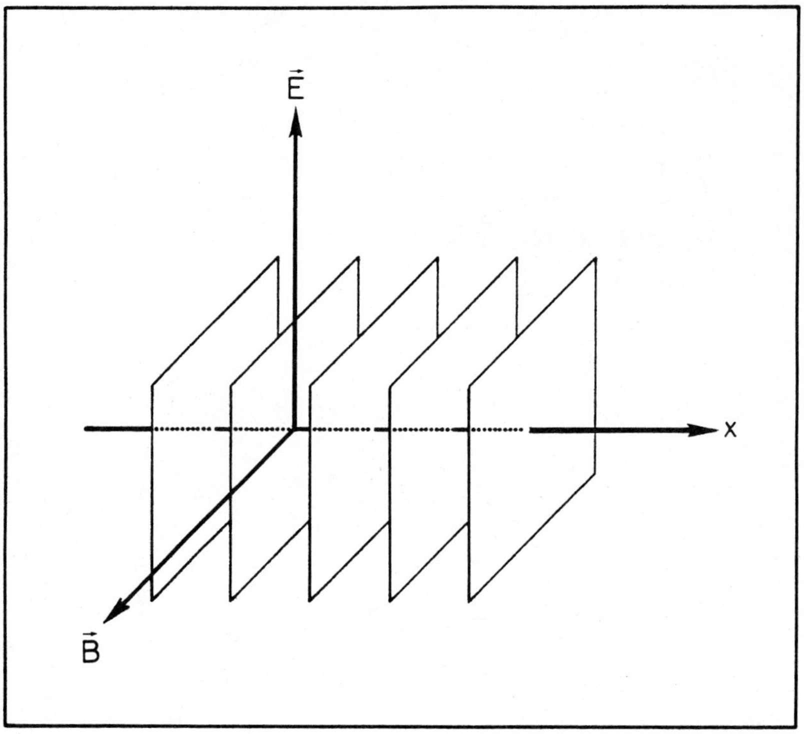

Figure 2.1 **In a plane wave the electric and magnetic field strengths are perpendicular to the direction of propagation. At any moment in time, the fields are constant within planes perpendicular to the direction of motion. As time advances, these planes move with constant velocity.**

such a point moves with the velocity

$$c = \frac{x}{t} = \frac{\omega}{k}$$

Since this velocity describes the speed of a point with a given phase, c is called the *phase velocity* of the wave. For light waves in a vacuum, it is independent of the wavelength. For positive, or negative, k the propagation is in the direction of the positive, or negative, x-axis, respectively.

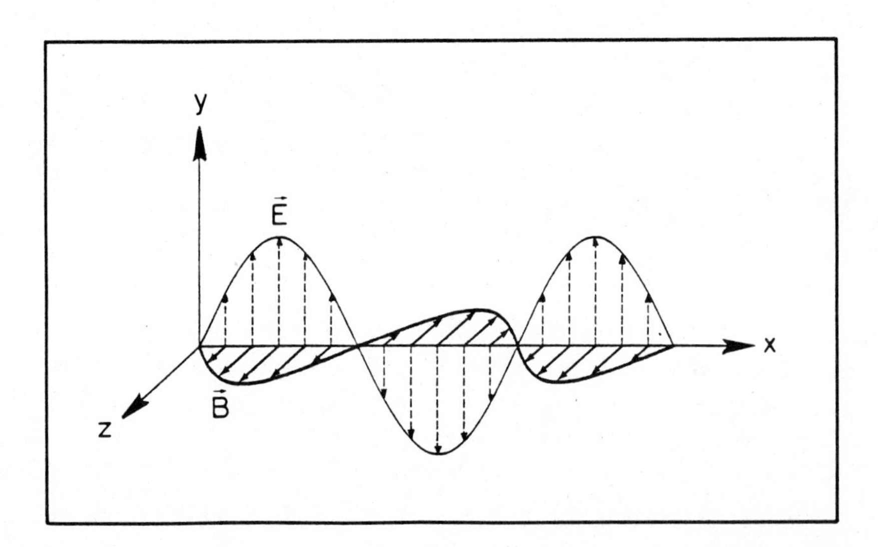

Figure 2.2 **For a given moment in time, the electric field strength E and the magnetic field strength B are shown along a line parallel to the direction of motion of the harmonic plane wave.**

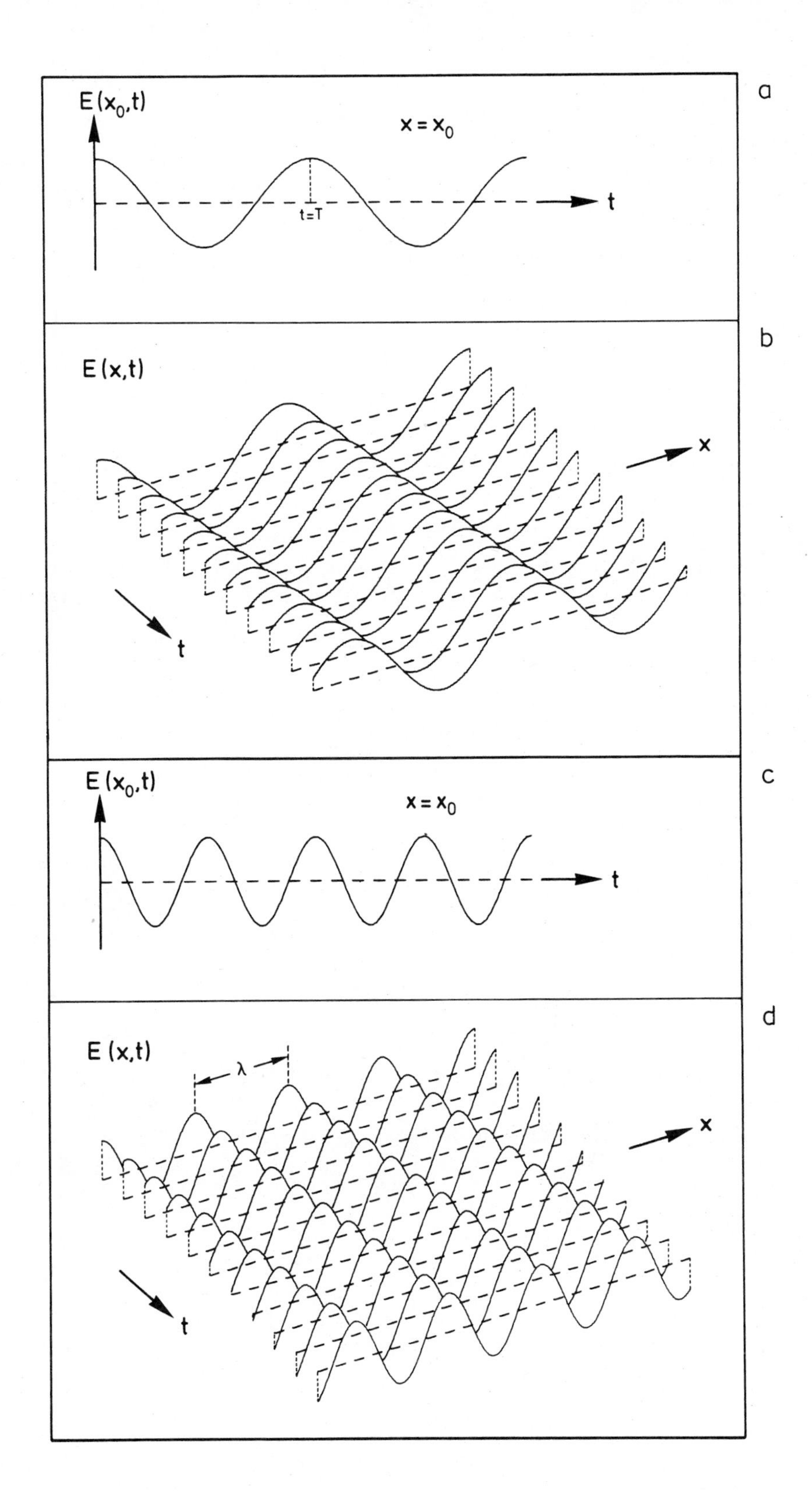

Figure 2.3 (a) Time dependence of the electric field of a harmonic wave at a fixed point in space. (b) Time development of the electric field of a harmonic wave. The field distribution along the x-direction is shown for several moments in time. Early moments are in the background, later moments in the foreground. (c, d) Here the wave has twice the frequency. We observe that the period T and the wavelength λ are halved, but that the phase velocity c stays the same. The time developments in parts b and d are drawn for the same interval of time.

At a fixed point in space, the field strengths E and B oscillate in time with the angular frequency ω (Figures 2.3a and c). The *period* of the oscillation is

$$T = \frac{2\pi}{\omega}$$

For fixed time the field strengths exhibit a periodic pattern in space with a spatial period, the *wavelength*

$$\lambda = \frac{2\pi}{|k|}$$

The whole pattern moves with velocity c along the x-direction. Figures 2.3b and d present the propagation of waves by a set of curves showing the field strength at a number of consecutive equidistant moments in time. Earlier moments in time are drawn in the background of the picture, later ones toward the foreground. We call such a representation a *time development*.

For our purpose it is sufficient to study only the electric field of a light wave,

$$E_y = E = E_0 \cos(\omega t - kx - \alpha)$$

We have included an additional phase α to allow for the fact that the maximum of E need not be at $x = 0$ for $t = 0$. To simplify many calculations, we now make use of the fact that cosine and sine are equal to the real and imaginary parts of an exponential,

$$\cos\beta + i\sin\beta = e^{i\beta}$$

that is,

$$\cos\beta = \operatorname{Re} e^{i\beta}, \qquad \sin\beta = \operatorname{Im} e^{i\beta}$$

The wave is then written as

$$E = \operatorname{Re} E_c$$

where E_c is the complex field strength:

$$E_c = E_0 e^{-i(\omega t - kx - \alpha)} = E_0 e^{i\alpha} e^{-i\omega t} e^{ikx}$$

It factors into a complex amplitude

$$A = E_0 e^{i\alpha}$$

and two exponentials containing the time and space dependences, respectively. As mentioned earlier, the wave travels in the positive or negative x-direction, depending on the sign of

k. Such waves with different amplitudes are

$$E_{c+} = Ae^{-i\omega t}e^{ikx}, \qquad E_{c-} = Be^{-i\omega t}e^{-ikx}$$

The factorization into a time- and a space-dependent factor is particularly convenient in solving Maxwell's equations. It allows the separation of time and space coordinates in these equations. If we divide by $\exp(-i\omega t)$, we arrive at the time-independent expressions

$$E_{s+} = Ae^{ikx}, \qquad E_{s-} = Be^{-ikx}$$

which we call *stationary waves*.

The energy density in an electromagnetic wave is equal to a constant, ϵ_0, times the square of the field strength,

$$w(x,t) = \epsilon_0 E^2$$

Because the plane wave has a cosine structure, the energy density varies twice as fast as the field strength. It remains always a positive quantity; therefore the variation occurs around a nonzero average value. This average taken over a period T of the wave can be written in terms of the complex field strength as

$$w = \frac{\epsilon_0}{2} E_c E_c^* = \frac{\epsilon_0}{2} |E_c|^2$$

Here E_c^* stands for the complex conjugate,

$$E_c^* = \operatorname{Re} E_c - i \operatorname{Im} E_c$$

of the complex field strength,

$$E_c = \operatorname{Re} E_c + i \operatorname{Im} E_c$$

For the average energy density in the plane wave, we obtain

$$w = \frac{\epsilon_0}{2} |A|^2 = \frac{\epsilon_0}{2} E_0^2$$

2.2 Light Wave Incident on a Glass Surface

The effect of glass on light is to reduce the phase velocity by a factor n called the *refractive index*,

$$c' = \frac{c}{n}$$

Although the frequency ω stays constant, wave number and wavelength are changed according to

$$k' = nk, \qquad \lambda' = \frac{\lambda}{n}$$

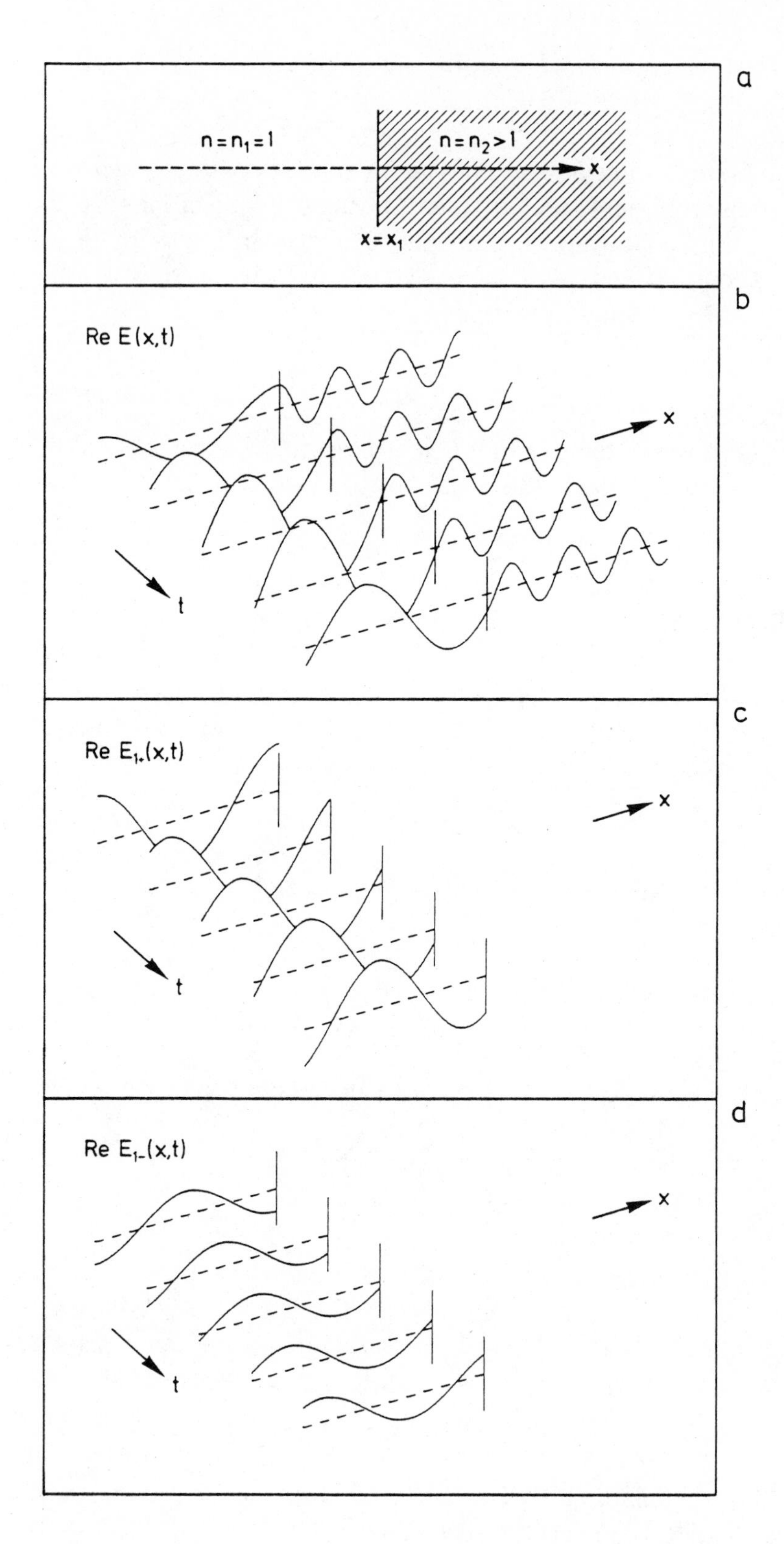

Figure 2.4 **(a) To the right of the plane $x = x_1$, a glass block extends with refractive index $n = n_2$; to the left there is empty space, $n = 1$. (b) Time development of the electric field strength of a harmonic wave which falls from the left onto a glass surface, represented by the vertical line, and is partly reflected by and partly transmitted into the glass. (c) Time development of the incoming wave alone. (d) Time development of the reflected wave alone.**

The Maxwell equations, which govern all electromagnetic phenomena, demand the continuity of the electric field strength and its first derivative at the boundaries of the regions with different refractive indices. We consider a wave traveling in the x-direction and encountering at position $x = x_1$ the surface of a glass block filling half of space (Figure 2.4a). The surface is oriented perpendicular to the direction of the light. The complex expression

$$E_{1+} = A_1 e^{ik_1 x}$$

describes the incident stationary wave to the left of the glass surface, that is, for $x < x_1$, where A_1 is the known amplitude of the incident light wave. At the surface only a part of the light wave enters the glass block; the other part will be reflected. Thus, in the region to the left of the glass block, $x < x_1$, we find in addition to the incident wave the reflected stationary wave

$$E_{1-} = B_1 e^{-ik_1 x}$$

propagating in the opposite direction. Within the glass the transmitted wave

$$E_2 = A_2 e^{ik_2 x}$$

propagates with the wave number

$$k_2 = n_2 k_1$$

altered by the refractive index n_2 of the glass. The waves E_{1+}, E_{1-}, and E_2 are called *incoming*, *reflected*, and *transmitted constituent waves*, respectively. The continuity for the field strength E and its derivative E' at $x = x_1$ means that

$$E_1(x_1) = E_{1+}(x_1) + E_{1-}(x_1) = E_2(x_1)$$

and

$$E_1'(x_1) = ik_1[E_{1+}(x_1) - E_{1-}(x_1)] = ik_2 E_2(x_1) = E_2'(x_1)$$

The two unknown amplitudes, B_1 of the reflected wave and A_2 of the transmitted, can now be calculated from these two continuity equations. The electric field in the whole space is determined by two expressions incorporating these amplitudes,

$$E_s = \begin{cases} A_1 e^{ik_1 x} + B_1 e^{-ik_1 x} & \text{for } x < x_1 \\ A_2 e^{ik_2 x} & \text{for } x > x_1 \end{cases}$$

The electric field in the whole space is obtained as a superposition of constituent waves physically existing in regions 1 and 2. By multiplication with the time-dependent phase $\exp(-i\omega t)$, we obtain the complex field strength E_c, the real part of which is the physical electric field strength.

Figure 2.4b gives the time development of this electric field strength. It is easy to see that in the glass there is a harmonic wave moving to the right. The picture in front of the glass is less clear. Figures 2.4c and d therefore show separately the time developments of the incoming and the reflected waves which add up to the total wave to the left of x_1 observed in Figure 2.4b.

2.3 Light Wave Traveling Through a Glass Plate

It is now easy to see what happens when light falls on a glass plate of finite thickness. When the light wave penetrates the front surface at $x = x_1$, again reflection occurs so that we have as before the superposition of two stationary waves in region $x < x_1$:

$$E_1 = A_1 e^{ik_1 x} + B_1 e^{-ik_1 x}$$

The wave moving within the glass plate suffers reflection at the rear surface at $x = x_2$, so that the second region, $x_1 < x < x_2$, also contains a superposition of two waves,

$$E_2 = A_2 e^{ik_2 x} + B_2 e^{-ik_2 x}$$

which now have the refracted wave number

$$k_2 = n_2 k_1$$

Only in the third region, $x_2 < x$, do we observe a single stationary wave,

$$E_3 = A_3 e^{ik_1 x}$$

with the original wave number k_1.

As a consequence of the reflection on both the front and the rear surface of the glass plate, the reflected wave in region 1 consists of two parts which interfere with each other. The most prominent phenomenon observed under appropriate circumstances is the destructive interference between these two reflected waves, so that no reflection remains in region 1. The light wave is completely transmitted into region 3. This phenomenon is called a *resonance of transmission*. It can be illustrated by looking at the *frequency dependence* of the stationary waves. Figure 2.5a shows the stationary waves for different fixed values of the angular frequency ω, with its magnitude rising from the background to the foreground. A resonance of transmission is recognized through a maximum in the amplitude of the transmitted wave, that is, in the wave to the right of the glass plate.

The signature of a resonance becomes even more prominent in the frequency dependence of the average energy density in the wave. As discussed in Section 2.1, in a vacuum the average energy density has the form

$$w = \frac{\epsilon_0}{2} E_c E_c^*$$

In glass, where the refractive index n has to be taken into account, we have

$$w = \frac{\epsilon\epsilon_0}{2} E_c E_c^* = n^2 \frac{\epsilon_0}{2} E_c E_c^*$$

where $\epsilon = n^2$ is the dielectric constant of glass. Thus, although E_c is continuous at the glass surface, w is not. It reflects the discontinuity of n^2. Therefore we prefer plotting the continuous quantity

$$\frac{2}{n^2 \epsilon_0} w = E_c E_c^*$$

This plot, shown in Fig. 2.5b, indicates a resonance of transmission by the maximum in the average energy density of the transmitted wave. Moreover, since there is no reflected wave at the resonance of transmission, the energy density is constant in region 1.

In the glass plate we observe the typical pattern of a resonance.

1. The amplitude of the average energy density is maximum.
2. The average energy density vanishes in a number of places called *nodes* because for a resonance a multiple of half a wavelength fits into the glass plate. Therefore different resonances can be distinguished by the number of nodes.

The ratio of the amplitudes of the transmitted and incident waves is called the *transmission coefficient* of the glass plate,

$$T = \frac{A_3}{A_1}$$

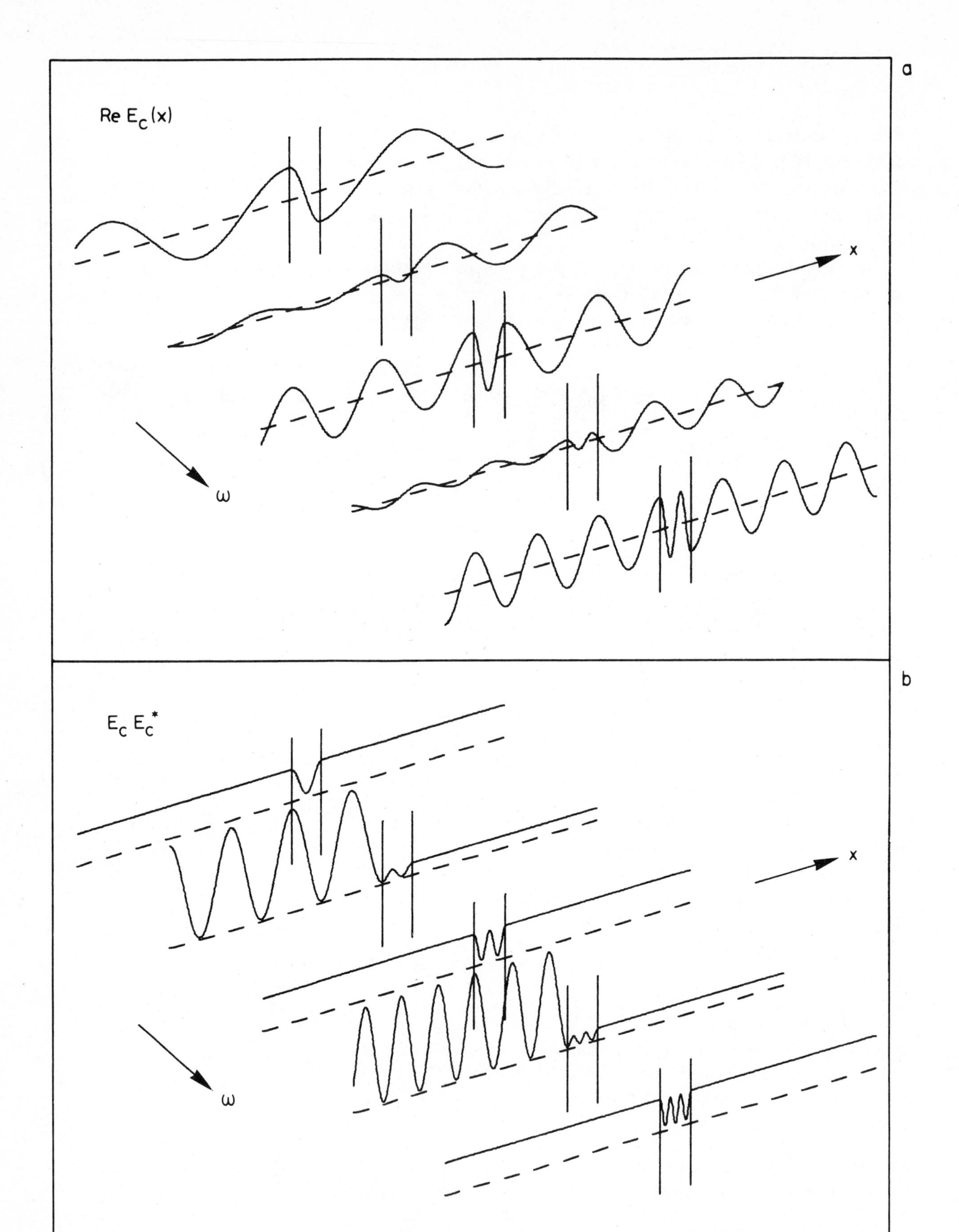
Re $E_c(x)$
a
x
ω
$E_c E_c^*$
b
x
ω

2.4 Free Wave Packet

The plane wave extends into all space, in contrast to any realistic physical situation in which the wave is localized in a finite domain of space. We therefore introduce the concept of a wave packet. It can be understood as a superposition, that is, a sum of plane waves of different frequencies and amplitudes. As a first step we concentrate the wave only in the x-direction. It still extends through all space in the y- and the z-direction. For simplicity we start with the sum of two plane waves with equal amplitudes, E_0:

$$E = E_1 + E_2 = E_0 \cos(\omega_1 t - k_1 x) + E_0 \cos(\omega_2 t - k_2 x)$$

For a fixed time this sum represents a plane wave with two periodic structures. The slowly varying structure is governed by a spatial period,

$$\lambda_- = \frac{4\pi}{|k_2 - k_1|}$$

the rapidly varying structure by a wavelength,

$$\lambda_+ = \frac{4\pi}{|k_2 + k_1|}$$

The resulting wave can be described as the product of a "carrier wave" with the short wavelength λ_+ and a factor modulating its amplitude with the wavelength λ_-:

$$E = 2E_0 \cos(\omega_- t - k_- x) \cos(\omega_+ t - k_+ x),$$
$$k_\pm = |k_2 \pm k_1|/2, \qquad \omega_\pm = ck_\pm$$

Figure 2.6 plots for a fixed moment in time the two waves E_1 and E_2 and the resulting wave E. Obviously the field strength is now concentrated for the most part in certain regions of space. These regions of great field strength propagate through

Figure 2.5 **(a) Frequency dependence of stationary waves when a harmonic wave is incident from the left on a glass plate. The two vertical lines indicate the thickness of the plate. Small values of the angular frequency ω are given in the background, large values in the foreground of the picture.**

(b) Frequency dependence of the quantity $E_c E_c^*$ (which except for a factor n^2 is proportional to the average energy density) of a harmonic wave incident from the left on a glass plate. The parameters are the same as in part a. At a resonance of transmission, the average energy density is constant in the left region, indicating through the absence of interference wiggles that there is no reflection.

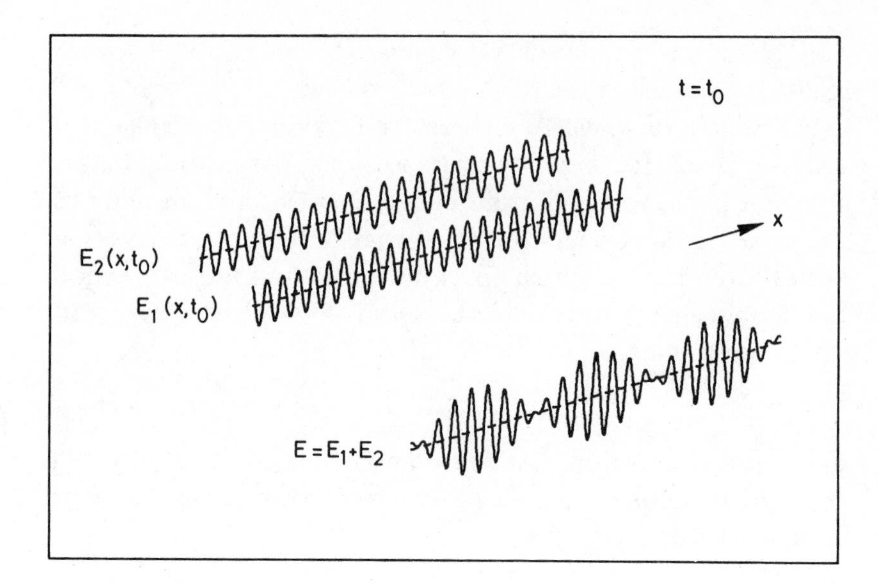

Figure 2.6 **Superposition of two harmonic waves of slightly different angular frequencies ω_1 and ω_2 at a fixed moment in time.**

space with the velocity

$$\frac{\Delta x}{\Delta t} = \frac{\omega_-}{k_-} = c$$

Now we use again complex field strengths. The superposition is written as

$$E_c = E_0 e^{-i(\omega_1 t - k_1 x)} + E_0 e^{-i(\omega_2 t - k_2 x)}$$

For the sake of simplicity, we have chosen in this example a superposition of two harmonic waves with equal amplitudes. By constructing a more complicated "sum" of plane waves, we can concentrate the field in a single region of space. To this end we superimpose a continuum of waves with different frequencies $\omega = c|k|$ and amplitudes:

$$E_c(x, t) = E_0 \int_{-\infty}^{+\infty} dk\, f(k) e^{-i(\omega t - kx)}$$

Such a configuration is called a *wave packet*. The *spectral function* $f(k)$ specifies the amplitude of the harmonic wave with wave number k and circular frequency $\omega = c|k|$. We now consider a particularly simple spectral function which is significantly different from zero in the neighborhood of the wave number k_0. We choose the *Gaussian* function

$$f(k) = \frac{1}{\sqrt{2\pi}\,\sigma_k} \exp\left[-\frac{(k - k_0)^2}{2\sigma_k^2}\right]$$

It describes a bell-shaped spectral function which has its the value of maximum at $k = k_0$; we assume the value of k_0

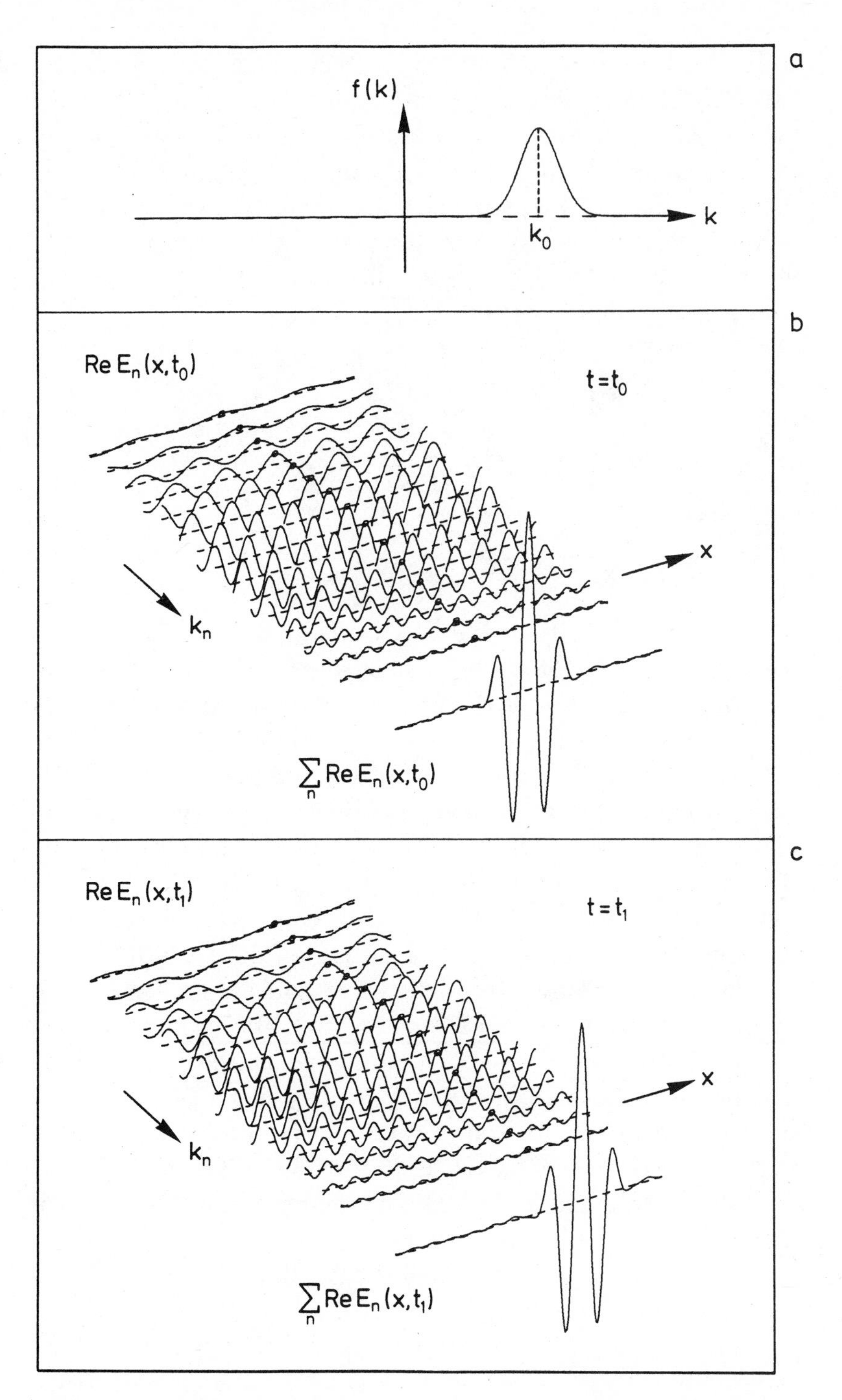

***Figure** 2.7* (a) Gaussian spectral function describing the amplitudes of harmonic waves of different wave numbers k. (b) Construction of a light wave packet as a sum of harmonic waves of different wavelengths and amplitudes. For time $t = 0$ the different terms of the sum are plotted, starting with the contribution of the longest wavelength in the background. Points $x = 0$ are indicated as circles on the partial waves. The resulting wave packet is shown in the foreground. (c) The same as part b, but for time $t_1 > 0$. The phases that were at $x = 0$ for $t = 0$ have moved to $x_1 = ct_1$ for all partial waves. The wave packet has consequently moved by the same distance and retained its shape.

to be positive, $k_0 > 0$. The width of the region in which the function $f(k)$ is different from zero is characterized by the parameter σ_k. In short, one speaks of a Gaussian with *width* σ_k. The Gaussian function $f(k)$ is shown in Figure 2.7a. The factors in front of the exponential are chosen so that the area under the curve equals one. We illustrate the construction of a wave packet by replacing the integration over k by a sum over a finite number of terms,

$$E_c(x,t) \approx \sum_{n=-N}^{N} E_n(x,t)$$

$$E_n(x,t) = E_0\,\Delta k f(k_n) e^{-i(\omega_n t - k_n x)}$$

where

$$k_n = k_0 + n\Delta k, \qquad \omega_n = c|k_n|$$

In Figure 2.7b the different terms of this sum are shown for time $t = 0$, together with their sum, which is depicted in the foreground. The term with the lowest wave number, that is, the longest wavelength, is in the background of the picture. The variation in the amplitudes of the different terms reflects the Gaussian form of the spectral function $f(k)$, which has its maximum, for $k = k_0$, at the center of the picture. On the different terms, the partial waves, the point $x = 0$ is marked by a circle. We observe that the sum over all terms is concentrated around a rather small region near $x = 0$.

Figure 2.7c shows the same wave packet, similarly made up of its partial waves, for later time $t_1 > 0$. The wave packet as well as all partial waves have moved to the right by the distance ct_1. The partial waves still carry marks at the phases that were at $x = 0$ at time $t = 0$. The picture makes it clear that all partial waves have the same velocity as the wave packet, which maintains the same shape for all moments in time.

If we perform the integral explicitly, the wave packet takes the simple form

$$E_c(x,t) = E_c(ct - x)$$
$$= E_0 \exp\left[-\frac{\sigma_k^2}{2}(ct - x)^2\right] \exp[-i(\omega_0 t - k_0 x)]$$

that is,

$$E(x,t) = \operatorname{Re} E_c = E_0 \exp\left[-\frac{\sigma_k^2}{2}(ct - x)^2\right] \cos(\omega_0 t - k_0 x)$$

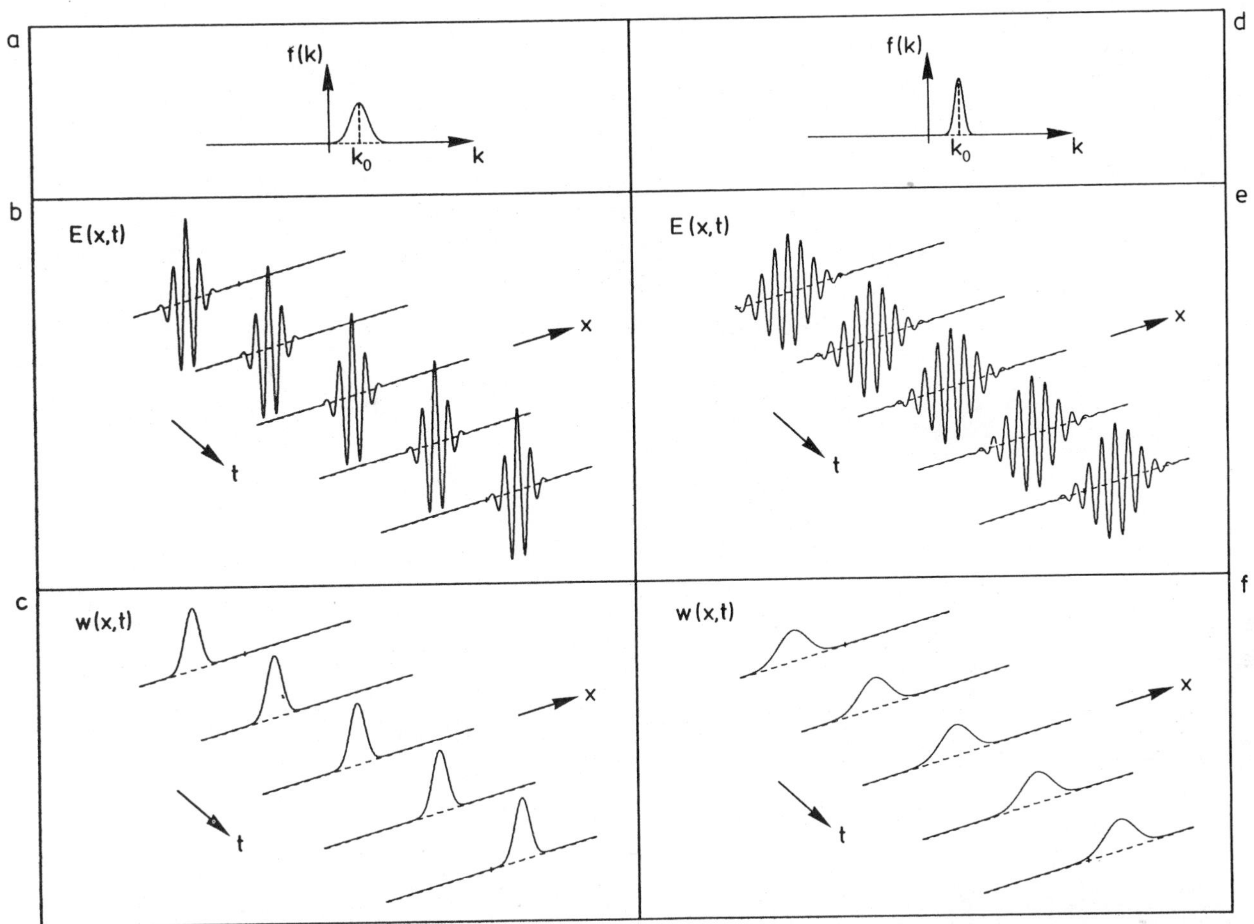

Figure 2.8 **(a, d) Spectral functions, (b, e) time developments of the field strength, and (c, f) time developments of the average energy density for two different Gaussian wave packets.**

It represents a plane wave propagating in the positive x-direction, with a field strength concentrated in a region of the spatial extension $1/\sigma_k$ around point $x = ct$. The time development of the field strength is shown in Figure 2.8b. Obviously the maximum of the field strength is located at $x = ct$; thus the wave packet moves with the velocity c of light. We call this configuration a *Gaussian wave packet* of spatial width

$$\Delta x = \frac{1}{\sigma_k}$$

and of wave number width

$$\Delta k = \sigma_k$$

We observe that a spatial concentration of the wave in the region Δx necessarily requires a spectrum of different wave numbers in the interval Δk so that

$$\Delta x \, \Delta k = 1$$

This is tantamount to saying that the sharper the localization of the wave packet in x-space, the wider is its spectrum in k-space. The original harmonic wave $E = E_0 \cos(\omega t - kx)$ was perfectly sharp in k-space ($\Delta k = 0$) and therefore not localized in x-space. The time development of average energy density w shown in Figure 2.8c appears even simpler than that of the field strength. It is merely a Gaussian traveling with the velocity of light along the x-direction. The Gaussian form is easily explained if we remember that

$$w = \frac{\epsilon_0}{2} E_c E_c^* = \frac{\epsilon_0}{2} E_0^2 e^{-\sigma_k^2 (ct - x)^2}$$

We demonstrate the influence of the spectral function on the wave packet by showing in Figure 2.8 spectral functions with two different widths σ_k. For both we show time development of the field strength and of the average energy density.

2.5 Wave Packet Incident on a Glass Surface

The wave packet, like the plane waves of which it is composed, undergoes reflection and transmission at the glass surface. Figure 2.9a shows the time development of the average energy density in a wave packet moving in from the left. As soon as it hits the glass surface, the already reflected part interferes with the incident wave packet, causing the wiggly structure at the top of the packet. Part of the packet enters the glass, moving with a velocity reduced by the refractive index. For this reason it is compressed in space. The remainder is reflected and moves to the left as a regularly shaped wave packet as soon as it has left the region in front of the glass where interference with the incident packet occurs.

We now demonstrate that the wiggly structure in the interference region is caused by the fast spatial variation of the carrier wave characterized by its wavelength. To this end let us examine the time development of the field strength in the packet, shown in Figure 2.9b. Indeed the spatial variation of the field strength has twice the wavelength of the average energy density in the interference region.

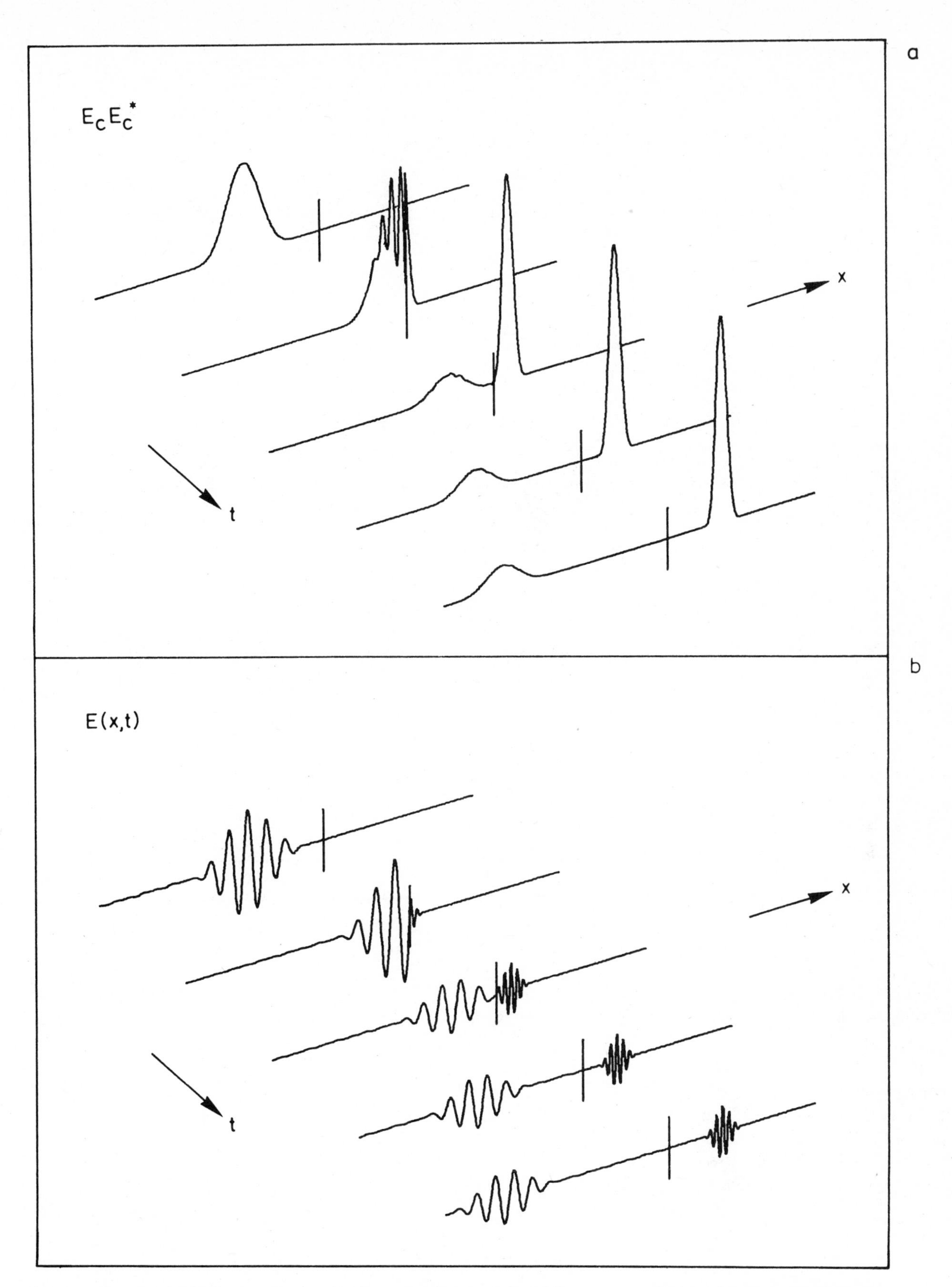

Figure 2.9 **Time developments of (a) the quantity $E_cE_c^*$ (which except for a factor n^2 is proportional to the average energy density) and of (b) the field strength in a wave packet of light falling onto a glass surface where it is partly reflected and partly transmitted through the surface. The glass surface is indicated by the vertical line.**

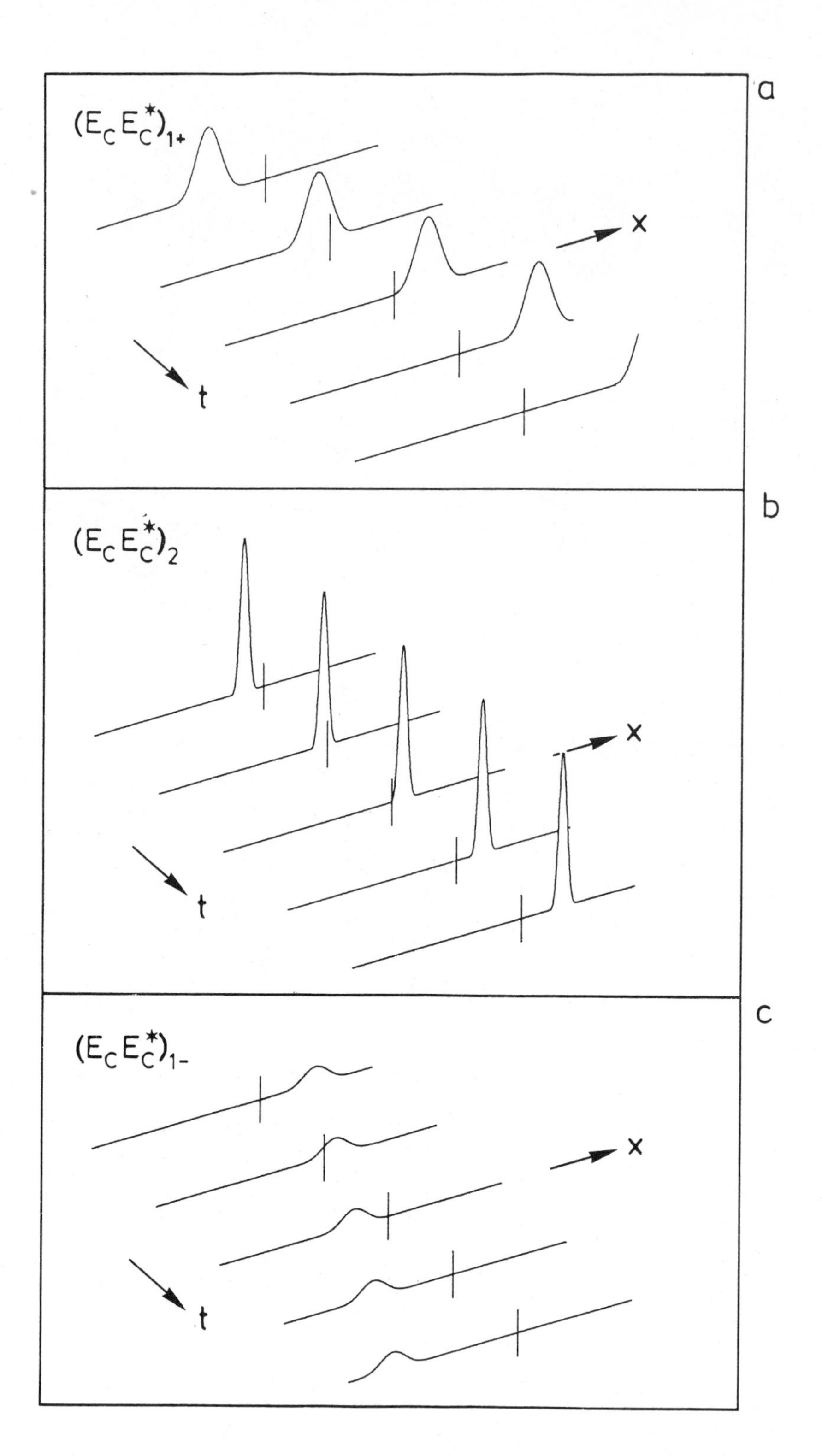

Figure 2.10 **Time developments of the quantity $E_cE_c^*$ (which except for a factor n^2 is proportional to the average energy density) of the constituent waves in a wave packet of light incident on a glass surface: (a) incoming wave, (b) transmitted wave, and (c) reflected wave.**

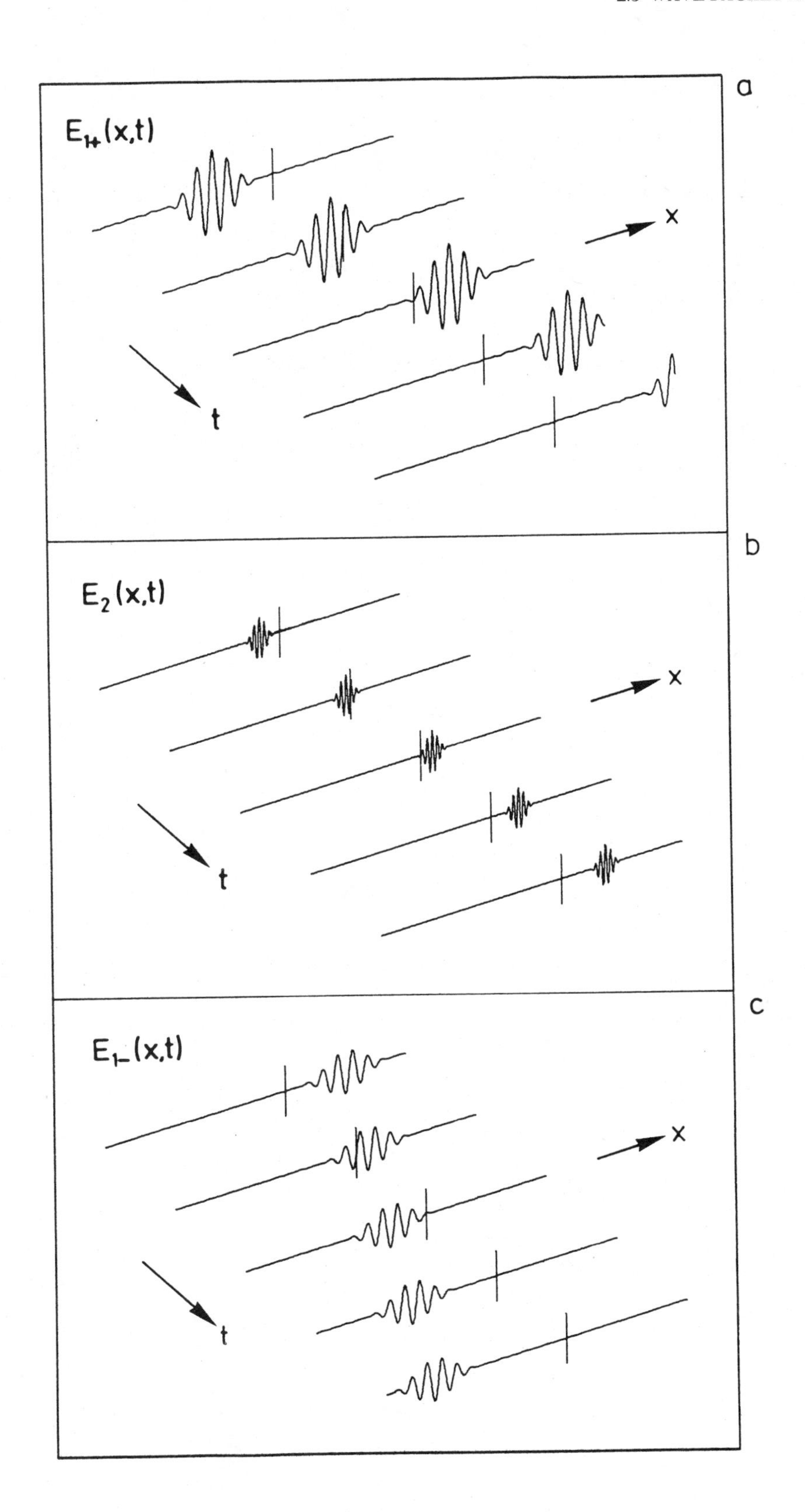

Figure 2.11 **Time developments of the electric field strengths of the constituent waves in a wave packet of light incident on a glass surface: (a) incoming wave, (b) transmitted wave, and (c) reflected wave.**

Another way of studying the reflection and transmission of the packet is to look separately at the average energy densities of the constituent waves, namely the incoming, transmitted, and reflected waves. We show these constituent waves in both regions 1, a vacuum, and 2, the glass, although they contribute physically only in either the one or the other. Figure 2.10 gives their time developments. All three have a smooth bell-shaped form and no wiggles, even in the interference region. The time developments of the field strengths of the constituent waves are shown in Figure 2.11. The observed average energy density of Figure 2.9a corresponds to the absolute square of the sum of the incoming and reflected field strengths in the region in front of the glass and, of course, not to the sum of the average energy densities of these two constituent fields. Their interference pattern shows half the wavelength of the carrier waves.

Figure 2.12 **Time development of the quantity $E_c E_c^*$ (which except for a factor n^2 is proportional to the average energy density) in a wave packet of light incident on a glass plate.**

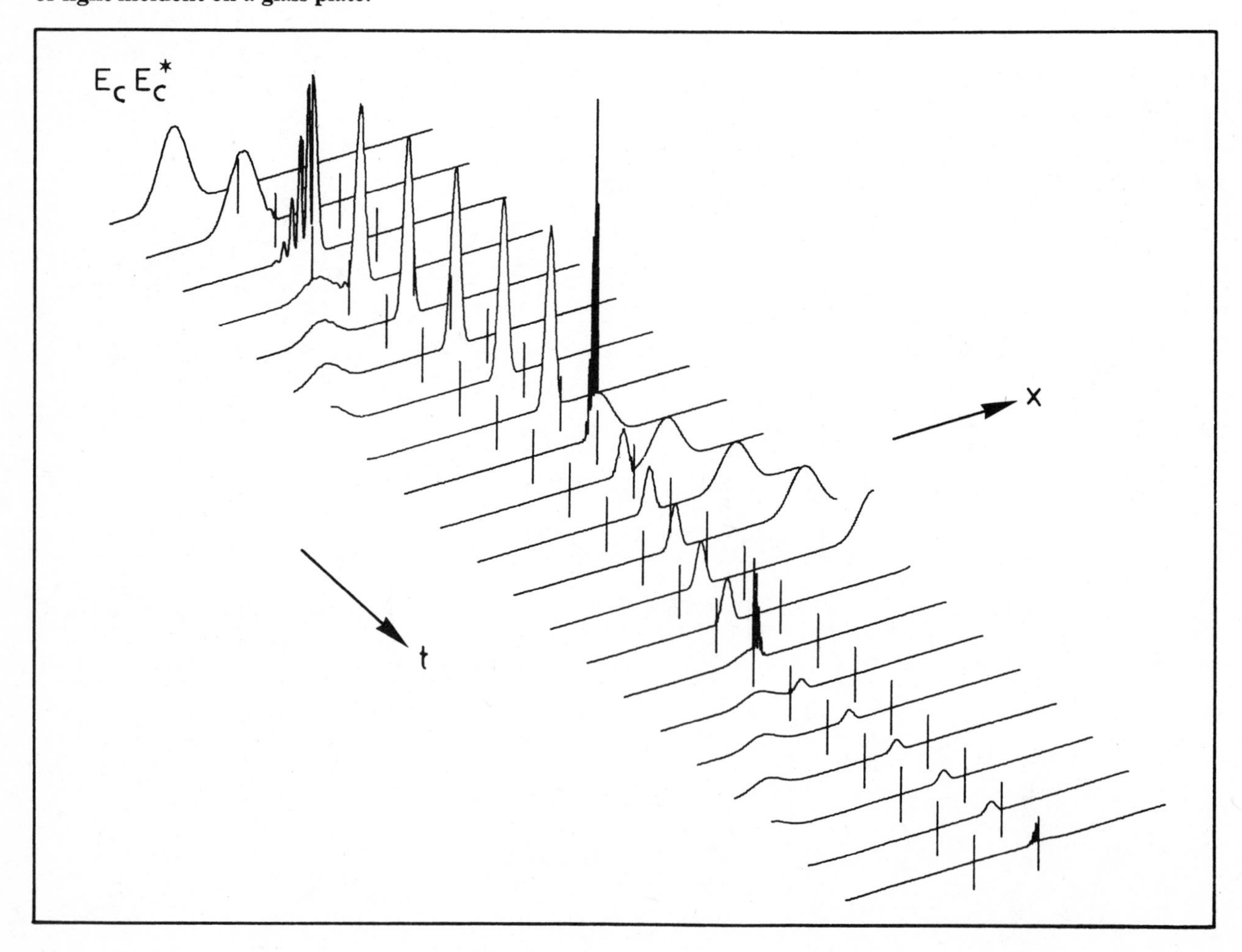

2.6 Wave Packet Traveling Through a Glass Plate

Let us study a wave packet that is relatively narrow in space, that is, one containing a wide range of frequencies. The time development of its average energy density (Figure 2.12) shows that, as expected, at the front surface of the glass plate part of the packet is reflected. Another part enters the plate, where it is compressed and travels with reduced speed. At the rear surface this packet is again partly reflected while another part leaves the plate, traveling to the right with the original width and speed. The small packet traveling back and forth in the glass suffers multiple reflections on the glass surfaces, each time losing part of its energy to packets leaving the glass.

2.7 The Photon

As we have seen in Chapter 1, there are quanta of electromagnetic energy called photons. They can be described by normalized wave packets of the mean angular frequency ω_0 and total energy $\hbar\omega_0$. A finite energy content can be attributed only to wave packets confined in all three dimensions. As the photoeffect indicates, a photon acts as a particle at a single location. Therefore a single photon cannot be understood as an object filling the space occupied by the wave packet. Nor can the wave representing the wave packet be interpreted as describing the electric field strength point by point. The same holds for the average energy density. Instead, one has to introduce the *probability interpretation* of quantum mechanics.

In Section 2.4 the spectral function

$$f(k) = \frac{1}{\sqrt{2\pi}\,\sigma_k} \exp\left[-\frac{(k-k_0)^2}{2\sigma_k^2}\right]$$

was introduced as the weight function specifying the amplitude of the harmonic wave with wave number k and angular frequency $\omega = c|k|$. For the description of a single photon, $|f(k)|^2$ has to be interpreted as a measure for the probability density $P(k)$ of the wave number of the photon. More explicitly, a wave packet with spectral function $f(k)$ and total energy $\hbar\omega_0$, with $\omega_0 = c|k_0|$, describes a photon. The probability interpretation of the spectral function states the following. For a given small interval, $k - \frac{1}{2}\Delta k$, $k + \frac{1}{2}\Delta k$, located symmetrically about wave number k, the product $|f(k)|^2\,\Delta k$

is proportional to the probability $P(k)\,\Delta k$ that the photon has a wave number within this interval, $k - \frac{1}{2}\Delta k$, $k + \frac{1}{2}\Delta k$. Since the probability that the photon possesses an arbitrary wave number equals one, the proportionality constant N is determined by the requirement

$$N\int_{-\infty}^{+\infty} |f(k)|^2\,dk = 1$$

which yields

$$N = 2\sqrt{\pi}\,\sigma_k$$

Thus

$$P(k)\,\Delta k = 2\sqrt{\pi}\,\sigma_k |f(k)|^2\,\Delta k$$

is the probability of finding the photon in the wave number interval $k - \frac{1}{2}\Delta k$, $k + \frac{1}{2}\Delta k$, and

$$P(k) = 2\sqrt{\pi}\,\sigma_k |f(k)|^2$$

is the probability density.

Conversely, a wave packet containing an energy much larger than $\hbar\omega_0$—for example, a wave packet produced by a radio transmitter which was switched on for a short time—contains a very large number of photons with a fixed phase relation. When there are many coherent photons, the wave represents an electromagnetic field strength.

As an example, let us apply the probability interpretation to the wave packet incident on a glass surface; assuming that the wave packet contains the energy $\hbar\omega_0$, that is, only one quantum of light. The fraction of the energy $\hbar\omega_0$ contained in a single reflected or transmitted wave packet is the probability that the incident photon will be reflected or transmitted. More generally, the fraction of the energy contained in a single wave packet is a measure indicating whether the photon is in the region of space inhabited by the packet.

The Planck-Einstein relation between energy E of the photon and its angular frequency,

$$E = \hbar\omega$$

necessitates a relation between the wave vector $\mathbf{k}$ of the photon and its momentum $\mathbf{p}$, the Compton relation

$$\mathbf{p} = \hbar\mathbf{k}$$

This is so since the energy and the momentum of a particle moving with the velocity of light are related by

$$E = c|\mathbf{p}|$$

The complex field strength E_c of a plane light wave of angular frequency ω and wave vector $\mathbf{k}$ can now be expressed in terms of energy E and momentum $\mathbf{p}$ of the photon:

$$E_c = E_0 \exp[-i(\omega t - \mathbf{k} \cdot \mathbf{x})] = E_0 \exp\left[-\frac{i}{\hbar}(Et - \mathbf{p} \cdot \mathbf{x})\right]$$

Problems

2.1 Estimate the refractive index n_2 of the glass plate in Figure 2.4b.

2.2 Calculate the energy density for the plane electromagnetic wave described by the complex electric field strength

$$E_c = E_0 e^{i(\omega t - kx)}$$

and show that its average over a temporal period T is $w = (\epsilon_0/2) E_c E_c^*$.

2.3 Give the qualitative reason why the resonance phenomena in Figure 2.5a occurs for the wavelengths

$$\lambda = l\frac{nd}{2}, \qquad l = 1, 2, 3 \ldots$$

Use the continuity condition of the electric field strength and its derivative. Here n is the refractive index of the glass plate of thickness d.

2.4 Calculate the ratio of the frequencies of the two electric field strengths, as they are plotted in Figure 2.6, from the beat in their superposition.

2.5 The one-dimensional wave packet of light does not show any dispersion, that is, spreading with time. What causes the dispersion of a wave packet of light confined in all three spatial dimensions?

2.6 Estimate the refractive index of the glass, using the change in width or velocity of the light pulse in Figure 2.9a.

2.7 Verify in Figure 2.12 that the stepwise reduction of the amplitude of the pulse within the glass plate proceeds with approximately the same reduction factor, thus following on the average an exponential decay law.

2.8 Calculate energy E and momentum p of a photon of blue ($\lambda = 450 \cdot 10^{-9}$ m), green ($\lambda = 530 \cdot 10^{-9}$ m), yellow ($\lambda = 580 \cdot 10^{-9}$), and red ($\lambda = 700 \cdot 10^{-9}$ m) light. Use Einstein's formula $E = Mc^2$ to calculate the relativistic mass of the photon. Give the results in SI units.

3. Probability Waves of Matter

3.1 De Broglie Waves

In Section 2.7 we learned that through the probability interpretation photons can be described by waves. We have made explicit use of the simple relation $E = c|\mathbf{p}|$ between energy and momentum of the photon, which holds only for particles moving with the velocity c of light. For particles with a finite rest mass m, which move with velocities v slow compared to the velocity of light, the corresponding nonrelativistic relation between energy and momentum is

$$E = \frac{p^2}{2m}, \qquad p = mv$$

Plane waves that are of the same type as those for photons, which were discussed at the end of Chapter 2, but have the nonrelativistic relation just given,

$$\psi_p(x,t) = \frac{1}{(2\pi\hbar)^{1/2}} \exp\left[-\frac{i}{\hbar}(Et - px)\right]$$
$$= \frac{1}{(2\pi\hbar)^{1/2}} \exp\left[-\frac{i}{\hbar}\left(\frac{p^2}{2m}t - px\right)\right]$$

are called *de Broglie waves* of matter. The phase velocity of these de Broglie waves is

$$v_p = \frac{E}{p} = \frac{p}{2m}$$

and is thus different from the particle velocity $v = p/m$.

3.2 Wave Packet, Dispersion

The harmonic de Broglie waves, like the harmonic electric waves, are not localized in space and therefore are not suited to describing a particle. To localize a particle in space, we again have to superimpose harmonic waves to form a wave packet. To keep things simple, we first restrict ourselves to discussing a one-dimensional wave packet.

For the spectral function we again choose a Gaussian function,[1]

$$f(p) = \frac{1}{(2\pi)^{1/4}\sqrt{\sigma_p}} \exp\left[-\frac{(p-p_0)^2}{4\sigma_p^2}\right]$$

The corresponding de Broglie wave packet is then

$$\psi(x,t) = \int_{-\infty}^{+\infty} f(p)\psi_p(x - x_0, t)\,dp$$

For the de Broglie wave packet, as for the light wave packet, we first approximate the integral by a sum,

$$\psi(x,t) = \sum_{n=-N}^{N} \psi_n(x,t)$$

where the $\psi_n(x,t)$ are harmonic waves for different values $p_n = p_0 + n\Delta p$ multiplied by the spectral weight $f(p_n)\Delta p$,

$$\psi_n(x,t) = f(p_n)\psi_{p_n}(x - x_0, t)\Delta p$$

Figure 3.1a shows the real parts $\mathrm{Re}\,\psi_n(x,t)$ of the harmonic waves $\psi_n(x,t)$ as well as their sum being equal to the real part $\mathrm{Re}\,\psi(x,t)$ of the wave function $\psi(x,t)$ for the wave packet at time $t = t_0 = 0$. The point $x = x_0$ is marked on each harmonic wave. In Figure 3.1b the real parts $\mathrm{Re}\,\psi_n(x,t)$ and their sum $\mathrm{Re}\,\psi(x,t)$ are shown at later time $t = t_1$. Because of their different phase velocities, the partial waves have moved by different distances $\Delta x_n = v_n(t_1 - t_0)$ where $v_n = p_n/(2m)$ is the phase velocity of the harmonic wave of momentum p_n. This effect broadens the extension of the wave packet.

The integration over p can be carried out so that the explicit expression for the wave packet has the form

$$\psi(x,t) = M(x,t)e^{i\phi(x,t)}$$

Here the exponential function represents the carrier wave with a *phase* ϕ varying rapidly in space and time. The

[1]We have chosen this spectral function to correspond to the square root of the spectral function that was used in Section 2.4 to construct a wave packet of light. Since the area under the spectral function $f(k)$ of Section 2.4 was equal to one, the area under $[f(p)]^2$ is now equal to one. This guarantees that the normalization condition of the wave function ψ in the next section will be fulfilled.

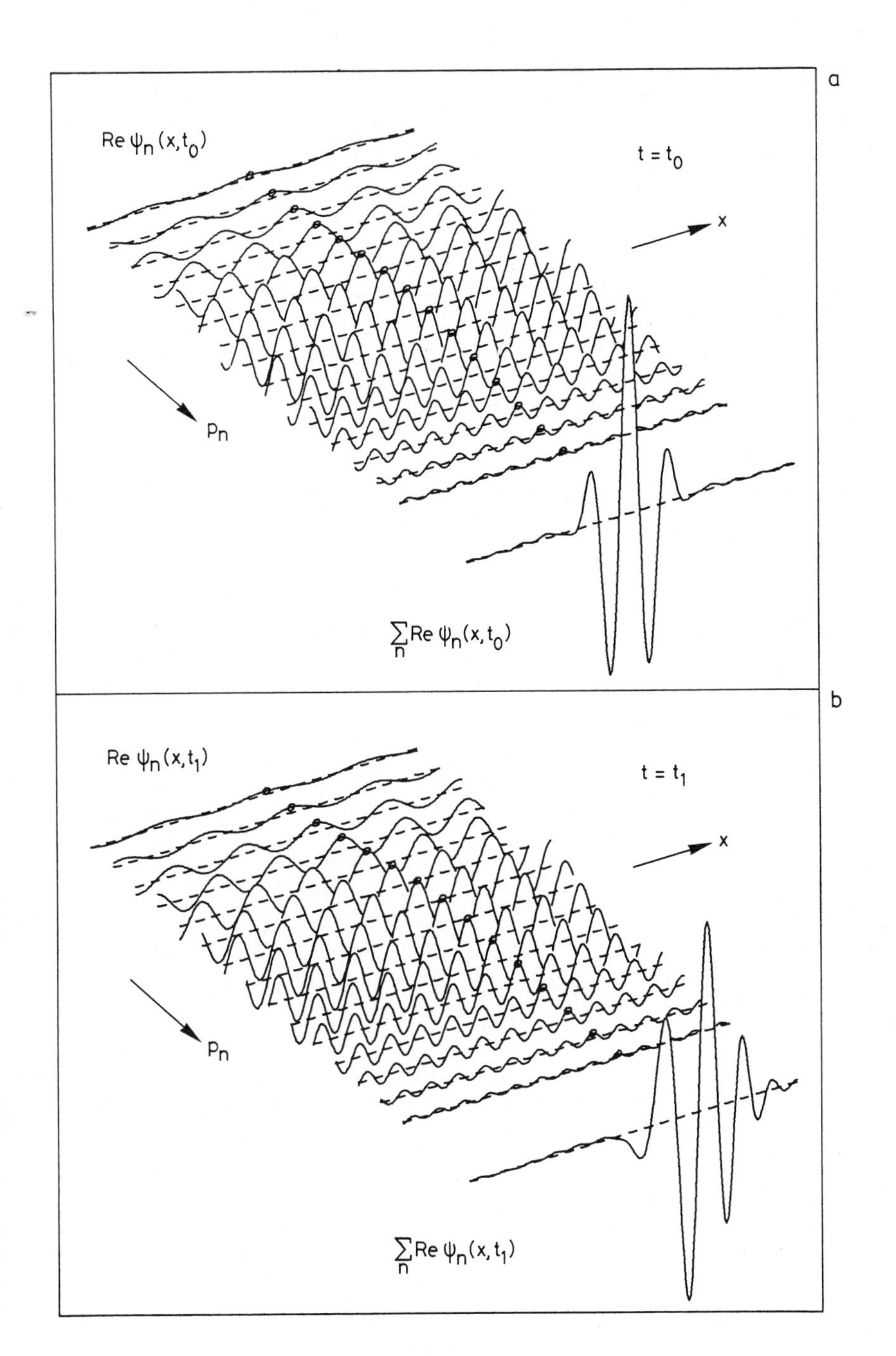

Figure 3.1 Construction of a wave packet as a sum of harmonic waves ψ_n of different momenta and consequently of different wavelengths. Plotted are the real parts of the wave functions. The terms of different momenta and different amplitudes begin with the one of longest wavelength in the background. In the foreground is the wave packet resulting from the summation.

(a) The situation for time $t = t_0$. All partial waves are marked by a circle at point $x = x_0$.

(b) The same wave packet and its partial waves at time $t_1 > t_0$. The partial waves have moved different distances $\Delta x_n = v_n(t_1 - t_0)$ because of their different phase velocities v_n, as indicated by the circular marks which have kept their phase with respect to those in part a. Because of the different phase velocities, the wave packet has changed its form and width.

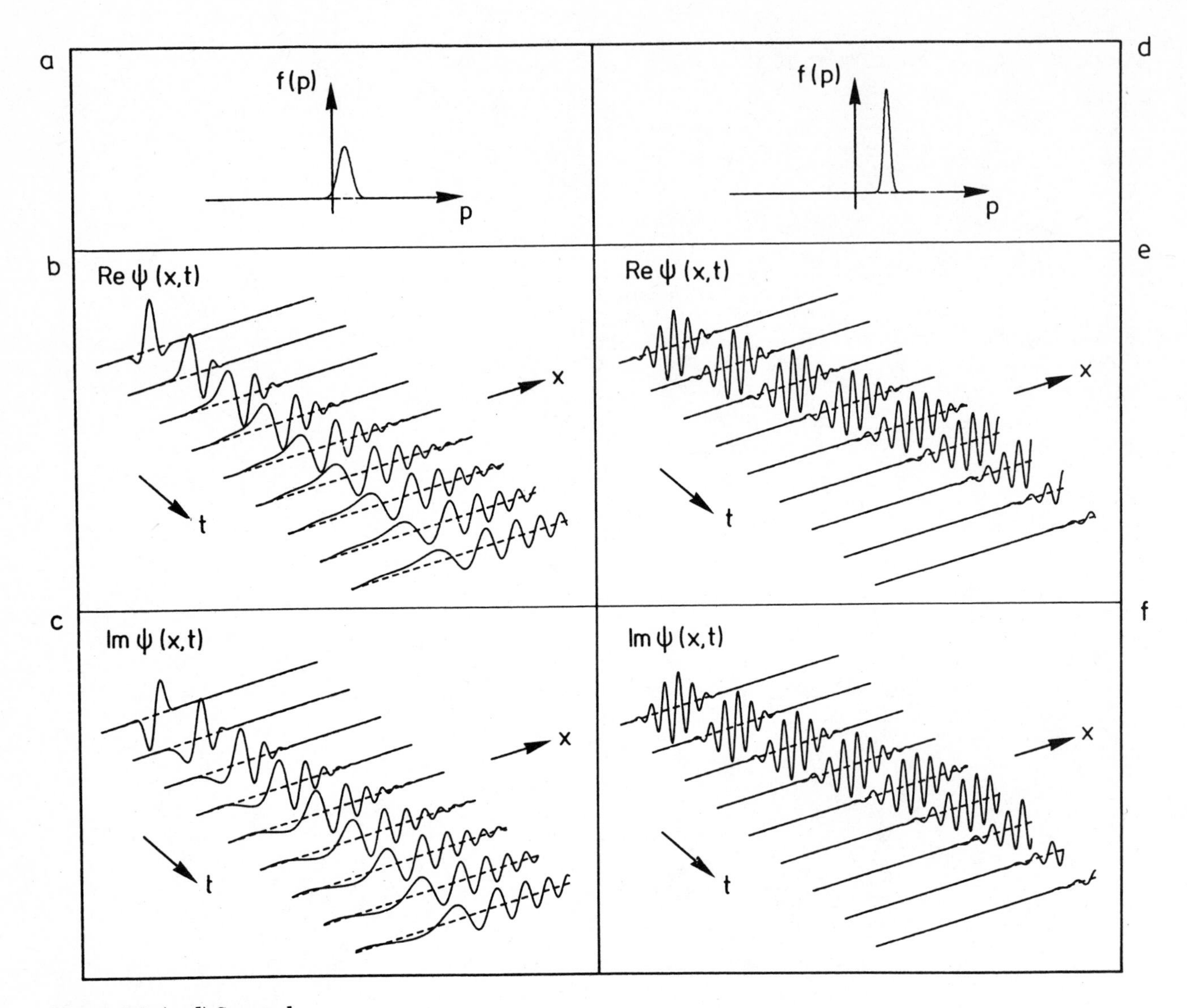

Figure 3.2 **(a, d) Spectral functions and time developments of (b, e) the real parts and (c, f) the imaginary parts of the wave functions for two different wave packets. The two packets have different group velocities and different widths and spread differently with time.**

bell-shaped amplitude function

$$M(x,t) = \frac{1}{(2\pi)^{1/4}\sqrt{\sigma_x}} \exp\left[-\frac{(x - x_0 - v_0 t)^2}{4\sigma_x^2}\right]$$

travels in x-direction with the group velocity

$$v_0 = \frac{p_0}{m}$$

The group velocity is indeed the particle velocity and different from the phase velocity. The localization in space is given by

$$\sigma_x^2 = \frac{\hbar^2}{4\sigma_p^2}\left(1 + \frac{4\sigma_p^4}{\hbar^2}\frac{t^2}{m^2}\right)$$

This formula shows that the spatial extension σ_x of the wave packet increases with time. This phenomenon is called *dispersion*. Figure 3.2 shows the time developments of the real and imaginary parts of two wave packets with different group velocities and widths. We easily observe the dispersion of the wave packets in time. The fact that a wave packet comprises a whole range of momenta is the physical reason why it disperses. Its components move with different velocities, thus spreading the packet in space.

The function $\phi(x,t)$ determines the phase of the carrier wave. It has the form

$$\phi(x,t) = \frac{1}{\hbar}\left[p_0 + \frac{\sigma_p^2}{\sigma_x^2}\frac{v_0 t}{2p_0}(x - x_0 - v_0 t)\right](x - x_0 - v_0 t) + \frac{p_0}{2\hbar}v_0 t + \frac{\alpha}{2}$$

with

$$\tan\alpha = \frac{2}{\hbar}\frac{\sigma_p^2}{m}(t - t_0)$$

For fixed time t it represents the phase of a harmonic wave modulated in wave number. The effective wave number k_{eff} is the factor in front of $x - x_0 - v_0 t$ and is given by

$$k_{\text{eff}}(x) = \frac{1}{\hbar}\left[p_0 + \frac{\sigma_p^2}{\sigma_x^2}\frac{v_0 t}{2p_0}(x - x_0 - v_0 t)\right]$$

At the value $x = \langle x \rangle$ corresponding to the maximum value of the bell-shaped amplitude modulation $M(x,t)$, that is, its position average,

$$\langle x \rangle = x_0 + v_0 t$$

the effective wave number is simply equal to the wave number that corresponds to the average momentum p_0 of the spectral function,

$$k_0 = \frac{1}{\hbar}p_0 = \frac{1}{\hbar}mv_0$$

For values $x > x_0 + v_0 t$ that is, in front of the average position $\langle x \rangle$ of the moving wave packet, the effective wave number increases,

$$k_{\text{eff}}(x > x_0 + v_0 t) > k_0$$

so that the local wavelength

$$\lambda_{\text{eff}}(x) = \frac{2\pi}{|k_{\text{eff}}(x)|}$$

decreases.

For values $x < x_0 + v_0 t$, that is, behind the average position $\langle x \rangle$, the effective wave number decreases,

$$k_{\text{eff}}(x < x_0 + v_0 t) < k_0$$

This decrease leads to negative values of k_{eff} of large absolute value, which, far behind the average position, makes the wavelengths $\lambda_{\text{eff}}(x)$ short again. This wave number modulation can easily be verified in Figures 3.1 and 3.2. For a wave packet at rest, that is, $p_0 = 0$, $v_0 = p_0/m = 0$, the effective wave number

$$k_{\text{eff}}(x) = \frac{1}{\hbar}\frac{\sigma_p^2}{\sigma_x^2}\frac{m}{2}t(x - x_0)$$

has the same absolute value to the left and to the right of the average position x_0. This implies a decrease of the effective wavelength that is symmetric on both sides of x_0. Figure 3.4 corroborates this statement.

3.3 Probability Interpretation, Uncertainty Principle

Following Max Born (1926), we interpret the wave function $\psi(x,t)$ as follows. Its absolute square

$$\rho(x,t) = |\psi(x,t)|^2 = M^2(x,t)$$

is identified with the probability density for observing the particle at position x and time t, that is, the probability of observing the particle at a given time t in the space region between x and $x + \Delta x$ is $P = \rho(x,t)\,\Delta x$. This is plausible since $\rho(x,t)$ is positive everywhere. Furthermore, its integral over all space is equal to one for every moment in time so that the *normalization condition*,

$$\int_{-\infty}^{+\infty} |\psi|^2\,dx = \int \psi(x,t)\psi^*(x,t)\,dx = 1$$

holds.

Notice, that there is a strong formal similarity between the average energy density $w(x,t) = |E_c(x,t)|^2$ of a light wave and the probability density $\rho(x,t)$. Because of the probability character, the wave function $\psi(x,t)$ is not a field strength, since the effect of a field strength must be measurable wherever the field is not zero. A probability density, however, determines the probability that a particle, which can be point-like, will be observed at a given position. This probability interpretation is, however, restricted to normalized wave functions. Since the integral over the absolute square of a harmonic plane wave is

$$\frac{1}{2\pi\hbar}\int_{-\infty}^{+\infty}\exp\left[\frac{i}{\hbar}(Et-px)\right]\exp\left[-\frac{i}{\hbar}(Et-px)\right]dx = \frac{1}{2\pi\hbar}\int_{-\infty}^{+\infty}dx$$

and diverges, the absolute square $|\psi(x,t)|^2$ of a harmonic plane wave cannot be considered a probability density. We shall call the absolute square of a wave function that cannot be normalized its *intensity*. Even though wave functions that cannot be normalized have no immediate physical significance, they are of great importance for the solution of problems. We have already seen that normalizable wave packets can be composed of these wave functions. This situation is similar to the one in classical electrodynamics in which the plane electromagnetic wave is indispensable for the solution of many problems. Nevertheless, a harmonic plane wave cannot exist physically, for it would fill all of space and consequently have infinite energy.

Figure 3.3 shows the time developments of the probability densities of the two Gaussian wave packets given in Figure 3.2. Underneath the two time developments the motion of a classical particle with the same velocity is presented. We see that the center of the Gaussian wave packet moves in the exact same way as the classical particle. But whereas the classical particle at every instant in time occupies a well-defined position in space, the quantum-mechanical wave packet has a finite width σ_x. It is a measure for the size of the region in space surrounding the classical position in which the particle will be found. The fact that the wave packet disperses in time means that the location of the particle becomes more and more uncertain with time.

The dispersion of a wave packet with zero group velocity is particularly striking. Without changing position it becomes wider and wider as time goes by (Figure 3.4a).

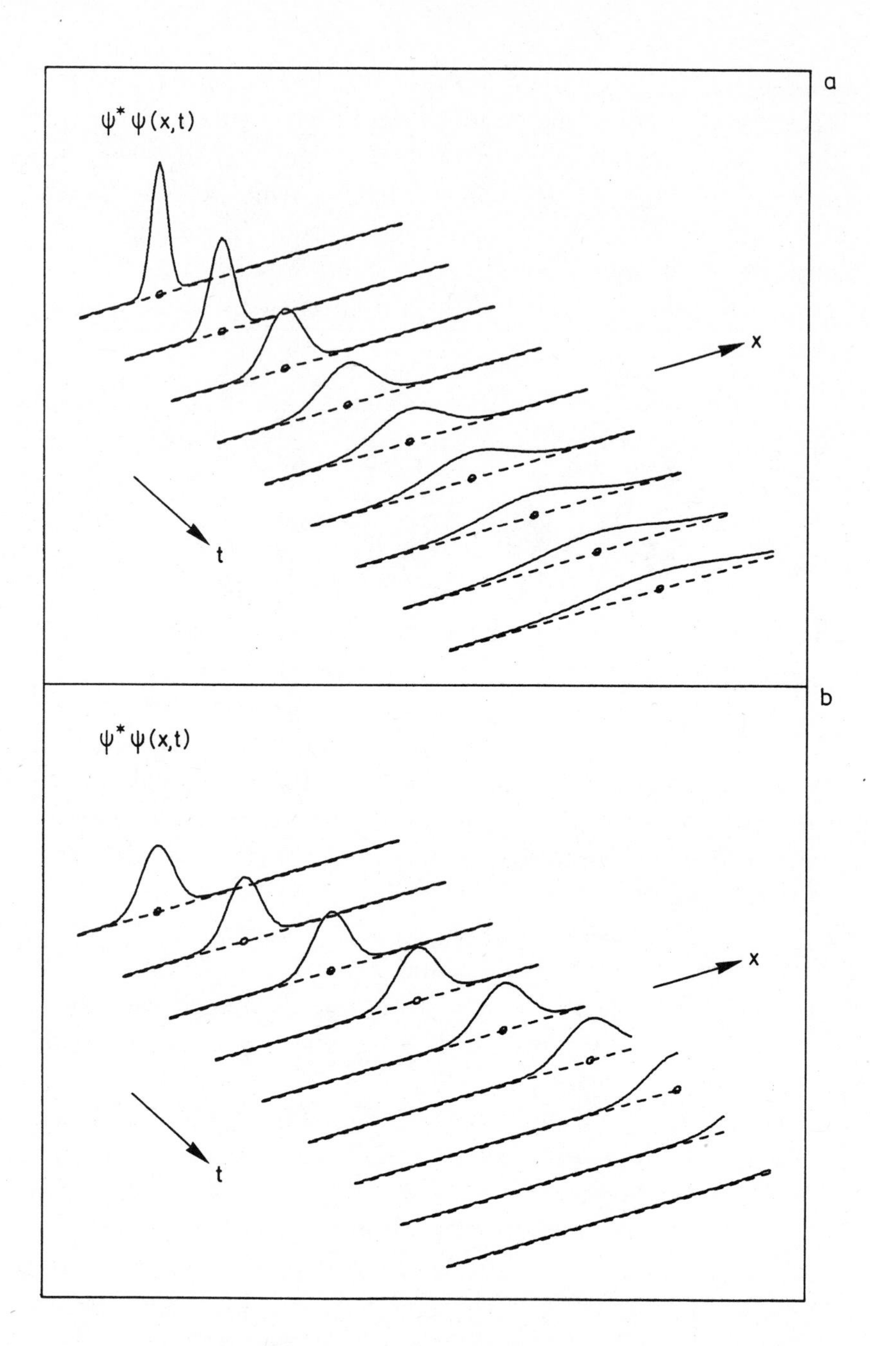

Figure 3.3 **Time developments of the probability densities for the two wave packets of Figure 3.2. The two packets have different group velocities and different widths. Also shown, by the small circles, is the position of a classical particle moving with a velocity equal to the group velocity of the packet.**

It is interesting to study the behavior of the real and imaginary parts of the wave packet at rest. Their time developments are shown in Figures 3.4b and 3.4c. Starting from a wave packet that at initial time $t = 0$ was chosen to be a real Gaussian packet, waves travel in both positive and negative x-directions. Obviously the harmonic waves with the highest phase velocities, those whose wiggles escape the most quickly from the original position $x = 0$, possess the shortest wavelengths. The spreading of the wave packet can be explained in

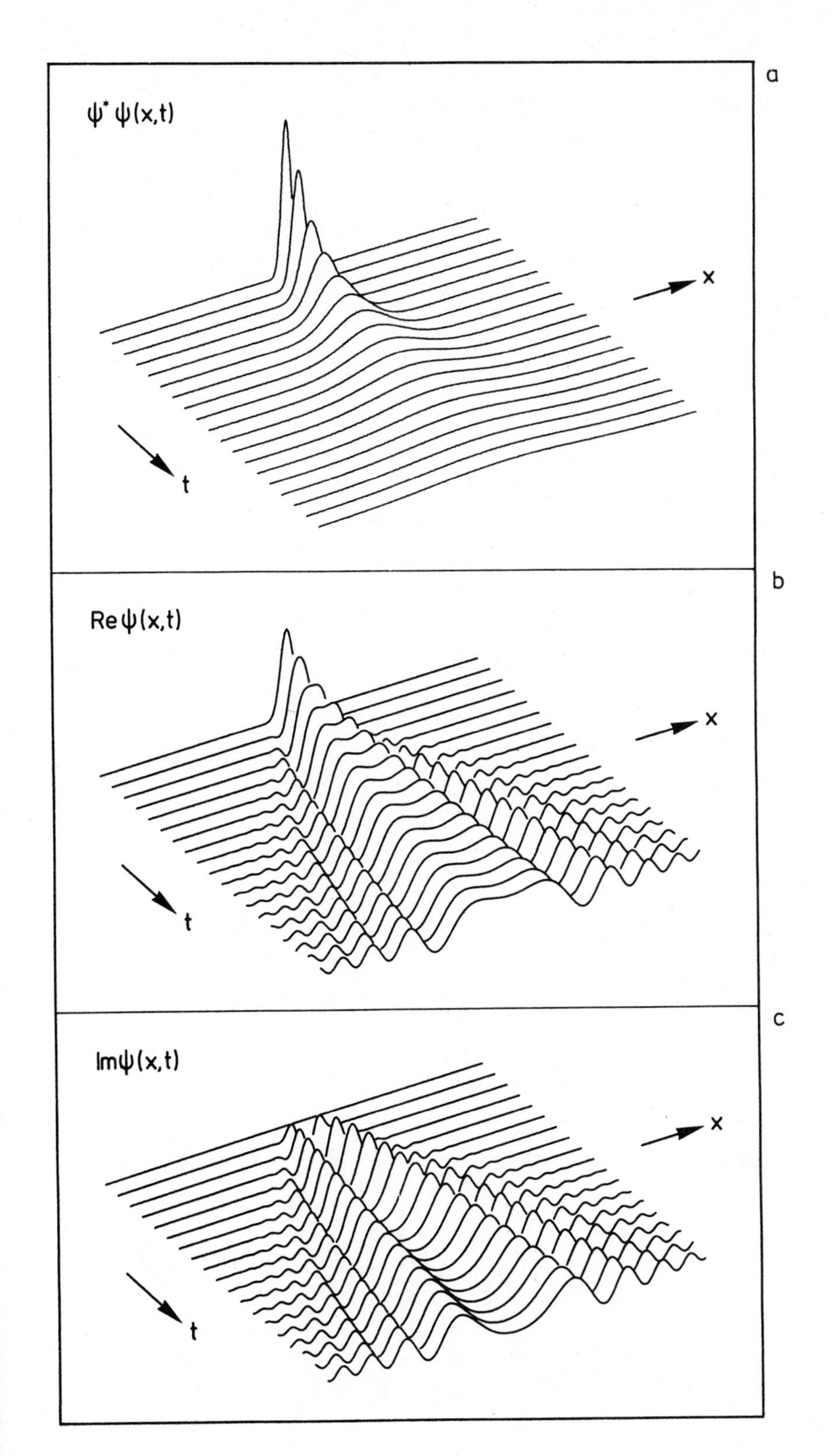

Figure 3.4 **Time developments of (a) the probability density for a wave packet at rest and of (b) the real part and (c) the imaginary part of its wave function.**

another way. Because the original wave packet at $t = 0$ contains spectral components with positive and negative momenta, it spreads in space as time elapses.

The probability interpretation of the wave function now suggests that we use standard concepts of probability calculus, in particular the expectation value and variance. The *expectation value* or *average value of the position* of a particle described by a wave function $\psi(x, t)$ is

$$\langle x \rangle = \int_{-\infty}^{+\infty} x\rho(x,t)\,dx = \int_{-\infty}^{+\infty} \psi^*(x,t)x\psi(x,t)\,dx$$

which, in general, remains a function of time. For a Gaussian wave packet the integration indeed yields

$$\langle x \rangle = x_0 + v_0 t, \qquad v_0 = \frac{p_0}{m}$$

corresponding to the trajectory of classical unaccelerated motion. We shall therefore interpret the Gaussian wave packet of de Broglie waves as a quantum-mechanical description of the unaccelerated *motion of a particle*, that is, a particle moving with constant velocity. Actually, the Gaussian form of the spectral function $f(k)$ allows the explicit calculation of the wave packet. With this particular spectral function, the wave function $\psi(x, t)$ can be given in closed form.

The *variance of the position* is the expectation value of the square of the difference between the position and its expectation:

$$\begin{aligned} \mathrm{var}(x) &= \left\langle (x - \langle x \rangle)^2 \right\rangle \\ &= \int_{-\infty}^{+\infty} \psi^*(x,t)(x - \langle x \rangle)^2 \psi(x,t)\,dx \end{aligned}$$

Again, for the Gaussian wave packet the integral can be carried out to give

$$\mathrm{var}(x) = \sigma_x^2 = \frac{\hbar^2}{4\sigma_p^2}\left(1 + \frac{4\sigma_p^4}{\hbar^2}\frac{t^2}{m^2}\right)$$

which agrees with the formula quoted in Section 3.2.

Calculation of the expectation value of the momentum of a wave packet

$$\psi(x,t) = \int_{-\infty}^{+\infty} f(p)\psi_p(x - x_0, t)\,dp$$

is carried out with the direct help of the spectral function $f(p)$, that is,

$$\langle p \rangle = \int_{-\infty}^{+\infty} p |f(p)|^2 \, dp$$

For the spectral function $f(p)$ of the Gaussian wave packet given at the beginning of Section 3.2, we find

$$\langle p \rangle = \int_{-\infty}^{+\infty} p \frac{1}{\sqrt{2\pi}\,\sigma_p} \exp\left[-\frac{(p-p_0)^2}{2\sigma_p^2} \right] dp$$

We replace the factor p by the identity

$$p = p_0 + (p - p_0)$$

Since the exponential in the integral above is an even function in the variable $p - p_0$, the integral

$$\int_{-\infty}^{+\infty} (p - p_0) \frac{1}{\sqrt{2\pi}\,\sigma_p} \exp\left[-\frac{(p-p_0)^2}{2\sigma_p^2} \right] dp = 0$$

vanishes, for the contributions in the intervals $-\infty < x < 0$ and $0 < x < \infty$ cancel. The remaining term is the product of the constant p_0 and the normalization integral,

$$\int_{-\infty}^{+\infty} |f(p)|^2 \, dp = 1$$

so that we find

$$\langle p \rangle = p_0$$

This result is not surprising, for the Gaussian spectral function gives the largest weight to momentum p_0 and decreases symmetrically to the left and right of this value. At the end of Section 3.2, we found $v_0 = p_0/m$ as the group velocity of the wave packet. Putting the two findings together, we have discovered that the momentum expectation value of a free, unaccelerated Gaussian wave packet is the same as the momentum of a free, unaccelerated particle of mass m and velocity v_0 in classical mechanics:

$$\langle p \rangle = p_0 = m v_0$$

The expectation value of momentum can also be calculated directly from the wave function $\psi(x, t)$. We have the simple relation

$$\frac{\hbar}{i}\frac{\partial}{\partial x}\psi_p(x - x_0, t) = \frac{\hbar}{i}\frac{\partial}{\partial x}\left\{\frac{1}{(2\pi\hbar)^{1/2}}\exp\left[-\frac{i}{\hbar}(Et - px)\right]\right\}$$
$$= p\psi_p(x - x_0, t)$$

This relation translates the momentum variable p into the *momentum operator*

$$p \to \frac{\hbar}{i}\frac{\partial}{\partial x}$$

The momentum operator allows us to calculate the expectation value of momentum from the following formula:

$$\langle p \rangle = \int_{-\infty}^{+\infty} \psi^*(x,t)\frac{\hbar}{i}\frac{\partial}{\partial x}\psi(x,t)\,dx$$

It is completely analogous to the formula for the expectation value of position given earlier. We point out that the operator appears between the functions $\psi^*(x,t)$ and $\psi(x,t)$, thus acting on the second factor only. To verify this formula, we replace wave function $\psi(x,t)$ by its representation in terms of the spectral function:

$$\langle p \rangle = \int_{-\infty}^{+\infty} \psi^*(x,t)\frac{\hbar}{i}\frac{\partial}{\partial x}\int_{-\infty}^{+\infty} f(p)\psi_p(x - x_0, t)\,dp\,dx$$
$$= \int_{-\infty}^{+\infty}\int_{-\infty}^{+\infty} \psi^*(x,t)\psi_p(x - x_0, t)\,dx\,p\,f(p)\,dp$$

The inner integral

$$\int_{-\infty}^{+\infty} \psi^*(x,t)\psi_p(x - x_0, t)\,dx$$
$$= \int_{-\infty}^{+\infty} \psi(x,t)\frac{1}{(2\pi\hbar)^{1/2}}\exp\left\{-\frac{i}{\hbar}[Et - p(x - x_0)]\right\}dx$$

is by Fourier's theorem the inverse of the representation

$$\psi^*(x,t) = \int f^*(p)\psi_p{}^*(x - x_0, t)\,dx$$
$$= \frac{1}{(2\pi\hbar)^{1/2}}\int f^*(p)\exp\left\{\frac{i}{\hbar}[Et - p(x - x_0)]\right\}dp$$

of the complex conjugate of the wave packet $\psi(x,t)$. Thus we have

$$\int_{-\infty}^{+\infty} \psi^*(x,t)\psi_p(x - x_0, t)\,dx = f^*(p)$$

Substituting this result for the inner integral of the expression for $\langle p \rangle$, we rediscover the expectation value of momentum in the form

$$\langle p \rangle = \int_{-\infty}^{+\infty} f^*(p) p f(p)\, dp = \int_{-\infty}^{+\infty} p |f(p)|^2\, dp$$

This equation justifies the identification of momentum p with the operator $(\hbar/i)(\partial/\partial x)$ acting on the wave function. The *variance of the momentum* for a wave packet is

$$\mathrm{var}(p) = \langle (p - \langle p \rangle)^2 \rangle = \int \psi^* \left(\frac{\hbar}{i} \frac{\partial}{\partial x} - p_0 \right)^2 \psi\, dx$$

For our Gaussian packet we have

$$\mathrm{var}(p) = \sigma_p^2$$

again independent of time because momentum is conserved.

The square root of the variance of the position,

$$\Delta x = \sqrt{\mathrm{var}(x)} = \sigma_x$$

determines the width of the wave packet in the position variable x and therefore is a measure of the *uncertainty* about where the particle is located. By the same token, the corresponding uncertainty about the momentum of the particle is

$$\Delta p = \sqrt{\mathrm{var}(p)} = \sigma_p$$

For our Gaussian wave packet we found the relation

$$\sigma_x = \frac{\hbar}{2\sigma_p} \left(1 + \frac{4\sigma_p^4}{\hbar^2} \frac{t^2}{m^2} \right)^{1/2}$$

For time $t = 0$ this relation reads

$$\sigma_x \sigma_p = \frac{\hbar}{2}$$

For later moments in time, the product becomes even larger so that, in general,

$$\Delta x \cdot \Delta p \geq \frac{\hbar}{2}$$

This relation expresses the fact that the product of uncertainties in position and momentum cannot be smaller than the fundamental Planck's constant h divided by 4π.

We have just stated the *uncertainty principle*, which is valid for wave packets of all forms. It was formulated by Werner Heisenberg in 1927. This relation says, in effect, that a small uncertainty in localization can only be achieved at the

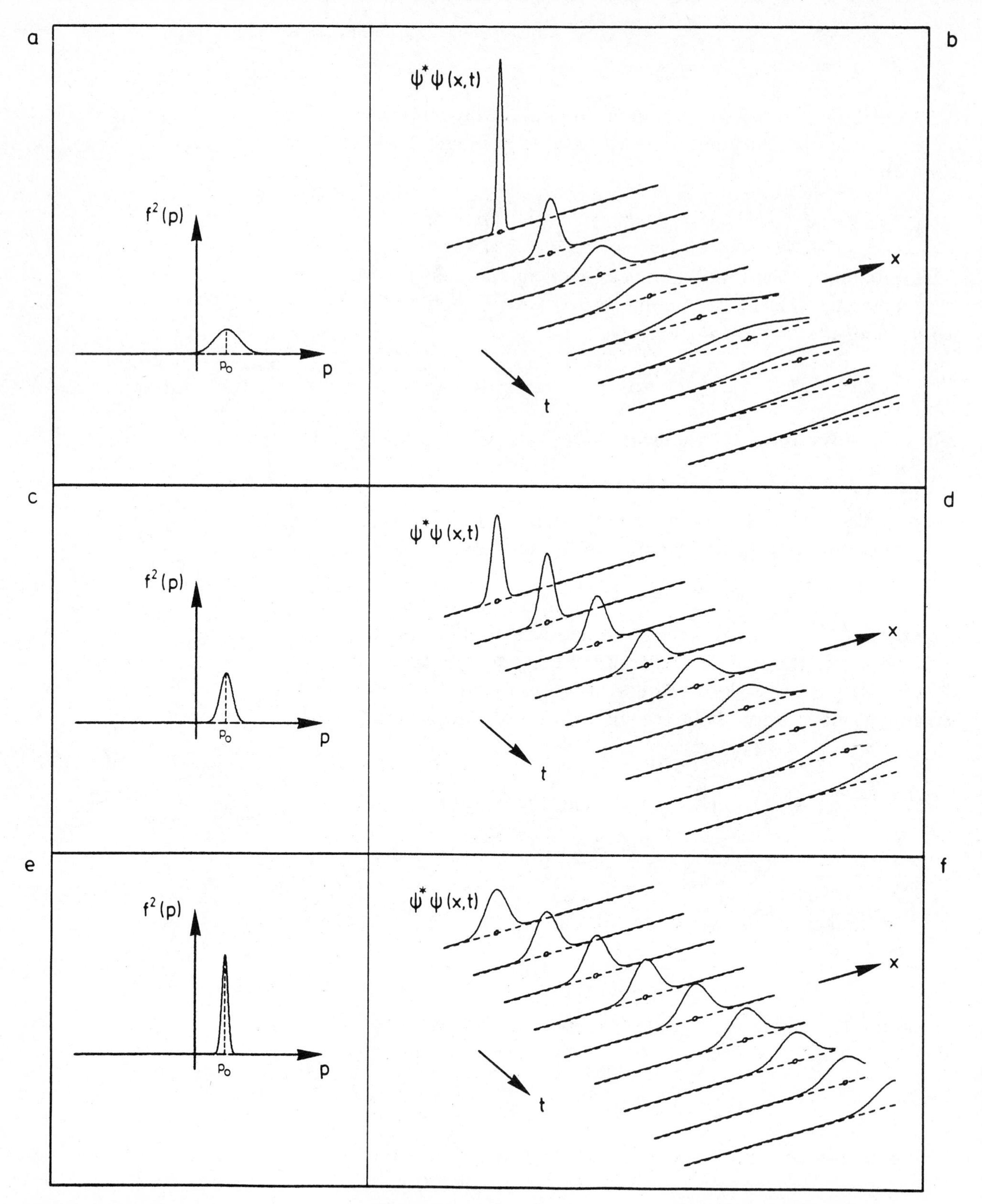

Figure 3.5 **Heisenberg's uncertainty principle. For three different Gaussian wave packets the square $f^2(p)$ of the spectral function is shown on the left, the time development of the probability density in space on the right. All three packets have the same group velocity but different widths σ_p in momentum. At $t = 0$ the widths σ_x in space and σ_p in momentum fulfill the equality $\sigma_x\sigma_p = \hbar / 2$. For later moments in time the wave packets spread in space so that $\sigma_x\sigma_p > \hbar / 2$.**

expense of a large uncertainty in momentum and vice versa. Figure 3.5 illustrates this principle by comparing the time development of the probability density $\rho(x,t)$ and the square of the spectral function $f^2(p)$. The latter, in fact, is the probability density in momentum. Looking at the spreading of the wave packets with time, we see that the initially narrow wave packet (Figure 3.5b) becomes quickly wide in space, whereas the initially wide wave packet (Figure 3.5f) spreads much more slowly. Actually, this behavior is to be expected. The spatially narrow wave packet requires a wide spectral function in momentum space. Thus it comprises components with a wide range of velocities. They, in turn, cause a quick dispersion of the packet in space compared to the initially wider packet with a narrower spectral function (Figures 3.5e and f).

At its initial time $t = 0$ the Gaussian wave packet discussed at the beginning of Section 3.2 has the smallest spread in space and momentum because Heisenberg's uncertainty principle is fulfilled in the equality form $\sigma_x \cdot \sigma_p = \hbar/2$. The wave function at $t = 0$ takes the simple form

$$\psi(x,0) = \frac{1}{(2\pi)^{1/4}\sqrt{\sigma_x}} \exp\left[-\frac{(x-x_0)^2}{4\sigma_x^2}\right] \exp\left[\frac{i}{\hbar} p_0 (x - x_0)\right]$$

$$= M(x,0)\exp[i\phi(x,0)]$$

The bell-shaped amplitude function $M(x,0)$ is centered around the position x_0 with the width σ_x; ϕ is the phase of the wave function at $t = 0$ and has the simple linear dependence

$$\phi(x,0) = \frac{1}{\hbar} p_0 (x - x_0)$$

This phase ensures that the wave packet at $t = 0$ stands for a particle with an average momentum p_0. We shall use this observation when we have to prepare wave functions for the initial state of a particle with the initial conditions $\langle x \rangle = x_0$, $\langle p \rangle = p_0$ at the initial moment of time $t = t_0$.

3.4 The Schrödinger Equation

Now that we have introduced the wave description of particle mechanics, we look for a *wave equation*, the solutions of which are the de Broglie waves. Starting from the harmonic wave

$$\psi_p(x,t) = \frac{1}{(2\pi\hbar)^{1/2}} \exp\left[-\frac{i}{\hbar}(Et - px)\right], \qquad E = \frac{p^2}{2m}$$

we compare the two expressions

$$i\hbar \frac{\partial}{\partial t} \psi_p(x,t) = E\psi_p(x,t)$$

and

$$-\frac{\hbar^2}{2m}\frac{\partial^2}{\partial x^2}\psi_p(x,t) = \frac{p^2}{2m}\psi_p(x,t) = E\psi_p(x,t)$$

Equating the two left-hand sides, we obtain the *Schrödinger equation* for a free particle,

$$i\hbar \frac{\partial}{\partial t} \psi_p(x,t) = -\frac{\hbar^2}{2m}\frac{\partial^2}{\partial x^2}\psi_p(x,t)$$

It was formulated by Erwin Schrödinger in 1925.

Since the solution ψ_p occurs linearly in this equation, an arbitrary linear superposition of solutions, that is, any wave packet, is also a solution of Schrödinger's equation. Thus this Schrödinger equation is the *equation of motion* for any free particle represented by an arbitrary wave packet $\psi(x,t)$:

$$i\hbar \frac{\partial}{\partial t} \psi(x,t) = -\frac{\hbar^2}{2m}\frac{\partial^2}{\partial x^2}\psi(x,t)$$

In the spirit of representing physical quantities by differential operators, as we did for momentum, we can now represent kinetic energy T, which is equal to the total energy of the free particle $T = p^2/(2m)$, by

$$T \to \frac{1}{2m}\left(\frac{\hbar}{i}\frac{\partial}{\partial x}\right)\left(\frac{\hbar}{i}\frac{\partial}{\partial x}\right) = -\frac{\hbar^2}{2m}\frac{\partial^2}{\partial x^2}$$

The equation can be generalized to describe the motion of a particle in a force field represented by a potential energy $V(x)$. This is done by replacing kinetic energy T with the total energy,

$$E = T + V \to -\frac{\hbar^2}{2m}\frac{\partial^2}{\partial x^2} + V(x)$$

With this substitution we obtain the *Schrödinger equation for the motion of a particle in a potential* $V(x)$:

$$i\hbar \frac{\partial}{\partial t} \psi(x,t) = -\frac{\hbar^2}{2m}\frac{\partial^2}{\partial x^2}\psi(x,t) + V(x)\psi(x,t)$$

We now denote the operator of total energy by the symbol

$$H = -\frac{\hbar^2}{2m}\frac{\partial^2}{\partial x^2} + V(x)$$

In analogy to the Hamilton function of classical mechanics, operator H is called the *Hamilton operator* or *Hamiltonian*. With its help the Schrödinger equation for the motion of a particle under the influence of a potential takes the form

$$i\hbar\frac{\partial}{\partial t}\psi(x,t) = H\psi(x,t)$$

At this stage we should point out that the Schrödinger equation, generalized to three spatial dimensions and many particles, is the fundamental law of nature for all of nonrelativistic particle physics and chemistry. The rest of this book will be dedicated to the pictorial study of the simple phenomena described by the Schrödinger equation.

Problems

3.1 Calculate the de Broglie wavelengths and frequencies of an electron and a proton that have been accelerated by an electric field through a potential difference of 100 V. What are the corresponding group and phase velocities?

3.2 An electron represented by a Gaussian wave packet with average energy $E_0 = 100$ eV was initially prepared to have momentum width $\sigma_p = 0.1p_0$ and position width $\sigma_x = \hbar/(2\sigma_p)$. How much time elapses before the wave packet has spread to twice the original spatial extension?

3.3 Show that the normalization condition $\int|\psi(x,t)|^2\,dx = 1$ holds true for any time if $\psi(x,t)$ is a Gaussian wave packet with a normalized spectral function $f(p)$.

3.4 Calculate the action of the commutator $[p,x] = px - xp$, $p = (\hbar/i)(\partial/\partial x)$ on a wave function $\psi(x,t)$. Show that it is equivalent to the multiplication of $\psi(x,t)$ by $\hbar/i$ so that we may write $[p,x] = \hbar/i$.

3.5 Express the expectation value of the kinetic energy of a Gaussian wave packet in terms of the expectation value of the momentum and the width σ_p of the spectral function.

3.6 Given a Gaussian wave packet of energy expectation value $\langle E \rangle$ and momentum expectation value $\langle p \rangle$, write its normalized spectral function $f(p)$.

3.7 A large virus may for purposes of this problem be approximated by a cube whose sides measure one micron and which has the density of water. Assuming as an upper estimate an uncertainty of one micron in position, calculate the uncertainty in velocity of the virus.

3.8 The radius of both the proton and the neutron is measured to be of the order of 10^{-15} m. A free neutron decays spontaneously into a proton, an electron, and a neutrino. The momentum of the emitted electron is typically 1 MeV/c. If the neutron were, as once thought, a bound system consisting of a proton and an electron, how large would be the position uncertainty of the electron and hence the size of the neutron? Take as the momentum uncertainty of the electron the value 1 MeV/c.

3.9 Show that the solutions of the Schrödinger equation satisfy the continuity equation,

$$\frac{\partial \rho(x,t)}{\partial t} + \frac{\partial j(x,t)}{\partial x} = 0$$

for the probability density

$$\rho(x,t) = \psi^*(x,t)\psi(x,t)$$

and the probability current density

$$j(x,t) = \frac{\hbar}{2mi}\left[\psi^* \frac{\partial}{\partial x}\psi - \psi \frac{\partial}{\partial x}\psi^*\right]$$

To this end, multiply the Schrödinger equation by $\psi^*(x,t)$ and its complex conjugate

$$i\hbar \frac{\partial \psi^*(x,t)}{\partial t} = \frac{\hbar^2}{2m}\frac{\partial^2}{\partial x^2}\psi^* - V(x)\psi^*$$

by $\psi(x,t)$, and add the two resulting equations.

3.10 Convince yourself with the help of the continuity equation that the normalization integral

$$\int_{-\infty}^{+\infty} \psi^*(x,t)\,\psi(x,t)\,dx$$

is independent of time if $\psi(x,t)$ is a normalized solution of the Schrödinger equation. To this end, integrate the continuity equation over all x and use the vanishing of the wave function for large $|x|$ to show the vanishing of the integral over the probability current density.

3.11 Calculate the probability current density for the free Gaussian wave packet as given at the end of Section 3.2. Interpret the result for $t = 0$ in terms of the probability density and the group velocity of the packet.

3.12 Show that the one-dimensional Schrödinger equation possesses spatial reflection symmetry, that is, is invariant under the substitution $x \to -x$ if the potential is an even function, that is, $V(x) = V(-x)$.

3.13 Show that the *ansatz* for the Gaussian wave packet of Section 3.2 fulfills the Schrödinger equation for a free particle.

4. Solution of the Schrödinger Equation in One Dimension

4.1 Separation of Time and Space Coordinates, Stationary Solutions

The simple structure of the Schrödinger equation allows a particular *ansatz* in which the time and space dependences occur in separate factors,

$$\psi_E(x,t) = \exp\left(-\frac{i}{\hbar}Et\right)\varphi_E(x)$$

As in the case of electromagnetic waves, we call the factor $\varphi(x)$ that is independent of time a *stationary solution*. Inserting our *ansatz* into the Schrödinger equation yields an equation for the stationary wave,

$$-\frac{\hbar^2}{2m}\frac{d^2}{dx^2}\varphi_E(x) + V(x)\varphi_E(x) = E\varphi_E(x)$$

which is often called the *time-independent Schrödinger equation*. It is characterized by the parameter E, which is called an *eigenvalue*. The left-hand side represents the sum of the kinetic and the potential energy, so that E is the total energy of the stationary solution. The solution $\varphi_E(x)$ is called an *eigenfunction* of the Hamilton operator

$$H = -\frac{\hbar^2}{2m}\frac{d^2}{dx^2} + V(x)$$

since the time-independent Schrödinger equation can be put into the form

$$H\varphi_E(x) = E\varphi_E(x)$$

We also say that the solution $\varphi_E(x)$ describes an *eigenstate* of the system specified by the Hamilton operator. This eigenstate is characterized by the eigenvalue E of the total energy. Often the stationary solution $\varphi_E(x)$ is also called a *stationary state* of the system.

The time-independent Schrödinger equation has a large manifold of solutions. It is supplemented by *boundary conditions* that have to be imposed on a particular solution. These boundary conditions must be abstracted from the physical process that the solution should describe. The boundary conditions on the solution for the elastic scattering in one dimension of a particle under the action of a force will be discussed in the next section. Because of the boundary conditions, solutions $\varphi_E(x)$ exist for particular values of the energy eigenvalues or for particular energy intervals only.

As a first example, we look at the de Broglie waves,

$$\psi_p(x - x_0, t) = \frac{1}{(2\pi\hbar)^{1/2}} \exp\left[-\frac{i}{\hbar}(Et - px + px_0)\right]$$

The function $\psi_p(x - x_0, t)$ factors into $\exp[-(i/\hbar)Et]$ and the stationary wave

$$\frac{1}{(2\pi\hbar)^{1/2}} \exp\left[\frac{i}{\hbar}p(x - x_0)\right]$$

It is a solution of the time-independent Schrödinger equation with a vanishing potential for the energy eigenvalue $E = p^2/2m$. A superposition of de Broglie waves fulfilling the normalization condition of Section 3.3 forms a wave packet describing an unaccelerated particle. Here x_0 is the position expectation value of the wave packet at time $t = 0$.

Since the momentum p is a real parameter, the energy eigenvalue of a de Broglie wave is always positive. Thus, for the case of de Broglie waves, we have found the restriction $E \geq 0$ for the energy eigenvalues.

The general solution of the time-dependent Schrödinger equation is given by a linear combination of waves of different energies. This is tantamount to stating that the various components of different energy E superimposed in the solution change independently of one another with time.

For initial time $t = 0$ the functions ψ_E and φ_E coincide. An initial condition prescribed at $t = 0$ determines the coefficients in the linear combination of spectral components of different energies. Therefore the procedure for solving the equation for a given initial condition has three steps. First, we determine the stationary solutions $\varphi_E(x)$ of the time-independent Schrödinger equation. Second, we superimpose them with appropriate coefficients to reproduce the initial condition $\psi(x, 0)$ at $t = 0$. Finally, we introduce into every term of this linear combination the time-dependent factor $\exp[-(i/\hbar)Et]$ corresponding to the energy of the stationary solution φ_E and sum them up to give $\psi(x, t)$, the solution of the time-dependent Schrödinger equation.

In the next section we study methods of obtaining the stationary solutions.

4.2 Stationary Scattering Solutions, Continuous Energy Spectrum

As in classical mechanics, the scattering of a particle by a force is called *elastic* if only its momentum is changed while its energy is conserved. A force is said to be of *finite range* if it is practically zero for distances from the center of force larger than a finite distance d. This distance d is called the *range* of the force. The elastic scattering of a particle through a force of finite range consists of three stages subsequent in time.

1. The incoming particle moves unaccelerated in a force-free region toward the range of the force.
2. The particle moves under the influence of the force. The action of the force changes the momentum of the particle.
3. After the scattering the outgoing particle moves away from the range of the force. Its motion in the force-free region is again unaccelerated.

In Section 3.3 we have seen that the force-free motion of a particle of mass m can be described by a wave packet of de Broglie waves,

$$\psi_p(x - x_0, t) = \frac{1}{(2\pi\hbar)^{1/2}} \exp\left[-\frac{i}{\hbar}(Et - px + px_0)\right],$$

$$E = \frac{p^2}{2m}$$

They can be factored into the time-dependent factor $\exp[-(i/\hbar)Et]$ and the stationary wave $(2\pi\hbar)^{-1/2}\exp[(i/\hbar)p(x - x_0)]$. This stationary wave is a solution of the time-independent Schrödinger equation with a vanishing potential.

If the spectral function $f(p)$ of the wave packet has values different from zero in a range of positive p values, the wave packet

$$\psi(x,t) = \int_{-\infty}^{+\infty} f(p)\psi_p(x - x_0, t)\, dx$$

$$= \int_{-\infty}^{+\infty} f(p) \exp\left(-\frac{i}{\hbar}Et\right) \frac{1}{(2\pi\hbar)^{1/2}} \exp\left[\frac{i}{\hbar}p(x - x_0)\right] dp$$

moves along the x-axis from left to right, that is, in the direction of increasing x-values.

Now we superimpose de Broglie waves of momentum $-p$,

$$\psi_{-p}(x - x_0, t) = \frac{1}{(2\pi\hbar)^{1/2}} \exp\left[-\frac{i}{\hbar}(Et + px - px_0)\right],$$

$$E = \frac{p^2}{2m}$$

with the same spectral function $f(p)$. A simple change of variables $p' = -p$ yields

$$\psi_-(x,t)$$

$$= \int_{-\infty}^{+\infty} f(-p') \exp\left(-\frac{i}{\hbar}Et\right) \frac{1}{(2\pi\hbar)^{1/2}} \exp\left[\frac{i}{\hbar}p'(x - x_0)\right] dp'$$

$$= \int_{-\infty}^{+\infty} f(-p)\psi_p(x - x_0, t)\, dp$$

We obtain a wave packet with a spectral function $f(-p)$ having its range of values different from zero at negative values of p. The wave packet $\psi_-(x,t)$ moves along the x-axis from right to left, that is, in the direction of decreasing x-values. Thus we learn that for a given spectral function, wave packets formed with $\psi_p(x-x_0,t)$ and $\psi_{-p}(x-x_0,t)$ move in opposite directions. This says that the sign of the exponent of the stationary wave $(2\pi\hbar)^{-1/2}\exp[\pm(i/\hbar)p(x-x_0)]$ decides the direction of motion. For a spectral function $f(p)$ different from zero at positive values of momentum p, a wave packet formed with the stationary wave

$$\exp\left[\frac{i}{\hbar}p(x-x_0)\right]=\exp[ik(x-x_0)]$$

moves in the direction of increasing x. A wave packet formed with the stationary wave

$$\exp\left[-\frac{i}{\hbar}p(x-x_0)\right]=\exp[-ik(x-x_0)]$$

moves in the direction of decreasing x.

Let us consider a particle moving from the left in the direction of increasing x. The force

$$F(x)=-\frac{d}{dx}V(x)$$

derived from the potential energy $V(x)$ has finite range d. This range is assumed to be near the origin $x=0$. The initial position x_0 of the wave packet is assumed to be far to the left from the origin at large negative values of the coordinate. As long as the particle is far to the left of the origin, the particle moves unaccelerated. In this region the solution is a wave packet of de Broglie waves $\psi_p(x-x_0,t)$. Thus the stationary solution $\varphi_E(x)$ of the time-independent Schrödinger equation for the eigenvalue E should contain a term approaching the function $\exp[(i/\hbar)p(x-x_0)]$ for negative x of large absolute value.

Through the scattering process in one dimension, the particle can only be transmitted or reflected. The transmitted particle will move force-free at large positive x. Here it will be represented by a wave packet of de Broglie waves of the form $\psi_{p'}(x-x_0,t)$. Therefore the solution of the stationary

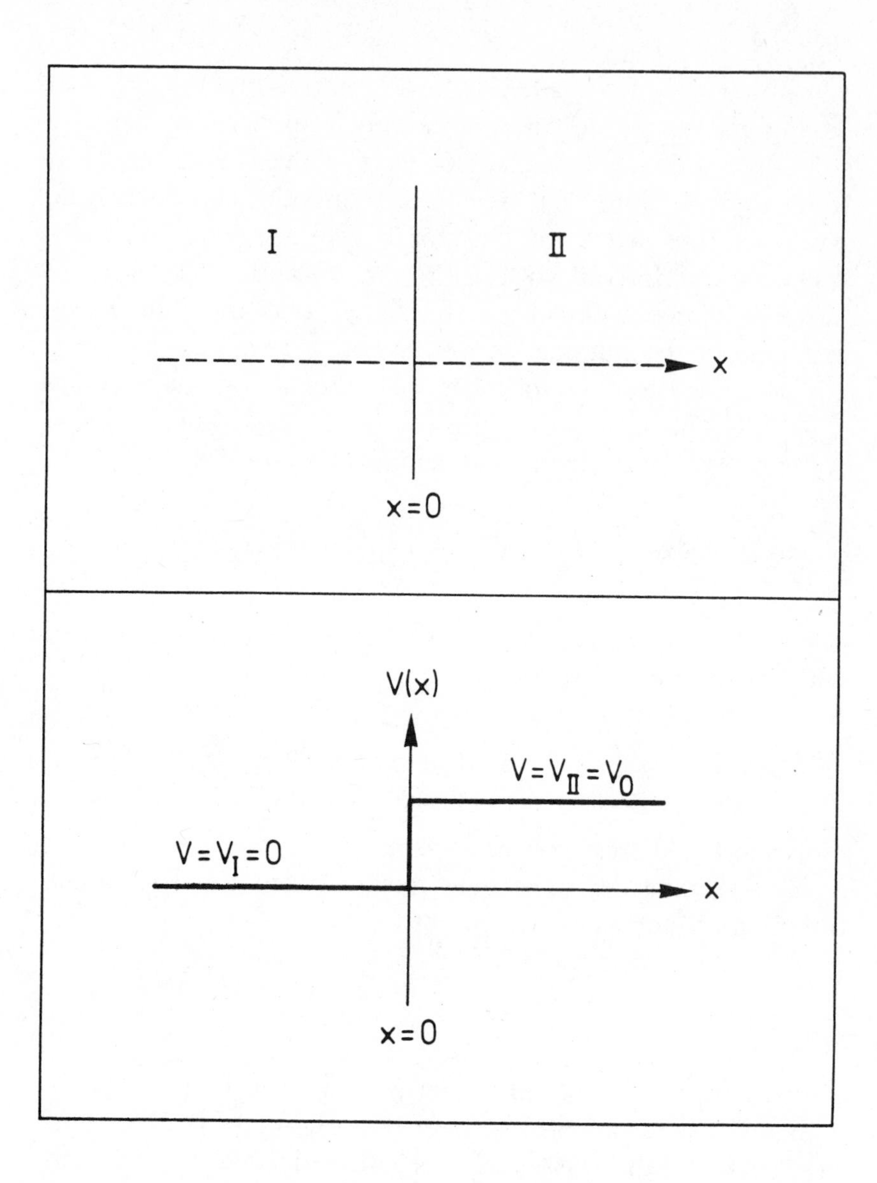

Figure 4.1 **Space is divided into region I, $x < 0$, and region II, $x > 0$. There is a constant potential in region II, $V = V_0$, whereas in region I there is no potential, $V = 0$.**

Schrödinger equation must approach the function $\exp[(i/\hbar)p'(x - x_0)]$ for large positive x. The value p' differs from p if the potential $V(x)$ assumes different values for large negative and large positive x. The reflected particle has momentum $-p$ and will leave the range of the potential to the left and thus return to the region of large negative x. In this region it will be represented by a wave packet of de Broglie waves $\psi_{-p}(x - x_0, t)$. Therefore the solution of the time-independent Schrödinger equation must also contain a contribution tending toward a function $\exp[-(i/\hbar)p(x - x_0)]$ for large negative x. The conditions for large positive and negative x just derived constitute the boundary condition that the stationary solution $\varphi_E(x)$, $E = p^2/2m$ must fulfill if its superpositions forming wave packets are to describe an elastic scattering process. We summarize the boundary conditions for stationary scattering solutions of the time-independent Schrödinger equation in the following statement:

$$\varphi_E(x) \xrightarrow[\text{approaches}]{} \begin{cases} \exp\left[\frac{i}{\hbar}p(x-x_0)\right] + B\exp\left[-\frac{i}{\hbar}p(x-x_0)\right] \\ \text{for large negative } x \\ A\exp\left[\frac{i}{\hbar}p'(x-x_0)\right] \\ \text{for large positive } x \end{cases}$$

Since there are no general methods for solving in closed form the Schrödinger equation for an arbitrary potential, we choose for our discussion particularly simple examples. We begin with a *potential step* of height $V = V_0$ at $x = 0$. The potential divides the space into two regions. In region I, that is, to the left of $x = 0$, the potential vanishes. To the right, region II, it has the constant value $V = V_0$ (Figure 4.1).

The time-independent Schrödinger equation has the form

$$-\frac{\hbar^2}{2m}\frac{d^2}{dx^2}\varphi + V_i\varphi = E\varphi$$

in both regions, with V_i assuming different but constant values in the two regions, $V_{\mathrm{I}} = 0$, $V_{\mathrm{II}} = V_0$. Thus the stationary solution for given energy E of the incoming wave is

$$\varphi_{\mathrm{I}} = \exp\left[\frac{i}{\hbar}p(x-x_0)\right] + B_{\mathrm{I}}\exp\left[-\frac{i}{\hbar}p(x-x_0)\right], \qquad x < 0$$

$$\varphi_{\mathrm{II}} = A_{\mathrm{II}}\exp\left[\frac{i}{\hbar}p'(x-x_0)\right], \qquad x > 0$$

Obviously, this solution fulfills the boundary conditions we have posed earlier in this section.

In region I the momentum is $p = \sqrt{2mE}$, in region II $p' = \sqrt{2m(E - V_0)}$. Since the potential is discontinuous at $x = 0$, the second derivative of φ has to reproduce the same discontinuity, reduced by the factor $-\hbar^2/(2m)$. Thus φ and $d\varphi/dx$ are continuous at $x = 0$. These conditions determine the complex coefficients B_{I} and A_{II}, which are as yet unknown. The coefficient of the incoming wave has been chosen equal to one, thus fixing the incoming amplitude. The phase of the wave function depends on the initial-position parameter x_0.

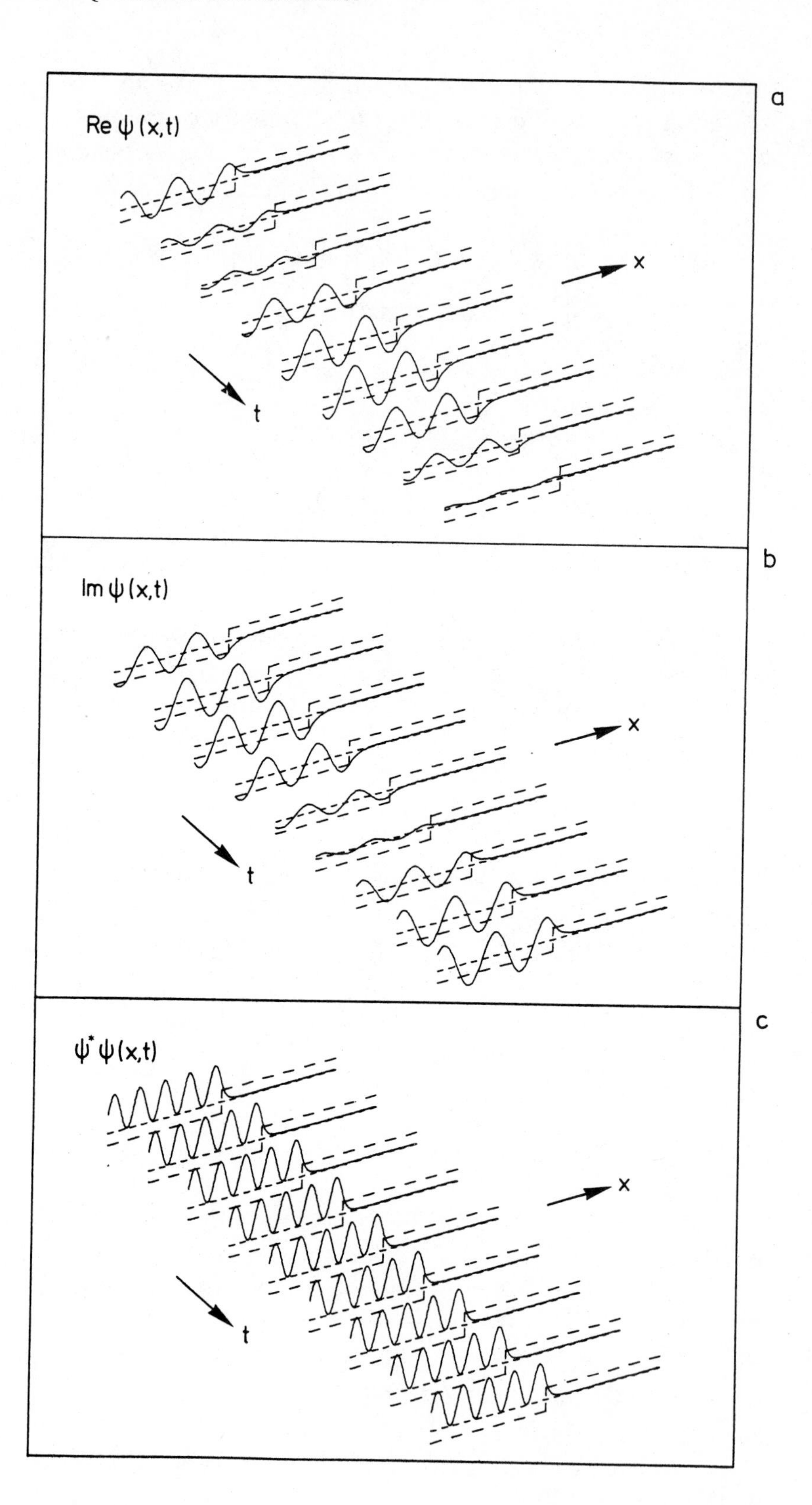

Figure 4.2 **Time developments of (a) the real part, (b) the imaginary part, and (c) the intensity of a harmonic wave of energy $E < V_0$ falling onto a potential step of height V_0. The form of the potential $V(x)$ is indicated by the line made up of long dashes, the energy of the wave by the short-dash horizontal line, which also serves as zero line for the functions plotted. To the left of the potential step is a standing wave pattern, as is apparent from the time-independent position of the nodes or zeroes of the functions Re $\psi(x,t)$ and Im $\psi(x,t)$. The absolute square $\psi^*\psi(x,t)$ is time-independent.**

As for light waves (Section 2.2), we denote the three members on the right-hand sides of the two expressions φ_{I} and φ_{II} as constituent waves. That is, we call

$$\varphi_{1+} = \exp[(i/\hbar)p(x - x_0)]$$

the incoming constituent wave

$$\varphi_{1-} = B_{\mathrm{I}} \exp[(-i/\hbar)p(x - x_0)]$$

the reflected constituent wave

$$\varphi_2 = A_{\mathrm{II}} \exp[(i/\hbar)p'(x - x_0)]$$

the transmitted constituent wave

As a first example, we choose a repulsive step, $V_0 > 0$, and an incoming wave of energy $E < V_0$, so that in classical mechanics the particle would be reflected by the potential step. The momentum of the transmitted wave in region II,

$$p' = \sqrt{2m(E - V_0)} = i\sqrt{2m(V_0 - E)}$$

is now imaginary so that the transmitted wave

$$\begin{aligned}\varphi_{\mathrm{II}} &= A_{\mathrm{II}} \exp\left[\frac{i}{\hbar} p'(x - x_0)\right] \\ &= A_{\mathrm{II}} \exp\left[-\frac{1}{\hbar}\sqrt{2m(V_0 - E)}\,(x - x_0)\right]\end{aligned}$$

becomes a real exponential function which falls off with increasing x in region II. We obtain the full solution of the time-dependent Schrödinger equation for a given energy by multiplying the stationary wave by the factor $\exp(-iEt/\hbar)$.

Figures 4.2a and b show the time developments of the real and imaginary parts of the wave function with fixed energy E. The real and imaginary parts behave in region I like standing waves, for they are superpositions of an incoming and a reflected wave of equal frequency and equal amplitude. We are easily convinced of this fact by looking at Figure 4.3, in which the time developments of the incoming and reflected constituent waves in region I are plotted separately. In Figure 4.2, region II, both the real and the imaginary parts are

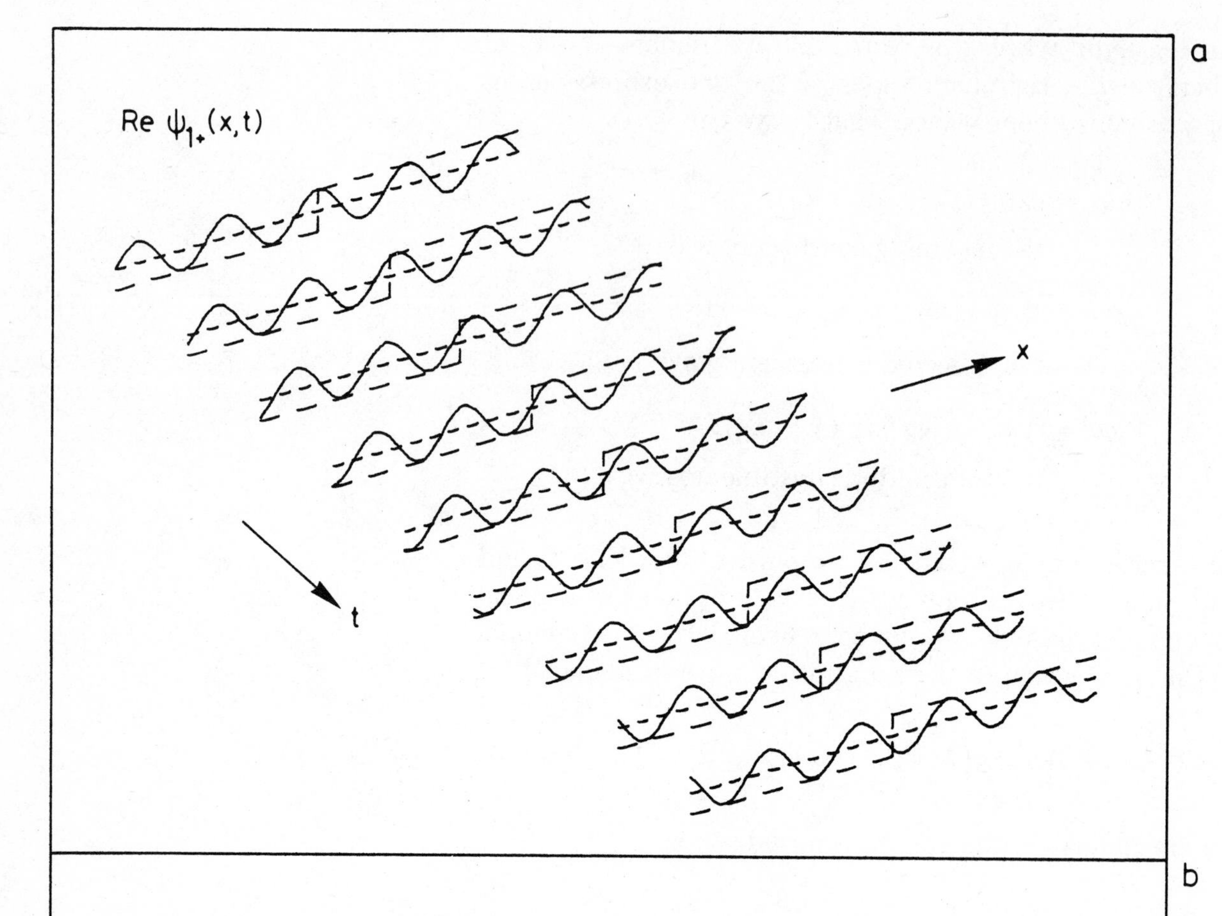
a
Re $\psi_{1+}(x,t)$
x
t

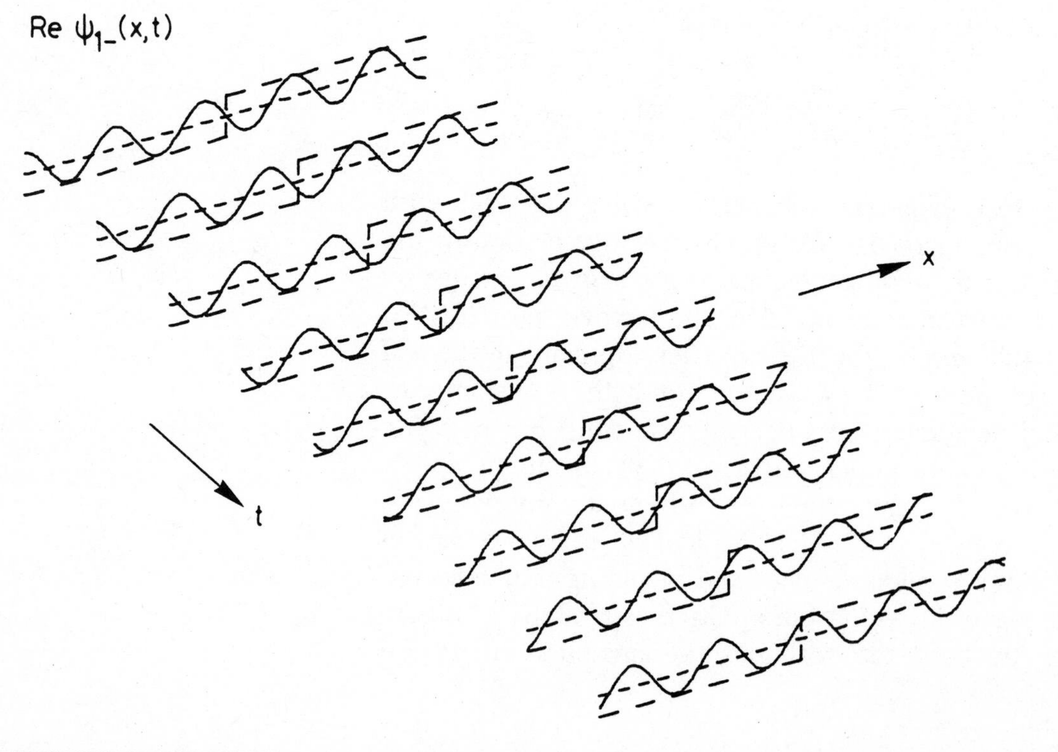
b
Re $\psi_{1-}(x,t)$
x
t

represented by exponentials oscillating in time. The time development of the absolute square of the wave function, which we shall call intensity (Figure 4.2c), shows no variation at all in time. In region I it is periodic in space, but in region II it shows an exponential falloff.

We now examine an incoming wave of energy $E > V_0$. Obviously, momentum $p' = \sqrt{2m(E - V_0)}$ in region II for $E > V_0$ is real. Therefore the stationary solution in this region, as in region I, is an oscillating function in space.

Figure 4.4a shows the energy dependence of the real parts of the stationary solutions. It includes both energies $E > V_0$ and energies $E < V_0$. For energies $E > V_0$ the wavelength in region II is longer than that in region I. For energies $E < V_0$ the stationary wave function has the exponential falloff just mentioned. The energy dependence of the intensity is given in Figure 4.4b. For $E > V_0$ the intensity is constant in region II, corresponding to the outgoing wave in this region. The periodic structure of the intensity in region I results from the superposition of the incoming and reflected waves.

For $V_0 < 0$ there is for all energies an oscillating transmitted wave in region II. Figure 4.5 shows the energy dependence of the real part and of the absolute square of the wave function. Since the potential is now attractive, the wavelength of the transmitted wave is decreased in region II.

Since for every energy $E > 0$ there is a stationary solution of the time-independent Schrödinger equation for a potential step, we say that the physical system has a *continuous energy spectrum*. For some types of potential, the Schrödinger equation has solutions only for certain particular values of energy. They form a *discrete energy spectrum*. The most general physical system has an energy spectrum composed of a discrete part and a continuous one (see Section 4.3).

We now turn to the example of a *potential barrier* of height $V = V_0$ between $x = 0$ and $x = d$. Outside this interval the potential vanishes. Here we have to study three

Figure 4.3 **Time developments of the real parts of (a) the incoming constituent wave and (b) the reflected constituent wave making up the harmonic wave of Figure 4.2.**

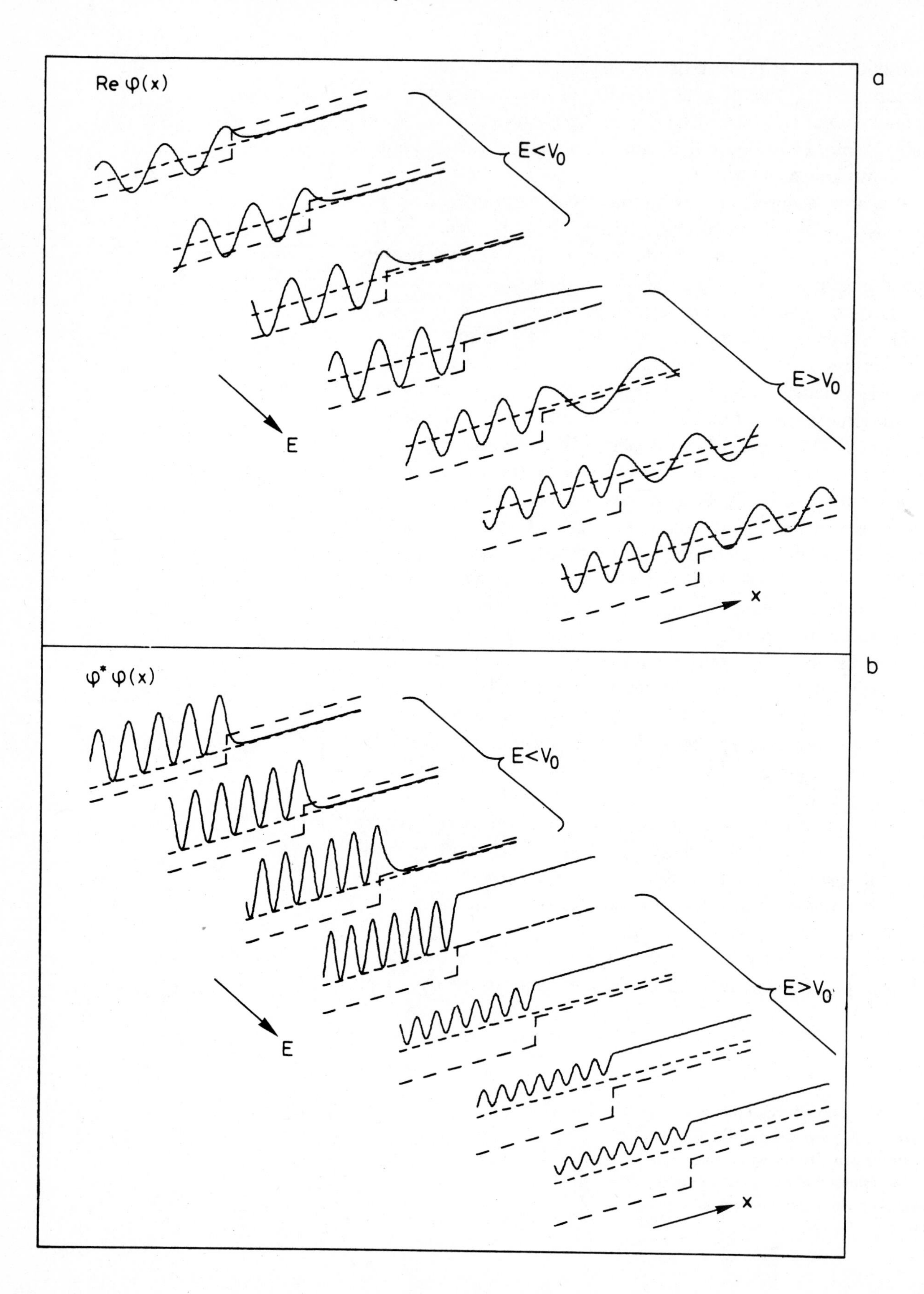

a
Re φ(x)
E<V0
E>V0
E
x
b
φ* φ(x)
E<V0
E>V0
E
x

different regions where the solution is given by

$$\varphi_{\mathrm{I}} = \exp\left[\frac{i}{\hbar}p(x - x_0)\right] + B_{\mathrm{I}} \exp\left[-\frac{i}{\hbar}p(x - x_0)\right], \qquad x < 0$$

$$\varphi_{\mathrm{II}} = A_{\mathrm{II}} \exp\left[\frac{i}{\hbar}p'(x - x_0)\right] + B_{\mathrm{II}} \exp\left[-\frac{i}{\hbar}p'(x - x_0)\right], \qquad 0 < x < d$$

$$\varphi_{\mathrm{III}} = A_{\mathrm{III}} \exp\left[\frac{i}{\hbar}p(x - x_0)\right], \qquad d < x$$

for a harmonic wave moving in from the left, that is, $x < 0$.

As before, the momentum in region II is $p' = \sqrt{2m(E - V_0)}$ and is real for $E > V_0$, imaginary for $E < V_0$. The complex coefficients A and B are again determined by continuity conditions for the wave function and its derivative $d\varphi/dx$ at the two boundaries of the barrier, $x = 0$ and $x = d$.

The energy dependence of the real part of the stationary solutions is presented in Figure 4.6a. Again transmission and reflection occur. The most striking feature, however, is the transmission of a wave into region III even for energies below the barrier height, $E < V_0$. The transmission of the wave corresponds to the penetration of a particle through the barrier. This remarkable quantum-mechanical phenomenon is called the *tunnel effect*. Within region II, of course, the wave does not have periodic structure since p' is imaginary, giving rise to a real exponential function.

A potential that is constant and negative in region II, that is, V_0 is less than zero for $0 < x < d$, and that vanishes in regions I and III is called a *square-well potential*. Here the waves keep their oscillating form in all three regions. Figure 4.6b shows the energy dependence of the real part of the

Figure 4.4 Energy dependence of stationary solutions for waves incident on a potential step of height $V_0 > 0$: (a) the real part of the wave function and (b) the intensity. Small energies are in the background.

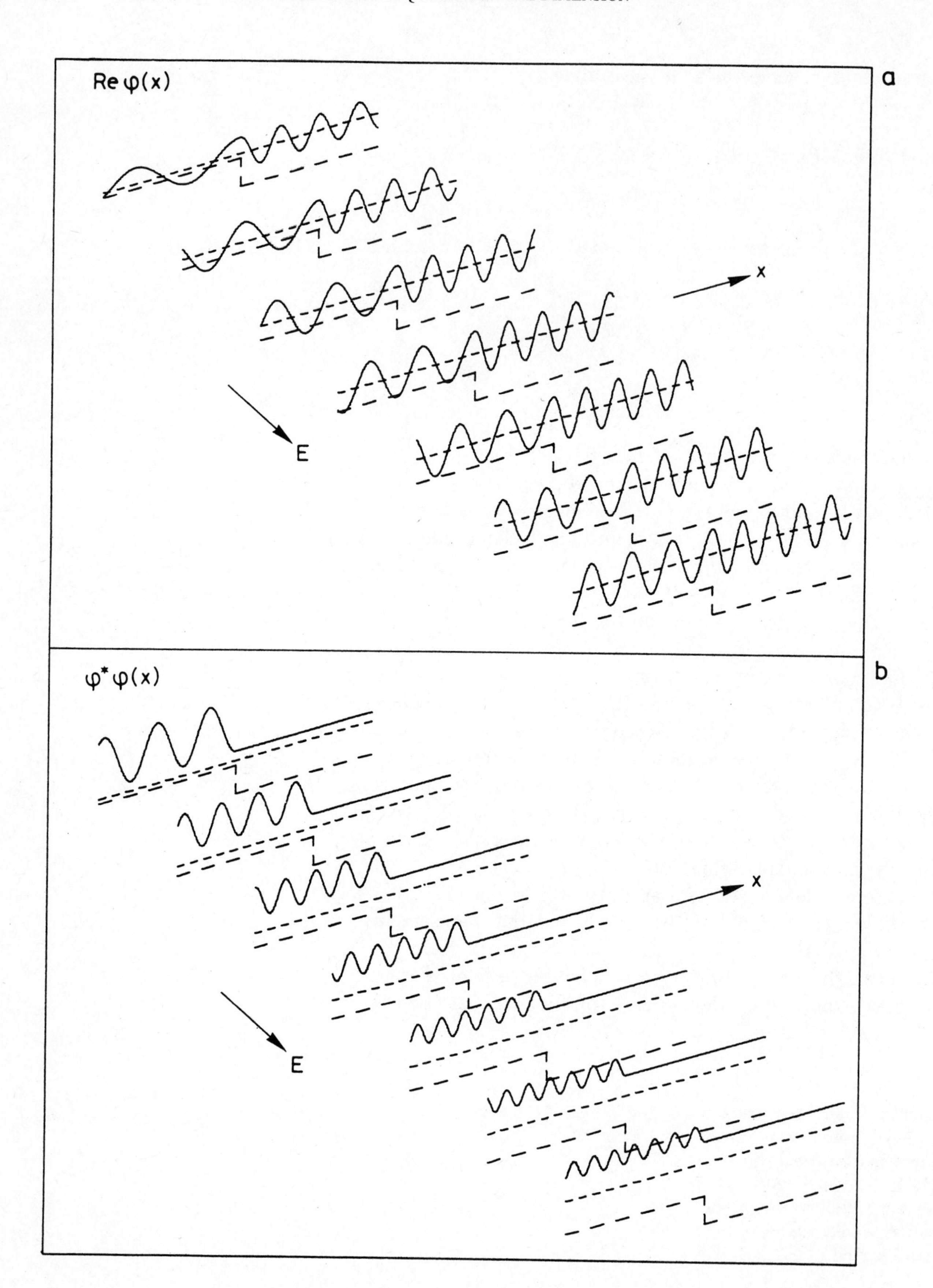
Re φ(x)
a
x
E
φ* φ(x)
b
x
E

stationary wave function. The wavelength is now decreased within the well, through the acceleration caused by the attractive potential. This effect is less obvious for the higher kinetic energies since the relative difference between the wave number k_{I} outside and $k_{\mathrm{II}} = \sqrt{2m(E - V_0)}/\hbar = \sqrt{2m(E + |V_0|)}/\hbar$ decreases with growing kinetic energy E.

4.3 Stationary Bound States

In classical mechanics the motion of a particle is also possible within the square-well potential at a negative total energy $E = E_{\mathrm{kin}} + V_0$, $E_{\mathrm{kin}} > 0$. Since the motion of the classical particle is restricted to region II, the quantum-mechanical analogue is described by stationary wave functions that fall off in regions I and III. Thus the stationary wave functions too are localized in the vicinity of the square well. Solutions with this property are called *bound states*. The exponential falloff in regions I and III guarantees the finiteness of the integral

$$\int_{-\infty}^{+\infty} |\varphi_{\mathrm{unnormalized}}(x)|^2 \, dx = N^2$$

in contrast to the stationary solutions of positive energy eigenvalues. Dividing the unnormalized solution by N yields the normalized stationary wave function $\varphi(x)$ fulfilling the normalization condition

$$\int |\varphi(x)|^2 \, dx = 1$$

which is analogous to the one for wave packets. The normalizability is a general feature of all bound-state wave functions. For negative total energy the Schrödinger equation admits as solutions real exponentials of the type $\exp(\pm ipx/\hbar)$ with

Figure 4.5 **Energy dependence of (a) the real part and (b) the intensity of stationary solutions for harmonic waves incident on a potential step of height $V_0 < 0$.**

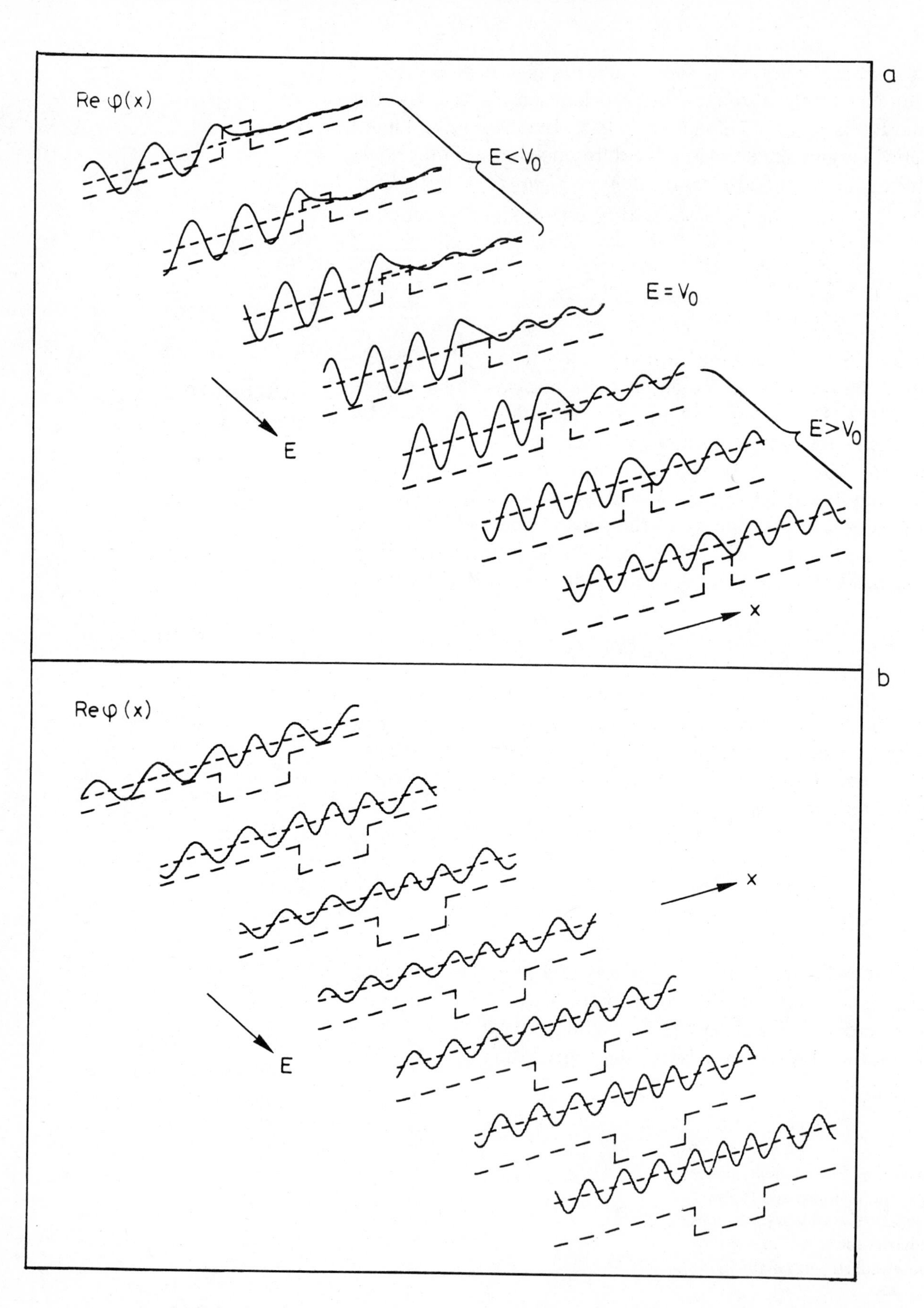
a
Re φ(x)
E<V0
E=V0
E>V0
E
x
b
Reφ (x)
x
E

$p = i\sqrt{2m|E|}$ in regions I and III. In order to guarantee the falloff of the exponential, only the negative sign is allowed in region I, only the positive sign in region III. In region II the solution is still oscillating since $p' = \sqrt{2m(E - V_0)}$ remains real. The continuity conditions at the boundaries of the potential must now be fulfilled with only one real exponential function in regions I and III. This is possible only for particular discrete values of the total energy. These values form the *discrete part* of the *energy spectrum*. The corresponding solutions can be chosen to be real. They are distinguished by the number of zeroes or *nodes* that they possess in region II.

The number of nodes increases as the energy of the bound states increases. This can be understood in the following way. For the ground state the wave function in region II is a cosine with half a wavelength slightly greater than the width of the square well. It is fitted into the square well in such a way that its slopes at the boundaries match those of the exponentials in regions I and III. The next bound state occurs at higher energy. As the energy increases, wavelength $\lambda = 2\pi\hbar[2m(E - V_0)]^{-1/2}$ in region II shortens. The slopes at the boundaries next match when approximately a full wavelength fits into the well, making the wave function a sine and thus giving rise to one node. As more and more wavelengths fit approximately into the well, more nodes appear.

Figure 4.7 shows the wave functions of the discrete energy spectrum and the corresponding probability densities. For a given width and depth of the square-well potential, there is only a finite number of bound states.

Figure 4.6 **Energy dependence of stationary solutions for waves incident onto (a) a positive potential barrier, $V_0 > 0$, and (b) a negative potential barrier, $V_0 < 0$, which is also called a square-well potential. The real part of the wave function is shown.**

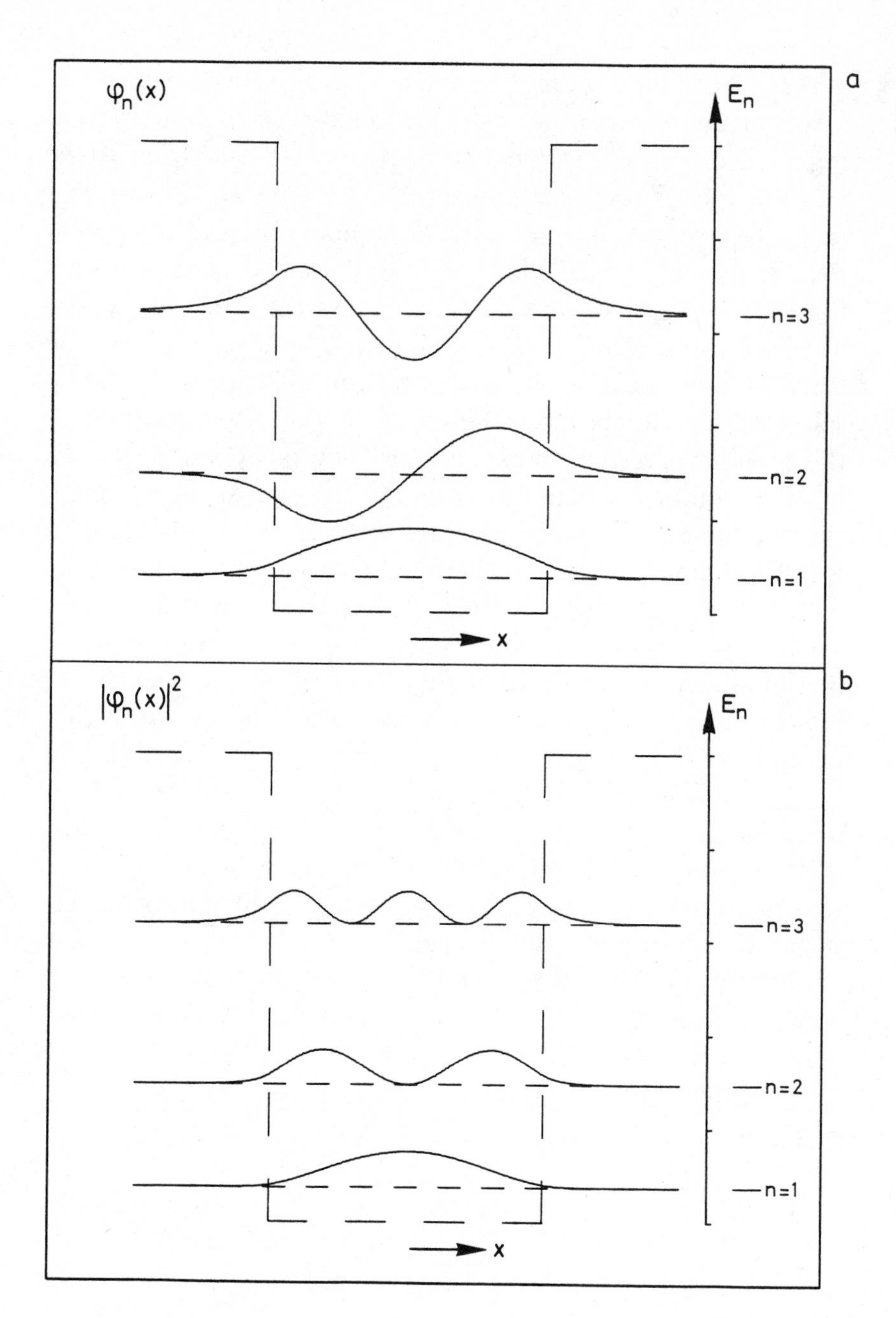

Figure 4.7 (a) Wave functions and (b) probability densities of the bound states in a square-well potential. On the right side of the picture an energy scale is shown with marks for the bound-state energies ($n = 1, 2, 3$). The form of the potential $V(x)$ is indicated by the long-dash line, the energy E_n of the bound states by the horizontal short-dash lines. The horizontal dashed lines also serve as zero lines for the functions shown.

Problems

4.1 Solve the stationary Schrödinger equation for energy E with a constant potential $V = V_0$.

4.2 Discuss the behavior of the solutions for energies $E > V_0$, $E < V_0$. Which solutions correspond to the particular energy $E = V_0$? These three cases play a role in the solution of the Schrödinger equation for stepwise constant potentials. Figures 4.4, 4.5, and 4.6 give examples.

4.3 Calculate the intensities $|\varphi|^2$ of the transmitted stationary wave and of the superposition of the incoming and reflected stationary waves φ_{I} of Section 4.2.

4.4 Calculate the probability current density

$$j(x) = \frac{\hbar}{2im}\left(\psi^* \frac{\partial}{\partial x}\psi - \psi\frac{\partial}{\partial x}\psi^*\right)$$

for the solution of the stationary Schrödinger equation. Consider a potential step of height V_0 as shown in Figure 4.1. Show that the current density is equal in the two spatial regions if the wave function and its derivative fulfill the continuity conditions at the boundaries between the different regions of the potential. Explain the result for $E \gtrless V_0$.

4.5 Show that the stationary bound-state wave functions can always be chosen to fulfill one of the following two relations,

$$\varphi(-x) = \varphi(x) \qquad \text{or} \qquad \varphi(-x) = -\varphi(x)$$

for an even potential $V(-x) = V(x)$. The function $\varphi(x)$ is said to have positive parity—also called natural or even parity—in the first relation and negative parity—unnatural or odd parity—in the second.

5. One-dimensional Quantum Mechanics: Scattering by a Potential

5.1 Sudden Acceleration and Deceleration of a Particle

We now study the motion of a wave packet incident on a potential step. As already discussed at the beginning of Section 4.2, the effect of the potential is the elastic scattering of the particle. In one-dimensional scattering the particle will be transmitted or reflected by the potential.

If we superimpose the stationary solutions of Section 4.2 with the spectral function that was used for the construction of the free wave packet in Section 3.2,

$$f(p) = \frac{1}{(2\pi)^{1/4}\sqrt{\sigma_p}} \exp\left[-\frac{(p-p_0)^2}{4\sigma_p^2}\right]$$

we obtain an initially Gaussian wave packet which is centered around $x = x_0$ for the values of x_0 that are far to the left of the potential step. Its time development is obtained by including the time-dependent factor $\exp(-iEt/\hbar)$, $E = p^2/2m$, in the superposition.

First, we discuss a repulsive potential, that is, a positive step, $V_0 > 0$, and a wave packet with $p_0 > \sqrt{2mV_0}$. Figure 5.1 presents the time developments of the real and imaginary parts of the wave function and of the probability density. Figure 5.1c also shows the position of a classical particle having the same momentum p_0 as the expectation value of the quantum-mechanical wave packet. Of course, the classical particle moves to the right in region I with velocity $v = p_0/m$. Entering region II, it is instantaneously decelerated to velocity

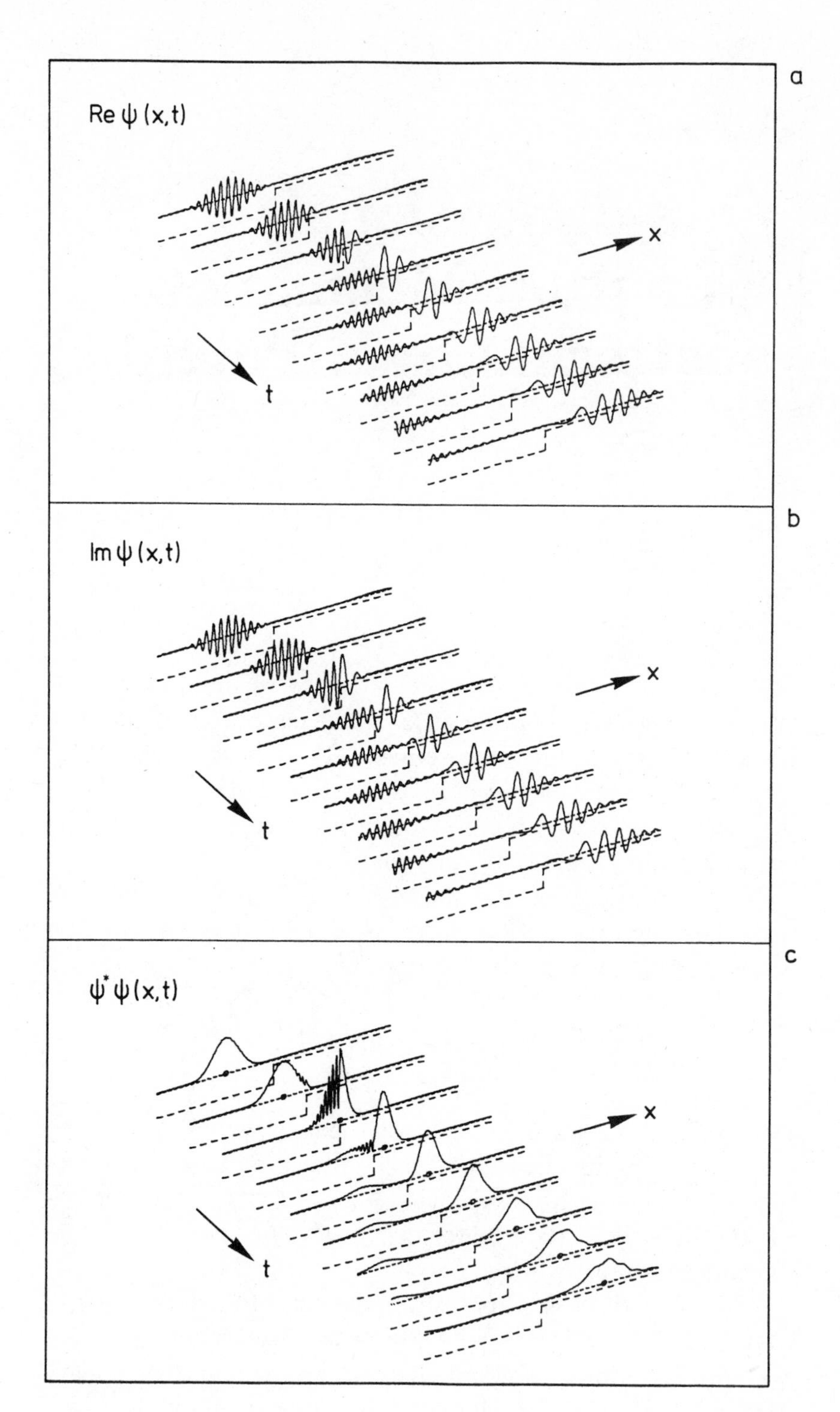

Figure 5.1 **Time developments of (a) the real part and (b) the imaginary part of the wave function and of (c) the probability density for a wave packet incident from the left on a potential step of height $V_0 > 0$. The form of the potential $V(x)$ is again indicated by the long-dash line, the expectation value of the energy of the wave packet by the short-dash line, which also serves as zero line for the functions plotted. The expectation value of the initial momentum is $p_0 > \sqrt{2mV_0}$. The small circles indicate the positions of a classical particle of the same initial momentum.**

$v' = p_0'/m = \sqrt{p_0^2 - 2mV_0}/m$. The most striking effect on the behavior of the wave packet is that it is partly reflected at the potential step. For large times we observe a wave packet moving to the right in region II and in addition a wave packet which is reflected at the step and is moving to the left in region I. The wiggly structure in the probability density that occurs close to the step in region I stems from the interference of the incoming and reflected wave packets. The wiggles are caused by the fast variation of the de Broglie wave function. It is interesting to compare the behavior of our quantum-mechanical wave packet incident on a potential step with that of the packet of light waves incident on a glass surface, which we studied in Section 2.5. The principal difference between the two phenomena is that the optical wave packet shows no dispersion, for its components with different wave numbers all move with the velocity of light.

We now use a wave packet with a lower initial-momentum expectation value p_0 so that the corresponding classical particle is reflected at the step, that is, $p_0 < \sqrt{2mV_0}$. The time developments of the wave function and the probability density (Figure 5.2) show that part of the wave packet penetrates for a short while, with an exponential falloff into region II that is forbidden for the classical particle. Eventually, the wave packet is also completely reflected. The penetration into region II is analogous to the reflection of light off a metal surface with finite conductivity.

For an attractive potential, that is, $V_0 < 0$, the picture is similar to Figure 5.1. The classical particle is now suddenly accelerated at the potential step and so is the transmitted part of the wave packet. Part of the wave packet is also reflected, however. The time developments of the wave function and the probability density are shown in Figure 5.3. The reflection is not too evident in Figure 5.3 but becomes apparent in Figure 5.4. Here the real parts of the incoming, transmitted, and reflected constituent waves are plotted separately. The constituent waves are shown in their mathematical form for the whole range of x-values. The physical significance of ψ_{1+} and ψ_{1-} is restricted to region I and that of ψ_2 to region II. Figure 5.4c shows that there is indeed a sizable reflected constituent wave moving to the left in region I.

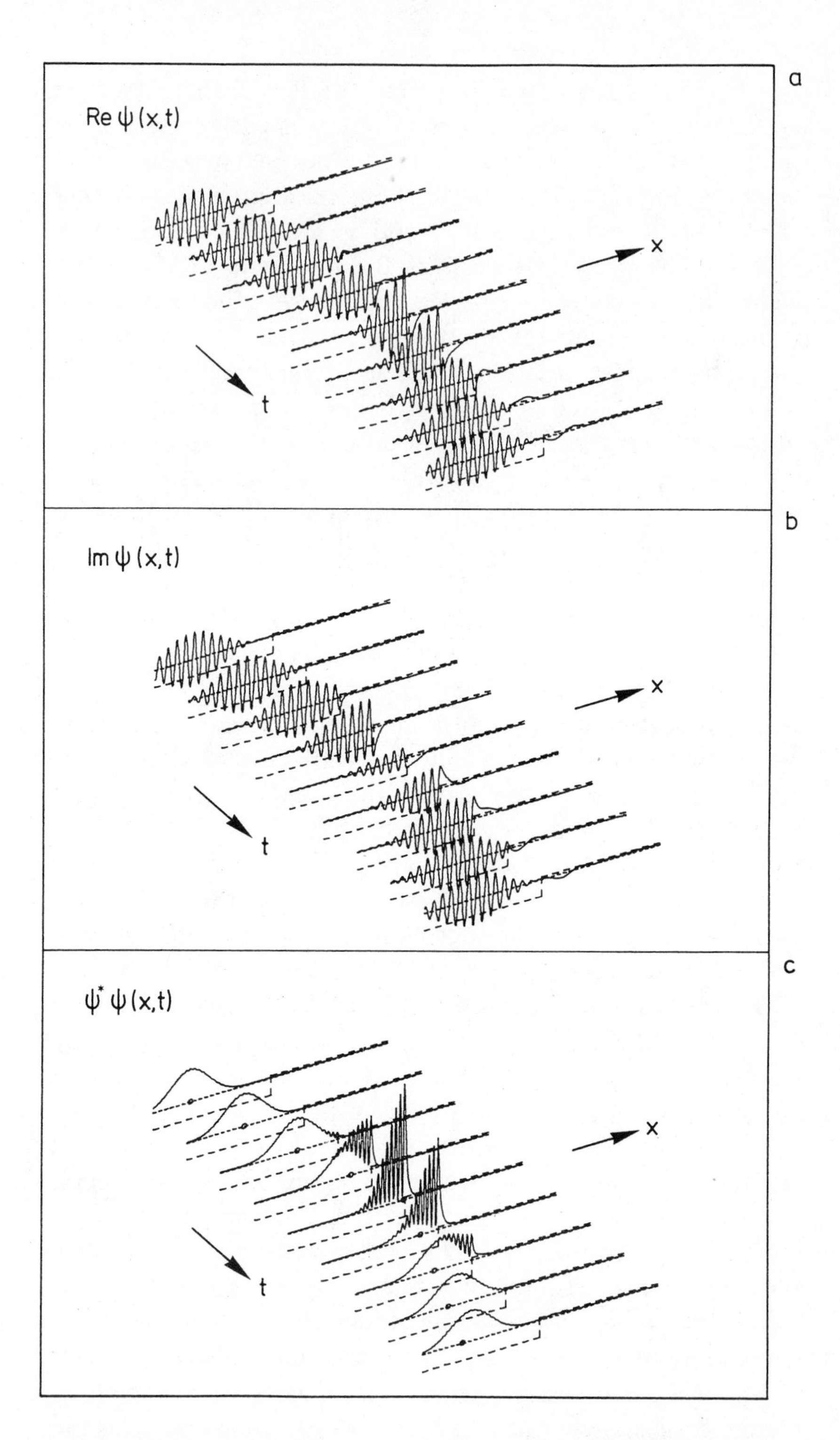

***Figure** 5.2* **Time developments of (a) the real part and (b) the imaginary part of the wave function and of (c) the probability density for a wave packet incident from the left on a potential step $V_0 > 0$. The initial momentum expectation value of the incident wave packet is $p_0 < \sqrt{2mV_0}$. The small circles indicate the positions of a classical particle of the same initial momentum.**

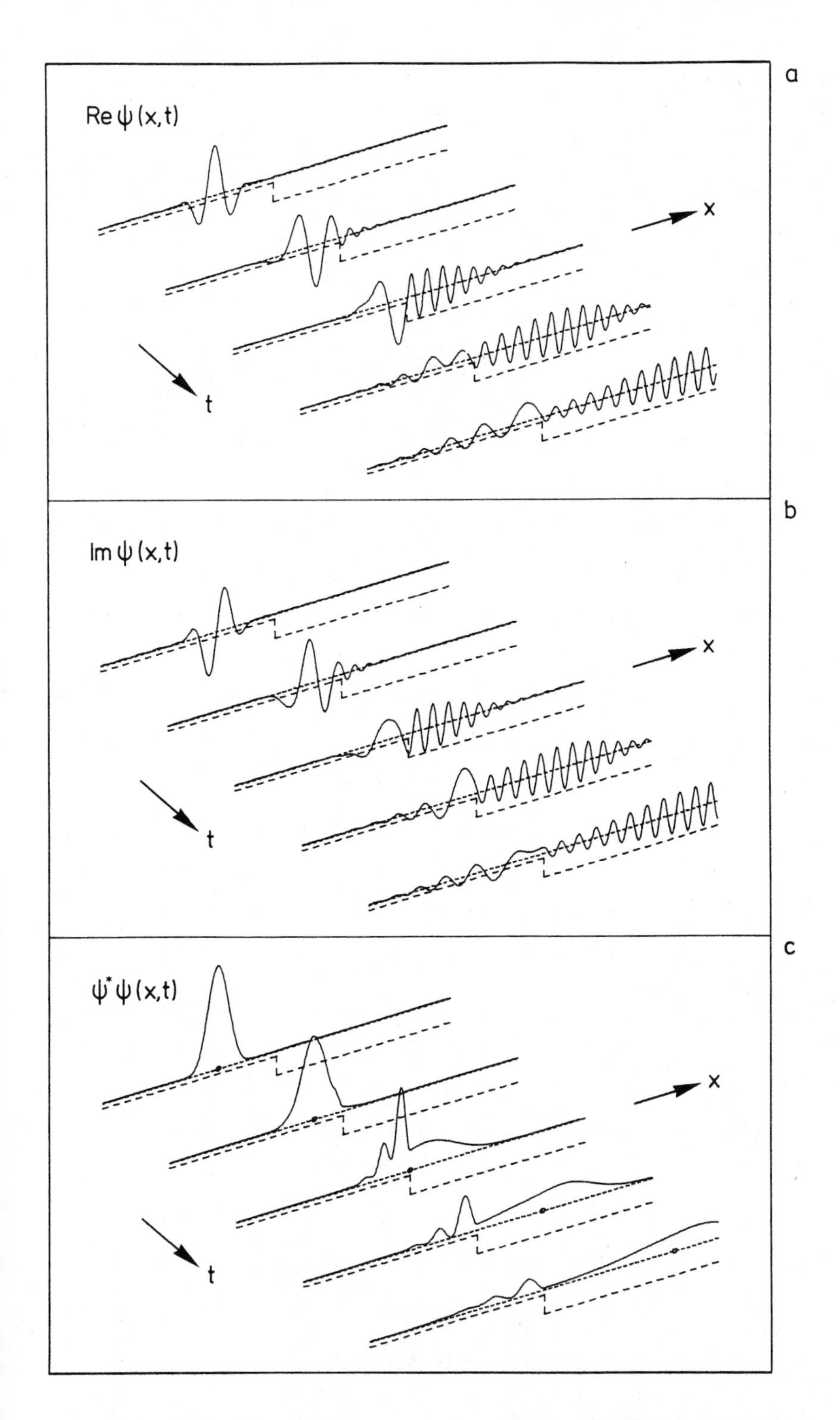

Figure 5.3 **Time developments of (a) the real part and (b) the imaginary part and of (c) the probability density for a wave packet incident from the left on a potential step of height $V_0 < 0$. The small circles in part c indicate the positions of a classical particle incident on the same potential step.**

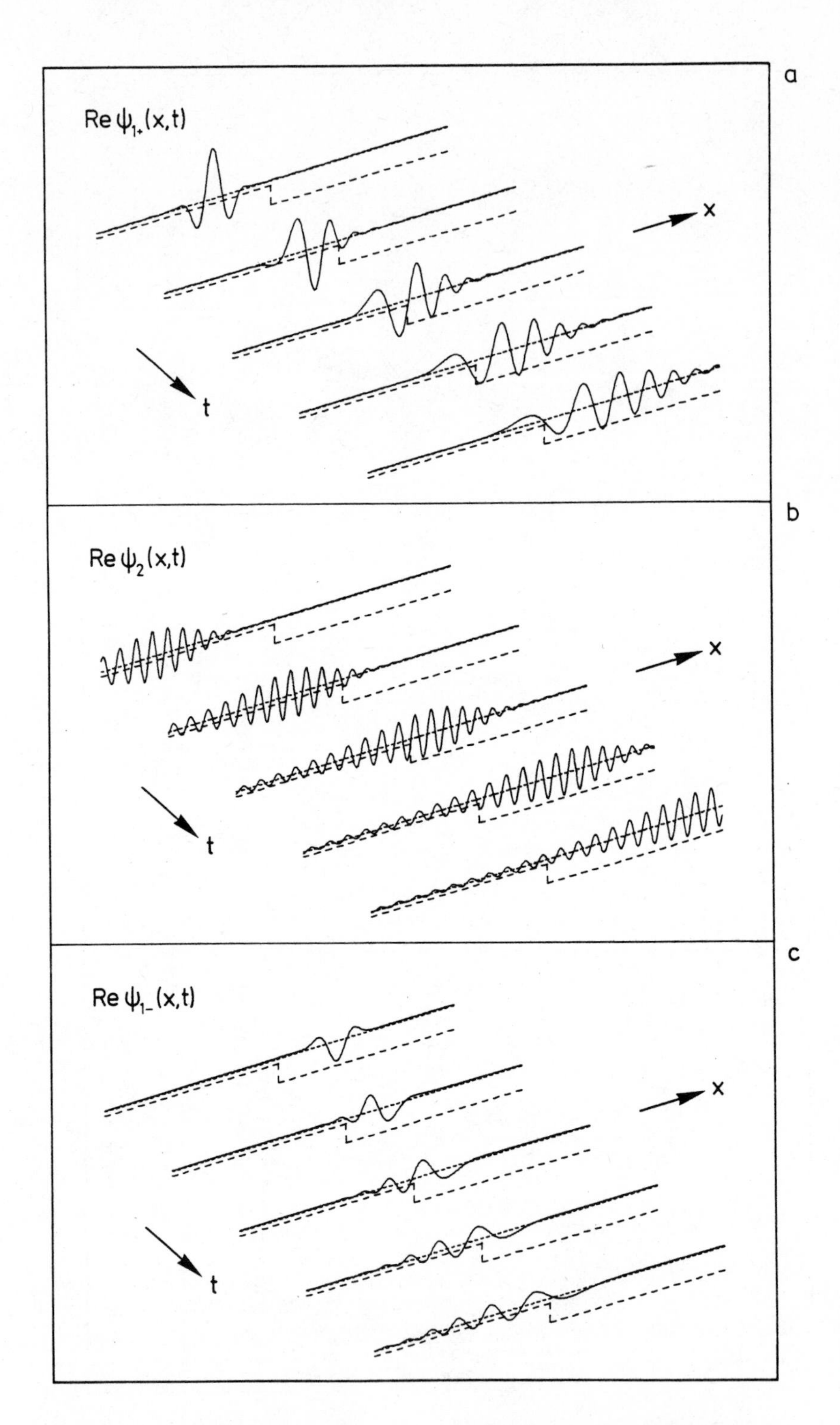

Figure 5.4 **A wave packet falls onto a potential step of height $V_0 < 0$, as in Figure 5.3. The time developments of the real parts of (a) the incident constituent wave, (b) the transmitted constituent wave, and (c) the reflected constituent wave.**

5.2 Tunnel Effect

In Figure 5.2 we studied the behavior of a wave packet that was reflected at a potential step higher than the average energy of the wave packet. We observed that, during the process of reflection, the wave packet penetrated to a certain extent into the region of high potential. It would be interesting now to see what happens if the region of high potential extends only over a distance comparable to the depth of penetration. We therefore study a wave packet which is under the influence of a potential barrier. The potential is constant, $V = V_0$, in a limited region of space, $0 < x < d$, called region II. It vanishes elsewhere, that is, in region I, $x < 0$, and in region III, $x > d$.

Figure 5.5a shows the time development of the probability density for a Gaussian wave packet incident in region I on such a potential barrier. At the upstep of the barrier at $x = 0$, we observe the expected pattern of a reflection. At the downstep we see a wave packet emerge and travel to the right in region III. According to our probability interpretation, this means that there is a finite probability that the particle described by the original Gaussian wave packet will pass the potential barrier, although it cannot do so under the laws of classical mechanics. Figure 5.5b shows that the probability of the particle's tunneling through the barrier increases when the barrier is narrower. Finally, from Figures 5.5b and c, we see that the probability of tunneling decreases as the barrier becomes higher. These general features have to be taken with caution, for there are discrete energy ranges in which the tunneling probability possesses maxima.

The tunnel effect just described is one of the most surprising in quantum mechanics. It is the basis for explaining a number of phenomena, including the radioactive decay of atomic nuclei through the emission of an α-particle. The surface region of the nucleus represents a potential barrier which with high probability keeps the α-particle from leaving the nucleus. The α-particle has only a small probability of penetrating the barrier through tunneling.

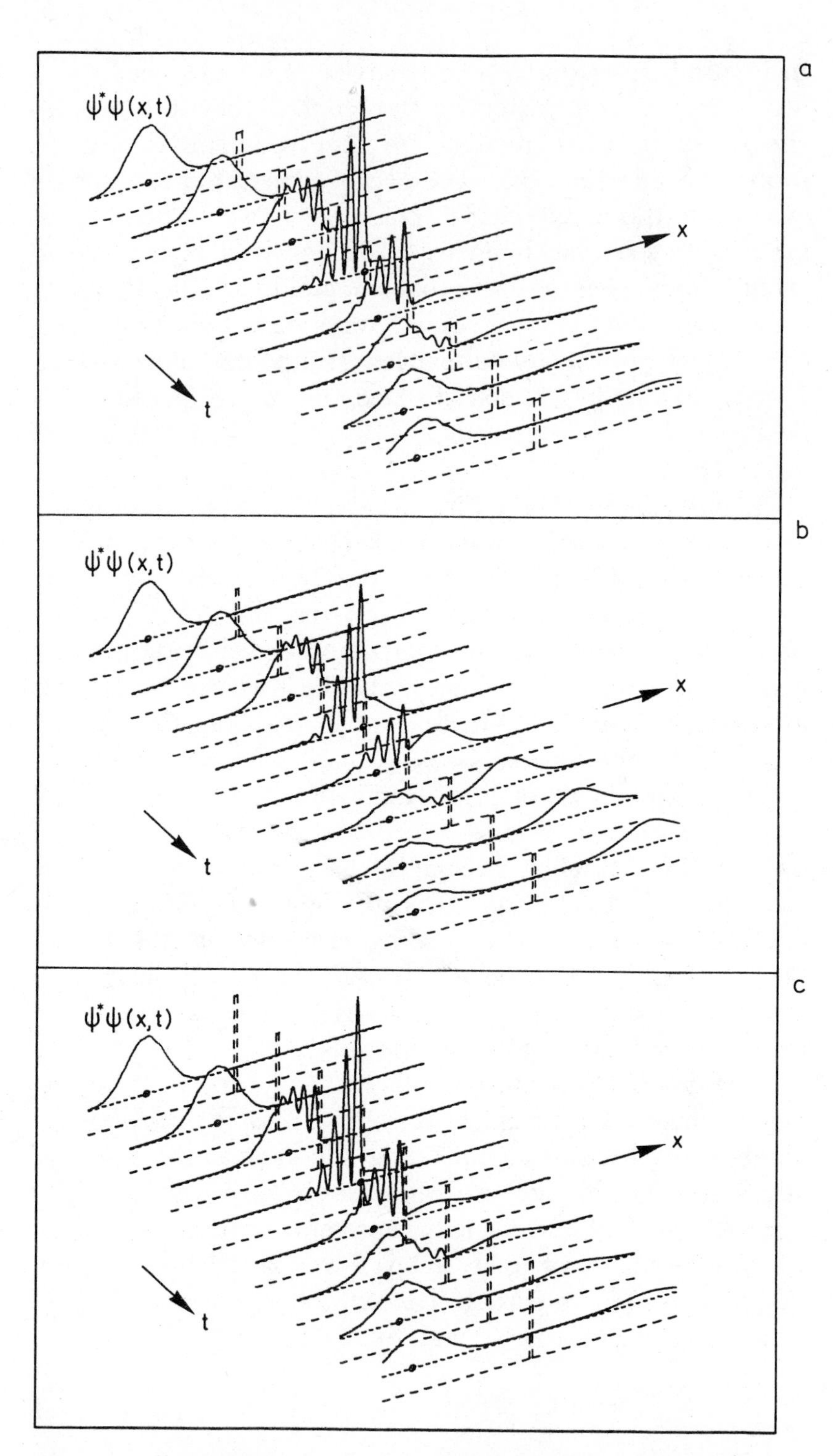

Figure 5.5 **Tunnel effect. (a) Time development of the probability density for a wave packet incident from the left onto a potential barrier of height V_0. The small circles indicate the positions of a classical particle incident onto the same potential barrier. (b) Same as for part a, but for a barrier of half the width. (c) Same as for part b, but for a barrier of double the height.**

5.3 Excitation and Decay of Metastable States

The scattering of a wave packet on two repulsive barriers that are far apart compared to the spatial width of the wave packet is a very interesting phenomenon. The width of the two barriers is chosen so that the tunnel effect allows a sizable fraction of the probability to pass through the two barriers. Figure 5.6 shows the time development of the packet entering from the left. We observe that although the major part of the probability is reflected at the first barrier, another part enters the region between the barriers and retains its bell shape at least while distant from the barriers. At a later moment in time the injected packet hits the barrier on the right, and again there is partial reflection and transmission. Later on in the process the particle is with a certain probability confined between the two walls, continuously bouncing back and forth and each time losing part of the probability to the outside region. Except for the continuous broadening of the particle wave packet, this situation is very similar to the analogous process in optics, namely a light wave packet falling onto a glass plate, which was shown in Figure 2.12.

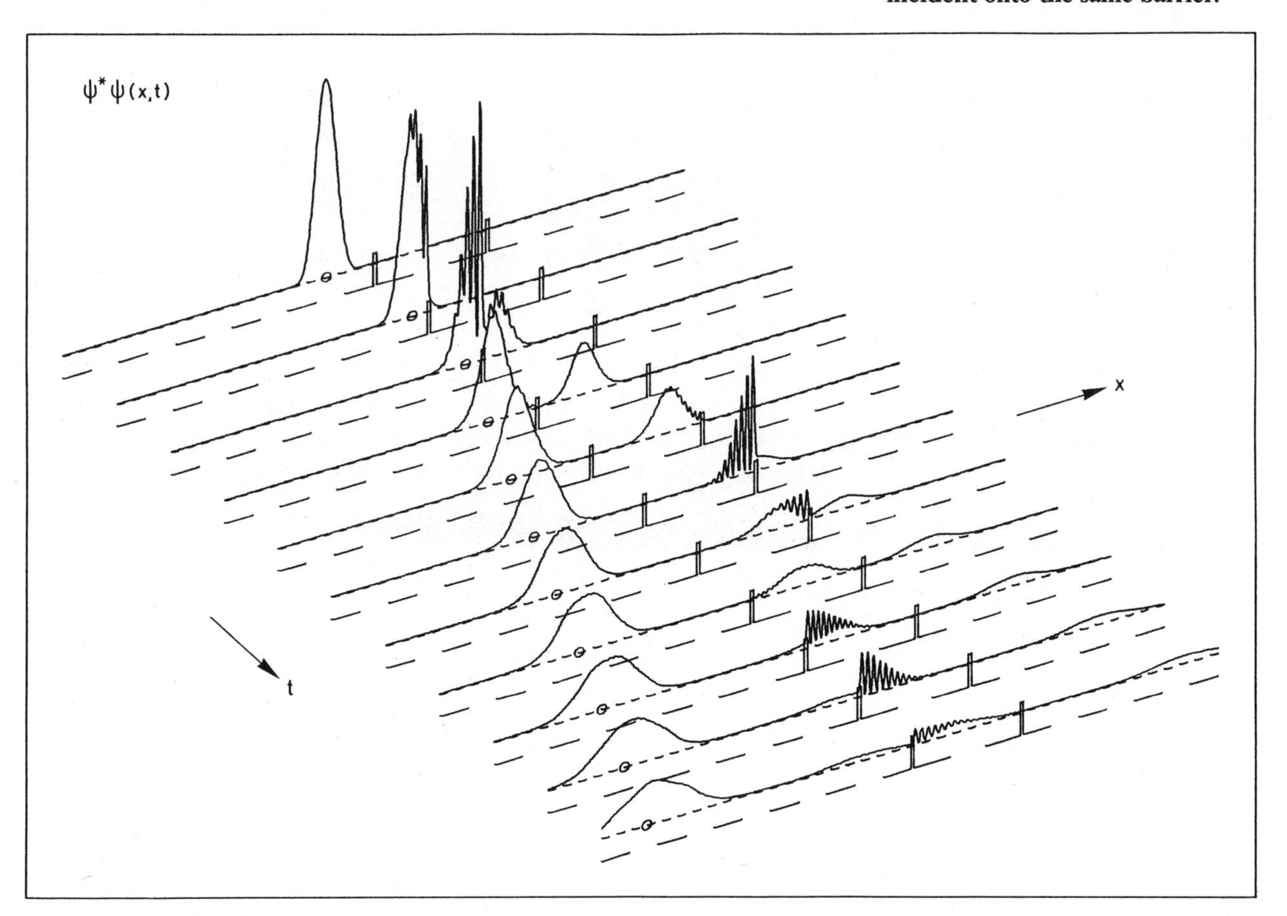

Figure 5.6 **Time development of the probability density for a wave packet incident from the left onto a double potential barrier. The small circles indicate the positions of a classical particle incident onto the same barrier.**

The situation in which a particle is partially confined to the region between the two barriers and the probability slowly leaks to the outside region is called a *metastable state*. The term was chosen to invoke the similarity of this state with the *stable state* or bound state, which we have already discussed briefly in Section 4.3. A particle in a bound state is permanently confined to a region of space.

In order to study metastable states more systematically, we now consider the situation in which the Gaussian wave packet is broad compared to the distance between the two barriers. Because of Heisenberg's uncertainty principle, the spatially wide Gaussian wave packet has a narrow momentum spread. The energy spectrum between zero and the top of the barrier can therefore be scanned in small intervals. For the two barriers of Figure 5.7, there are various energy, and thus momentum, values for which a fraction of the probability enters the region between the barriers and stays there for quite some time, even though the wave packet has traveled rather far away from the reflecting barrier. Figure 5.7 shows the time developments of the probability densities for wave packets of the three average energies that correspond to the three lowest-lying metastable states in this system of two barriers. The probability densities of metastable states between the two walls are distinguishable by the number of nodes they possess. This number increases as the energy of the state increases. When the potential between the two barriers is not less than zero—in our case it is exactly zero—the lowest metastable state has no node.

If the potential is sufficiently negative between the walls, the lowest-lying metastable state, which, of course, has positive energy, may have one or more nodes. Then the states with a smaller number of nodes have negative energy. Therefore no probability can leave the region between the walls, for no particle with negative energy can exist outside the barriers. Thus these states are stable or bound. To complicate matters, the behavior of the wave packets discussed so far depends not only on their average energy but also on their spectral function in momentum space, that is, on their spatial form. In order to rid ourselves of this complication, we shall study wave packets with a smaller and smaller momentum spread. They are of course very wide in space.

Figure 5.8 shows the time development of the probability density for a wave packet whose average energy is equal to that of the third metastable state. It corresponds to Figure

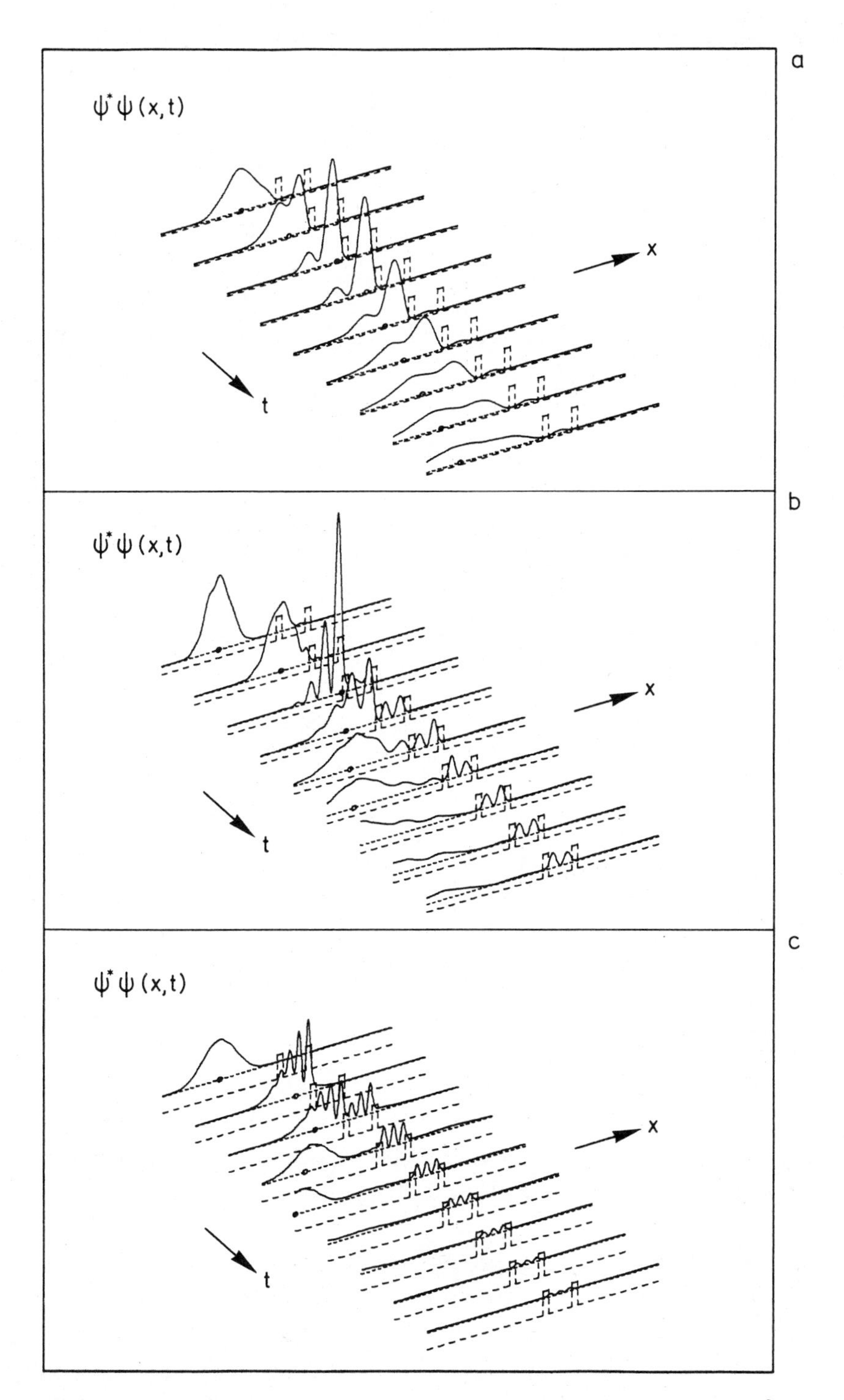

Figure 5.7 Time developments of the probability densities for wave packets of mean energies corresponding to (a) the first, (b) the second, and (c) the third metastable states in a system of two barriers. The wave packets, which are rather wide in space and thus possess a small momentum width, are incident from the left onto the double potential barrier. The small circles indicate the positions of a classical particle incident on the same barrier.

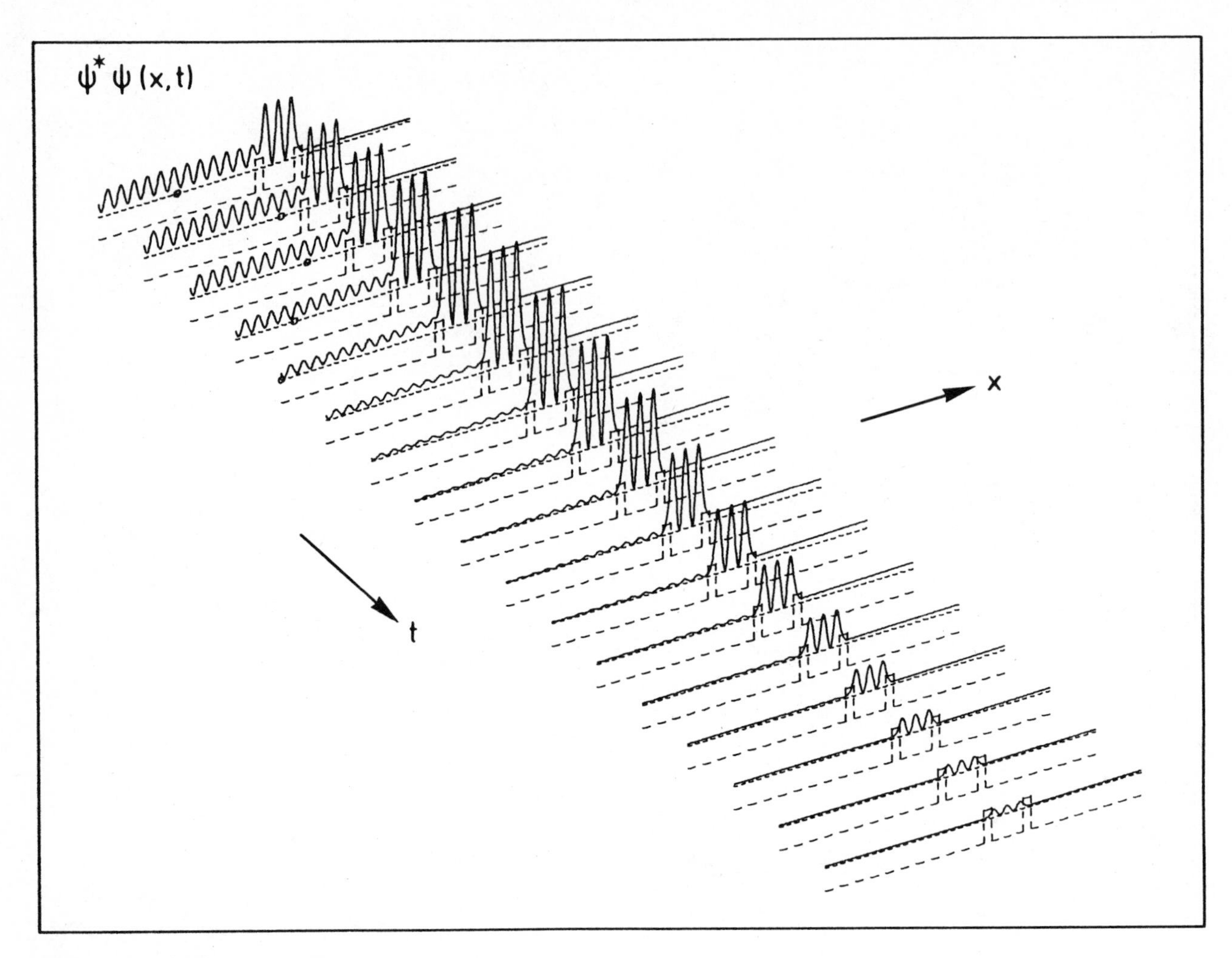

Figure 5.8 **Time development of the probability density for a wave packet that has the same mean energy as that of Figure 5.7c but is ten times as wide. Again, the wave packet is incident onto a double potential barrier. The small circles indicate the positions of a classical particle incident onto the same barrier.**

5.7c, except that the wave packet now has ten times the spatial width; its width exceeds by far the dimension of the figure. In the region to the left of the barriers, we observe the wiggly pattern typical for the interference between the incoming and the reflected wave packet. Between the two barriers the probability density keeps increasing with time up to the maximum amplitude, which is reached when the bulk of the wave packet has been reflected and has moved to the left. From then on the metastable state with two nodes decays slowly, in fact exponentially. The excitation of the metastable state in Figure 5.8 is much greater than the excitation of that in Figure 5.7c. The greater width of the wave packet in Figure 5.8 implies a narrower spectral function, which therefore contains more probability within the energy range of the metastable state.

To study the lifetime of metastable states, we observe their excitation and decay, as shown in Figure 5.8, over a longer

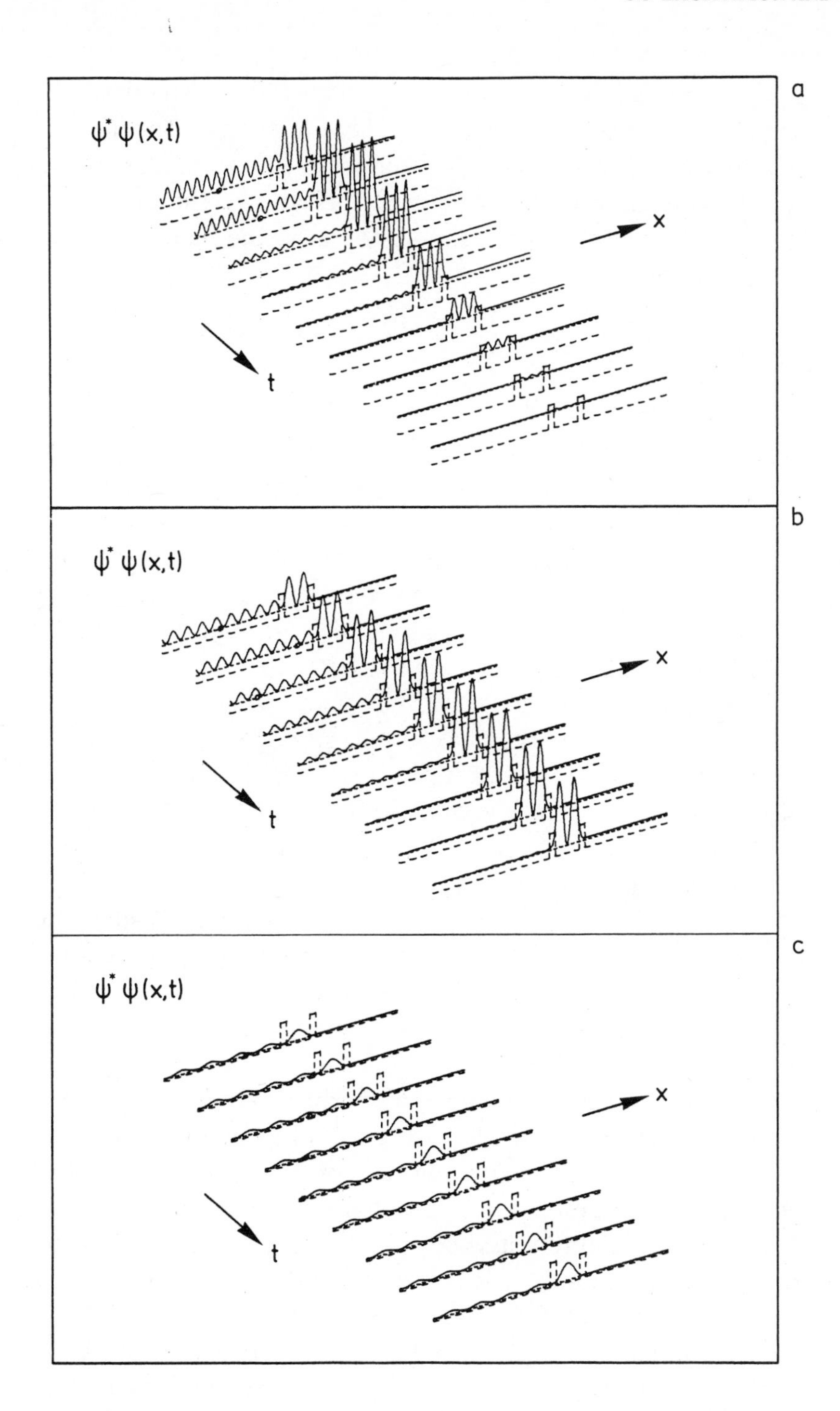

Figure 5.9 (a) Time development of the process shown in Figure 5.8 but observed over a longer period of time. Once the bulk of the wave packet has been reflected, the metastable state decays like an exponential function in time. Parts b and c are the same as part a but for the two metastable states that lie lower in energy. Parts a, b, and c of this figure correspond to parts c, b, and a of Figure 5.7. The wave packets, however, are much broader, and the time interval shown is much longer.

period of time. In Figure 5.9a it is easy to see that the amplitude in the region between the two barriers drops exponentially with time. We can measure the lifetime by determining the time in which the amplitude decreases by a factor of two. This we call the *half-life* of the state. Figure 5.9b shows the excitation and decay of the metastable state with only one node, corresponding to a lower energy, in the same time scale as the metastable state with two nodes. The half-life is now considerably longer. Even longer is the lifetime of the metastable state with no nodes. In Figure 5.9c, which is in the same time scale as Figures 5.9a and b, the amplitude has not decreased yet; the time interval of the figure is still in the excitation phase.

5.4 Stationary States of Sharp Momentum

We have just discussed the one-dimensional scattering of wave packets of narrow momentum spread and large extension in space. By reducing the momentum spread further, we obtain as the limiting case a harmonic wave $\psi_E(x, t)$ of fixed energy and momentum. After separating off the energy-dependent phase factor $\exp(-iEt/\hbar)$, we are left with a stationary state $\varphi_E(x)$, which was discussed in Chapter 4. The intimate relation between wide wave packets and stationary states allows a direct physical interpretation, in terms of particle mechanics, of the characteristic features of stationary states. A stationary state can be thought of as a limiting description of a particle with sharp momentum.

We are able to understand important details about metastable states through the study of stationary states in our potential with two barriers. We recall that within the barriers, that is, in regions II and IV, the potential is constant and positive, $V = V_0 > 0$. Outside the barriers, in regions I, III, and V, it vanishes, $V = 0$. Figure 5.10 shows the energy dependence of the stationary state $\varphi_E(x)$ in the potential with two barriers. That is, the solution of the time-independent Schrödinger equation for this potential is energy-dependent. The quantity plotted in Figure 5.10 is the intensity, introduced in Section 4.2, of this stationary solution. The range of energies shown in the figure comprises the energy of the metastable state with two nodes, which we discussed earlier and showed in Figures 5.7c, 5.8, and 5.9a. When the energy is lower than that of the metastable state—in the background of the picture—only a small fraction of the intensity is transmitted through the barriers into region V. There is a prominent interference pattern in region I from the superposition of the incoming and the reflected wave. As the energy approaches that of the metastable state, the reflection decreases to zero, the interference pattern vanishes, and the full inten-

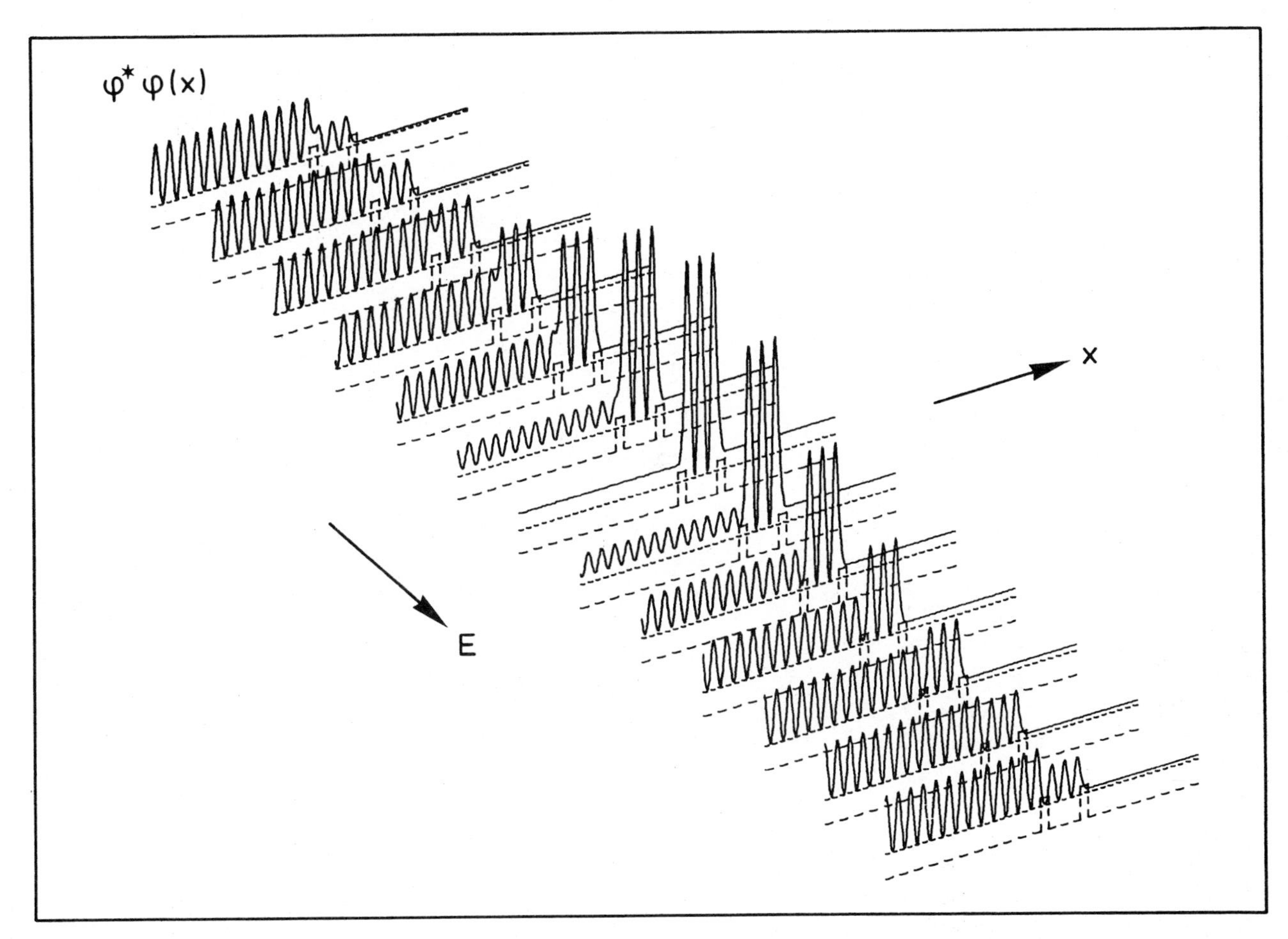

Figure 5.10 **Energy dependence, over a small range of energies, of the intensity of a harmonic wave incident onto a double potential barrier. The middle line corresponds to a resonance energy.**

sity of the incoming wave is transmitted through both barriers into region V. At the energy of the metastable state, the intensity in region III, between the barriers, reaches its maximum and assumes the two-node structure that is characteristic of the metastable state. This phenomenon is called a *resonance* of the system. As the energy increases further, the intensity in region III decreases as does the transmission into region V. The interference pattern in region I reappears as reflection grows.

Resonance phenomena are well known in many branches of physics. The best-known example from classical physics is the resonance of a pendulum excited to forced oscillation of a particular frequency. Our example of a quantum-mechanical resonance has a striking similarity to optical resonances. In Section 2.3 we saw that light at particular frequencies is transmitted through a glass plate without reflection. In the terminology of quantum mechanics, the words *metastable state* and *resonance* are often used synonymously.

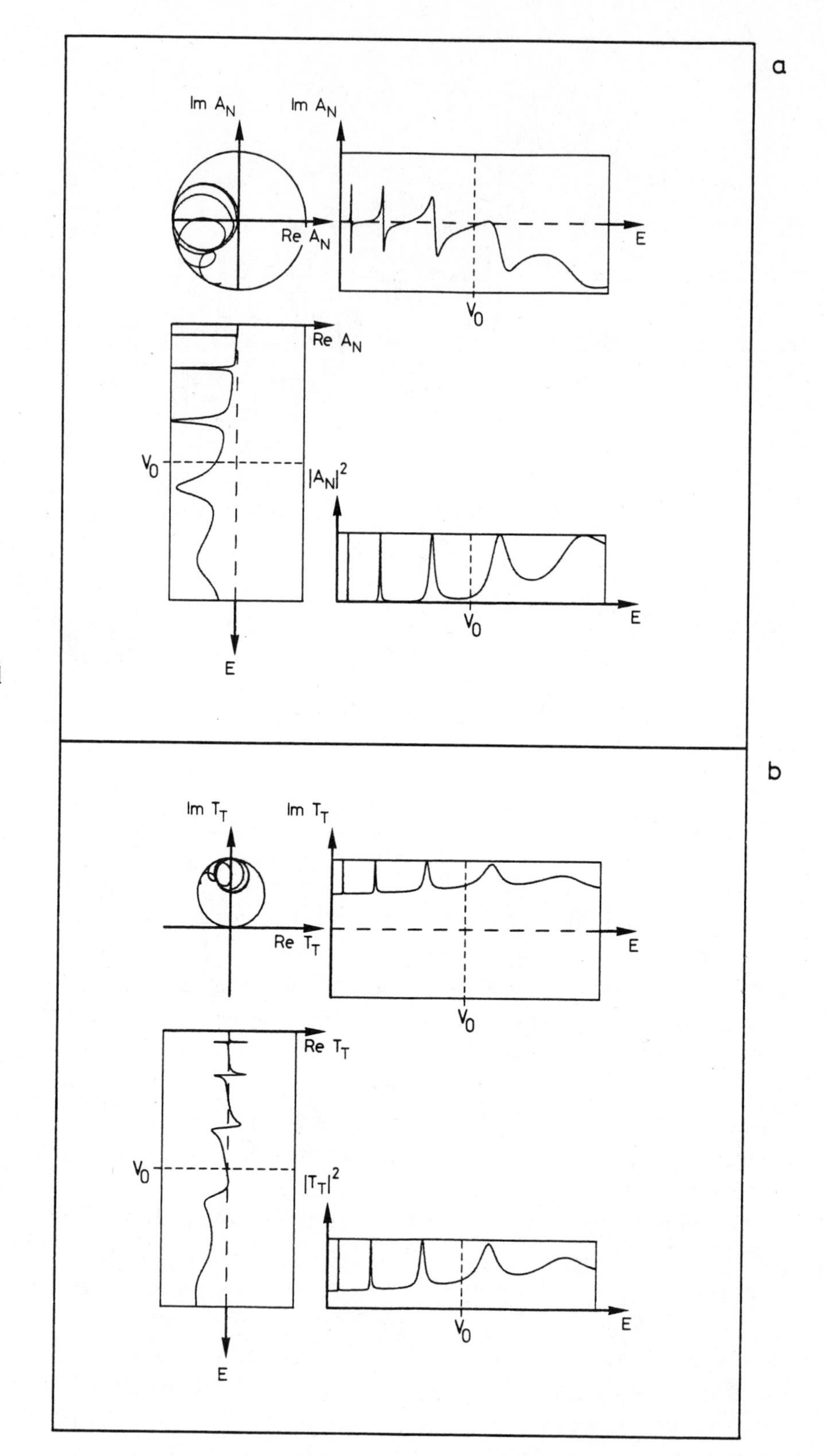

Figure 5.11 **(a) Energy dependence of the complex amplitude A_n of the part of a harmonic wave that is transmitted through the system with a double potential barrier. The energy ranges from zero to a value twice the barrier height. The energy dependence of A_n is shown as a line, starting from the origin, in the complex plane at upper left. The circle around the origin indicates the maximally allowed region for A_N. The energy dependence of the real part, projection onto the real axis, is shown below, that of the imaginary part, projection onto the imaginary axis, to the right. The lower right of the figure shows the energy dependence of $|A_N|^2$.**

(b) The parts of this figure are the same as those of part a, but they are for the transmission matrix element $T_T = (A_N - 1)/(2i)$. The line starts at point $i/2$ in the complex plane. The circle around point $i/2$ indicates the maximally allowed region for T_T.

As long as we are not interested in the details of the propagation and deformation of a wave packet with definite initial shape, but only in the fraction of probability with which reflection or transmission occurs, knowledge of the complex amplitudes of the stationary waves in the far left and far right regions—in our example regions I and V, in general regions I and N—is entirely sufficient. The stationary waves in these two regions are

$$\varphi_{\mathrm{I}} = \exp\left[\frac{i}{\hbar}p(x - x_0)\right] + B_{\mathrm{I}}\exp\left[-\frac{i}{\hbar}p(x - x_0)\right]$$

$$\varphi_N = A_N \exp\left[\frac{i}{\hbar}p(x - x_0)\right]$$

The probability for transmission is $|A_N|^2$, for reflection $|B_{\mathrm{I}}|^2$. The fact that we are dealing with a particle that can only be reflected or transmitted obviously requires that

$$|A_N|^2 + |B_{\mathrm{I}}|^2 = 1$$

This relation, which expresses the conservation of the total probability of observing the particle, is called the *unitarity relation* for the scattering amplitudes A_N and B_{I}. For vanishing reflection, $B_{\mathrm{I}} = 0$, the unitarity relation allows a circle of radius 1 in the complex plane for A_N. Figure 5.11a shows the energy dependence of the complex number A_N, again for the problem with two potential barriers. The upper left part of the figure is an *Argand diagram*. A plane is spanned by the axes $\mathrm{Re}\,A_N$ and $\mathrm{Im}\,A_N$. For a fixed energy value the complex number A_N is given by point $A_N = \mathrm{Re}\,A_N, \mathrm{Im}\,A_N$ in the Argand diagram. The line in this figure shows the variation with energy of A_N as a complex number. The outer circle corresponds to $|A_N| = 1$. Obviously, A_N always stays within this circle, indicating the energy dependence of $\mathrm{Im}\,A_N$ and $\mathrm{Re}\,A_N$ respectively. We follow the energy dependence from $E = 0$ to $E = 2V_0$, where V_0 is the height of both potential barriers. The imaginary part of A_N stays near zero for almost all energies below the barrier height V_0, slowly deviating from that value for larger energies. For resonance regions of the energy, $\mathrm{Im}\,A_N$ rises quickly, then falls even more steeply to negative values in order to rise quickly again to zero. The real part of A_N displayed below the circle also shows for most energies lower than V_0 very little deviation from zero. With increasing energy it drops to negative values. For the resonance regions it has negative peaks which become wider with increasing energies. Finally, the absolute square of A_N, shown in the lower right corner, again has peaks of increasing width in the resonance regions. $|A_N|^2$ has a tendency to approach one for energies far above the barrier height. For these energies total transmission is expected.

Returning to the Argand diagram, we are now able to recognize the typical signatures of resonances. Outside the resonance region A_N varies very slowly with energy; for energies lower than the barrier height, it stays near the origin of the complex plane. In the resonance region it passes quickly and counter-clockwise through a circle and causes the typical resonance patterns in the real and imaginary parts. For energies above the barrier height, the circles no longer return to the origin of the complex plane, for the transmission outside the resonance regions is sizable.

Yet another set of parameters is used to characterize the effect of the potential on the particle waves,

$$A_N = 1 + 2iT_T$$

$$B_{\mathrm{I}} = 2iT_R$$

The *transition matrix elements* T_T and T_R describe the deviation of the parameters A_N and B_{I} from the situation in which the wave travels without a potential being present, that is, the deviation from $A_N = 1$, $B_{\mathrm{I}} = 0$. The factors $2i$ are introduced for convenience.

Inserting these expressions into the unitarity relation for the scattering amplitudes, $|A_N|^2 + |B_{\mathrm{I}}|^2 = 1$, we find the unitarity relation for the T-matrix elements:

$$\operatorname{Im} T_T = T_T T_T^* + T_R T_R^*$$

This equation can be rewritten in terms of real and imaginary parts of T_T:

$$(\operatorname{Re} T_T)^2 + (\operatorname{Im} T_T - \tfrac{1}{2})^2 = \tfrac{1}{4} - T_R T_R^*$$

For $T_R T_R^* = 0$ this relation describes complex numbers T_T on a circle of radius $\frac{1}{2}$ centered around the point $i/2$. Because of $|B_{\mathrm{I}}|^2 \leq 1$, we have $|T_R|^2 \leq \frac{1}{4}$ so that the right-hand side of the equation remains positive or zero. For nonvanishing T_R the complex numbers T_T therefore fall within the circle. Figure 5.11b shows in the upper left the Argand diagram of T_T, with the circle of radius $\frac{1}{2}$ centered around $i/2$ limiting its possible values. It also contains the projections $\operatorname{Im} T_T(E)$, $\operatorname{Re} T_T(E)$ as well as $|T_T(E)|^2$. Because of the simple relation between T_T and A_N, the features of these diagrams are in one-to-one correspondence with those of Figure 5.11a.

In elementary particle physics Argand diagrams of the type given in Figure 5.11 are used to study the complex scattering amplitude. This amplitude describes the collision probability of two particles. Detection of characteristic circular features is equivalent to the discovery of metastable states. Such states are considered to be elementary particles with very short lifetimes.

Problems

5.1 Figure 5.1c shows the probability density and the classical position of a particle moving toward and beyond a potential step. Why is the wave packet narrower immediately after passing the positive potential step than before passing it? Predict the behavior of the wave packet at a negative potential step and verify this in Figure 5.3c.

5.2 Determine the ratios of the amplitudes of the metastable state at successive equidistant moments in time. Use a ruler to measure the amplitudes in Figure 5.8. For later moments in time, the ratios tend to a constant value, indicating that the decay is becoming exponential. Why is the decay slower earlier in time?

5.3 Plot the amplitudes of the probability densities in the region between the two potential barriers for the energies E_i corresponding to the thirteen situations shown in Figure 5.10. The energies are equidistant, that is, $E_{i+1} - E_i = \Delta E =$ constant. Fit the result to a Breit-Wigner distribution,

$$f(E) = A\frac{\Gamma^2/4}{(E - E_r)^2 + \Gamma^2/4}$$

For E_r use the energy of the maximum amplitude and give the width Γ of the distribution in units of ΔE.

5.4 Figure 5.11 shows the energies of the resonances in the double potential barrier. Calculate the ratios of the energies of the three lowest resonance peaks as they are given in Figure 5.11b. Compare the ratios to the corresponding ones of the bound-state energies of Figure 4.7. Compare both sets of ratios to Figure 6.1 and the formula for the deep square well given at the beginning of Section 6.1.

6. One-Dimensional Quantum Mechanics: Motion Within a Potential, Stationary Bound States

So far we have dealt with the motion of particles with a total energy $E = E_{\text{kin}} + V$ that is positive at least in region I, the region of the incoming particle. Of course, classical motion inside a finite region where the potential is negative is also possible for negative total energies, as long as kinetic energy $E_{\text{kin}} = E - V$ is positive. We now study this system from the point of view of quantum mechanics.

6.1 Spectrum of a Deep Square Well

As a particularly simple system, let us consider the force-free motion in a region of zero potential between two infinitely high potential walls at $x = -d/2$ and $x = d/2$. Since the potential outside this region is infinite, the solutions of the time-independent Schrödinger equation vanish there. Within this region they have the simple forms

$$\varphi_n(x) = \sqrt{\frac{2}{d}}\cos\left(n\pi\frac{x}{d}\right), \qquad n = 1, 3, 5, \ldots$$

or

$$\varphi_n(x) = \sqrt{\frac{2}{d}}\sin\left(n\pi\frac{x}{d}\right), \qquad n = 2, 4, 6, \ldots$$

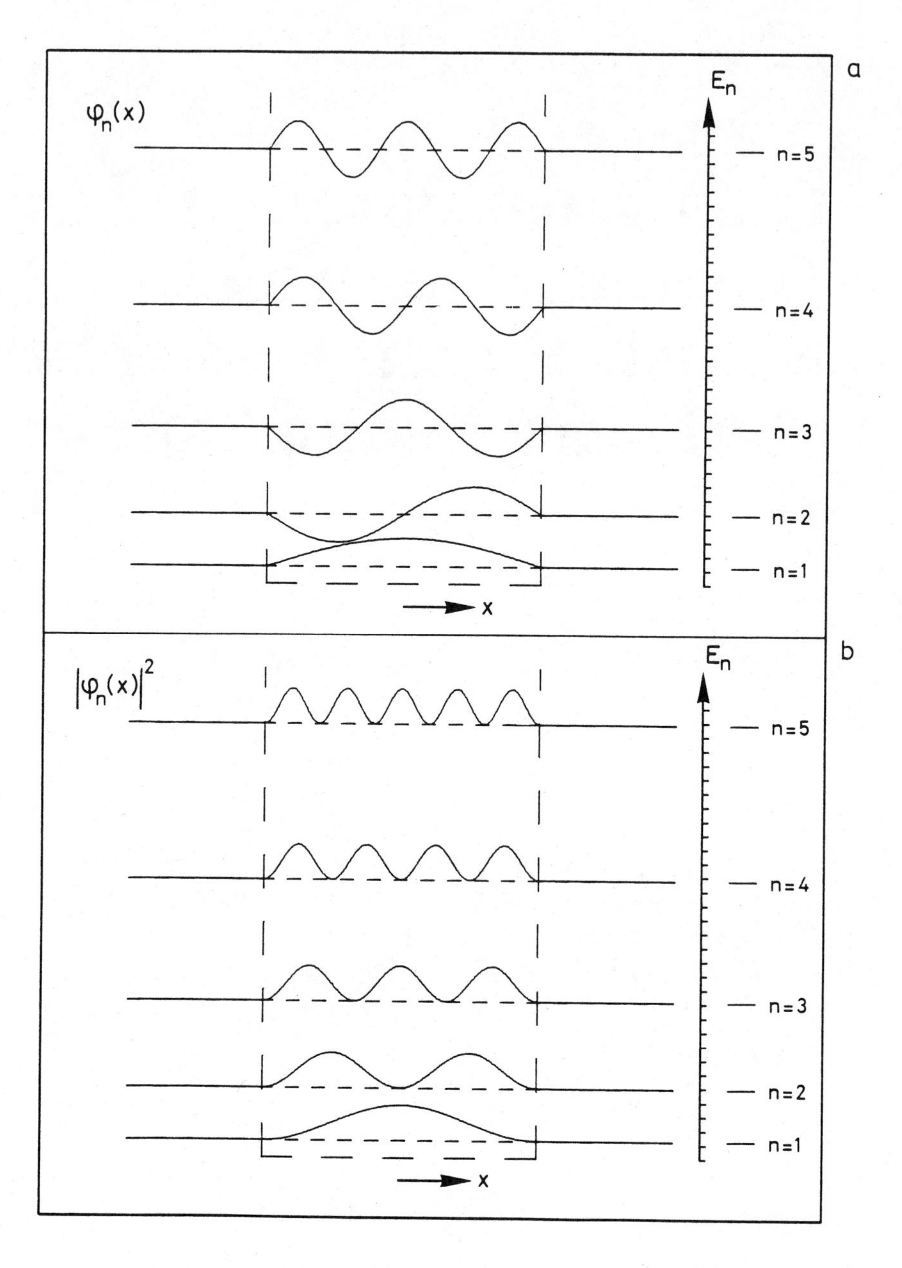

Figure 6.1 **Bound states in an infinitely deep square well. The long-dash line indicates the potential energy $V(x)$. It vanishes for $-d/2 < x < d/2$ and is infinite elsewhere. Points $x = \pm d/2$ are indicated as vertical walls. On the right side an energy scale is drawn, and to the right of it the energies E_n of the lower-lying bound states are indicated by horizontal lines. These lines are repeated as short-dash lines on the left. They serve as zero lines for (a) the wave functions and (b) the probability densities of the bound states.**

The energies of these bound states are

$$E_n = \frac{1}{2m}\left(\frac{\hbar n\pi}{d}\right)^2 , \qquad n = 1, 2, 3, \ldots$$

as we easily verify by inserting φ_n into the time-independent Schrödinger equation,

$$-\frac{\hbar^2}{2m}\frac{d^2}{dx^2}\varphi_n = E_n\varphi_n$$

which is valid between the two walls. Figure 6.1 presents the wave function, the probability density, and the energy spectrum. The lowest-lying state at E_1, called the *ground state*, has a finite energy $E_1 > 0$, which implies a kinetic energy $E_{\text{kin}} > 0$ since the potential energy V is zero by construction. Already this situation differs from that in classical mechanics, where the state of least energy is of course the state of rest with $E = E_{\text{kin}} = 0$. The higher states increase in energy proportionally to n^2. The *quantum number n* is equal to one plus the number of nodes of the wave function in region $-d/2 < x < d/2$; that is, boundaries $x = \pm d/2$ are excluded. The wave function has even or odd symmetry with respect to point $x = 0$, depending on whether n is odd or even, respectively. Even wave functions, here the cosine functions, are said to possess *even* or *natural parity*, odd wave functions *odd* or *unnatural parity*. Obviously, wave functions with an even number of nodes have even parity, those with an odd number odd parity. This property also holds for other one-dimensional potentials that are mirror-symmetric.

6.2 Particle Motion in a Deep Square Well

In Section 6.1 we found the spectrum of eigenvalues E_n and the wave functions describing the corresponding eigenstates $\varphi_n(x)$ for the deep square well. The solutions of the time-dependent Schrödinger equation are obtained by multiplying $\varphi_n(x)$ with a factor $\exp(-iE_n t/\hbar)$. Through a suitable superposition of such time-dependent solutions, we form a moving wave packet which at the initial time $t = 0$ is bell-shaped with a momentum average p_0. Its wave function is

$$\psi(x,t) = \sum_{n=1}^{\infty} a_n(p_0, x_0)\varphi_n(x)\exp\left[-\frac{i}{\hbar}E_n t\right]$$

where the coefficients $a_n(p_0, x_0)$ have been chosen to ensure a bell shape around location x_0 for $t = 0$ and the momentum average p_0.

Figure 6.2 shows the time development of the probability density $|\psi(x,t)|^2$ for such a wave packet. We observe that for $t = 0$ the wave packet is well localized about initial position $x_0 = 0$ of the classical particle. It moves toward one wall of the well, where it is reflected. Here it shows the pattern typical of interference between incident and reflected waves. The pattern is very similar to that caused by a free wave packet incident on a sharp potential step, shown in Figure 5.2c. On its way back through the well, the packet resumes its

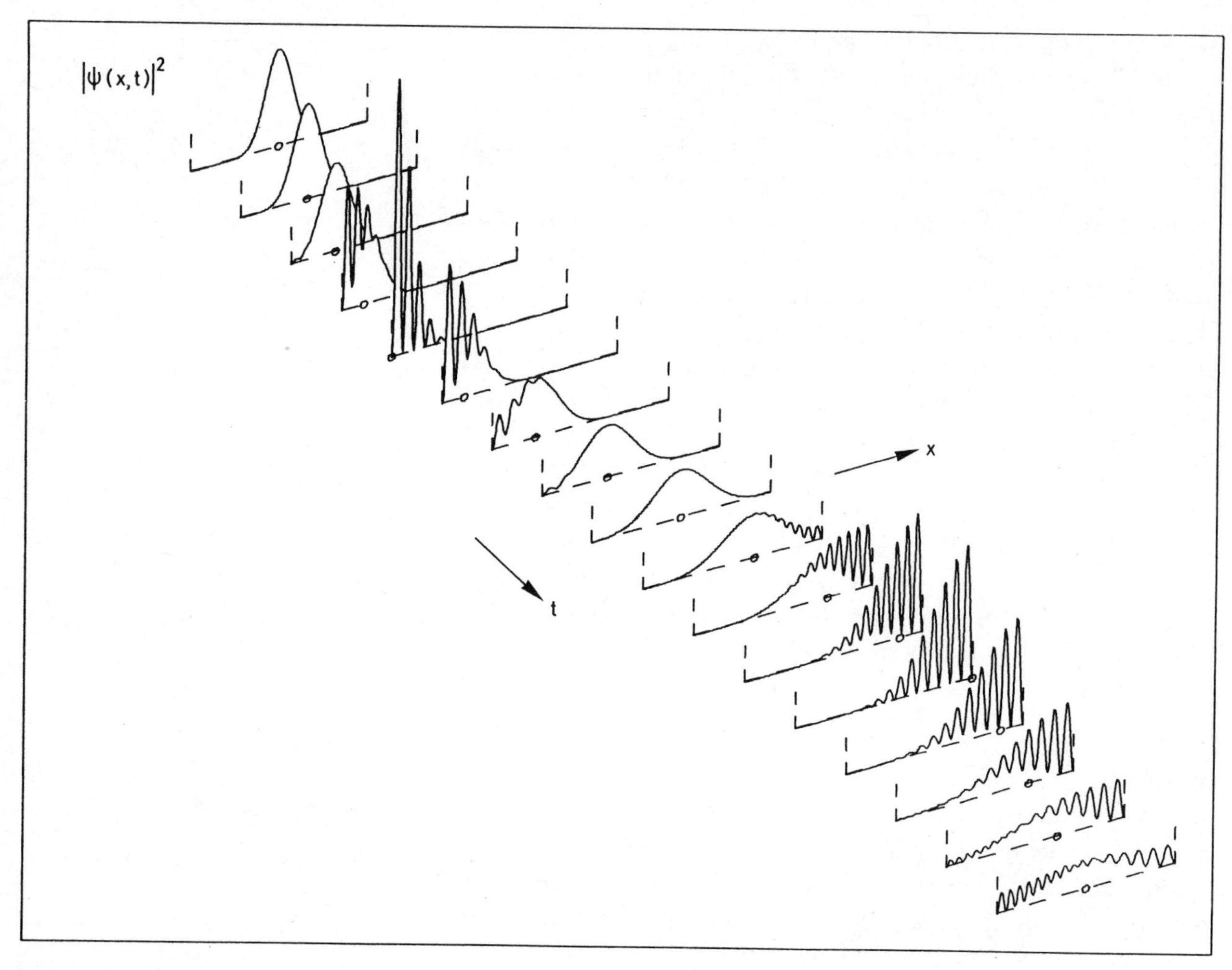

Figure 6.2 **Time development of a wave packet moving in an infinitely deep square well. At $t = 0$, in the background, the packet is well concentrated in the middle of the well. Its initial momentum makes it bounce back and forth between the two walls. The characteristic interference pattern of the reflection process, as well as the dispersion of the packet with time, is apparent. Once the packet is wider than the well itself, interference occurs simultaneously at both walls.**

smooth bell shape, but it is wider than it was initially. It continues to bounce between the two walls and is soon so wide that the packet touches both walls simultaneously. It would be wrong, however, to conclude that the packet dissolves completely to form a uniform distribution in the well. The packet always maintains sufficient structure so that the classical position of the particle coincides with the expectation value $\langle x \rangle$ of the position coordinate (see Section 3.3). The classical motion describing the reflection of a particle between two walls has a period in time given by $T_c = 2d/v_0 = 2md/p_0$. The quantum-mechanical wave function possesses another period, in general a much longer one, given by

$$T_1 = \frac{2\pi}{\omega_1}$$

where ω_1 is the frequency of the ground-state wave function,

$$\omega_1 = \frac{E_1}{\hbar} = \frac{\hbar}{2m}\left(\frac{\pi}{d}\right)^2$$

Since all energies E_n, $n = 2, 3, \ldots$, are integer multiples of E_1, the period T_1 of the ground state is also the period of the superposition $\psi(x,t)$ that describes the wave packet. Because of this periodicity of the wave function in time, the original wave packet will be restored after time T_1 has elapsed.

6.3 Spectrum of the Harmonic Oscillator Potential

The particle in a deep square well experiences a force only when hitting the wall. A simple, continuously acting force $F(x)$ can be thought of as the force of a spring, which follows Hooke's law,

$$F(x) = -kx, \qquad k > 0$$

This force, also called a *harmonic force*, is proportional to the displacement x from equilibrium position $x = 0$. A physical system in which a particle moves under the influence of a harmonic force is called a *harmonic oscillator*. The proportionality constant k gives the stiffness of the spring. The potential energy stored in the spring is

$$V(x) = \frac{k}{2}x^2$$

A classical particle of mass m performs *harmonic oscillations* of angular frequency

$$\omega = \sqrt{k/m}$$

so that $V(x)$ can be equivalently expressed by $V(x) = (m/2)\omega^2 x^2$. Introducing this expression into the time-independent Schrödinger equation yields

$$\left(-\frac{\hbar^2}{2m}\frac{d^2}{dx^2} + \frac{m}{2}\omega^2 x^2\right)\varphi(x) = E\varphi(x)$$

With the help of the dimensionless variable

$$\xi = \frac{x}{\sigma_0}, \qquad \sigma_0 = \sqrt{\hbar/m\omega}$$

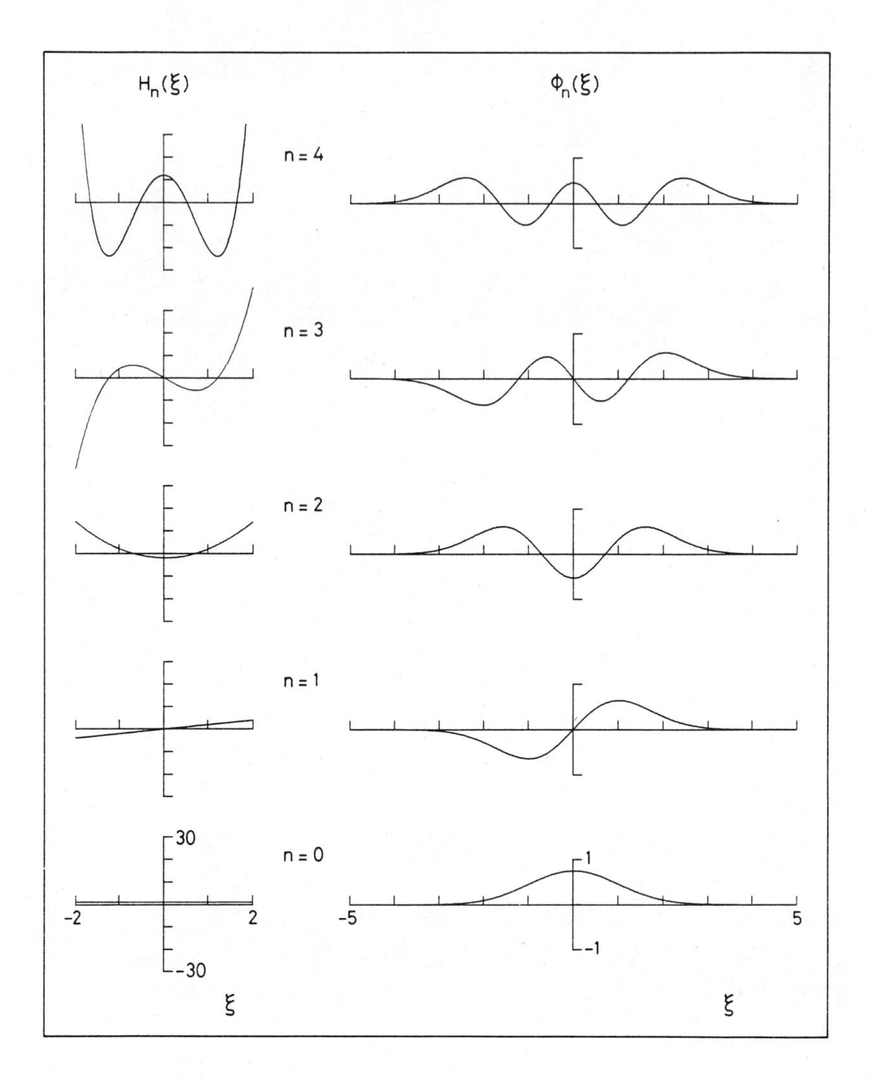

Figure 6.3 **Hermite polynomials $H_n(\xi)$ and eigenfunctions $\phi_n(\xi)$ of the harmonic oscillator for low values of n.**

the equation above simplifies to the reduced form

$$\frac{1}{2}\left(-\frac{d^2}{d\xi^2}+\xi^2\right)\phi(\xi)=\epsilon\phi(\xi),\qquad \phi(\xi)=\varphi(\sigma_0\xi)$$

The dimensionless eigenvalue $\epsilon = E/\hbar\omega$ measures the energy of the oscillator in multiples of Planck's quantum of energy $\hbar\omega$.

The solutions of the Schrödinger equation for the harmonic oscillator can be normalized (see Section 4.3) for the eigenvalues

$$\epsilon_n = n + \tfrac{1}{2},\qquad n = 0, 1, 2, \ldots$$

thus determining the energy eigenvalues of the harmonic oscillator,

$$E_n = \epsilon_n\hbar\omega = \left(n + \tfrac{1}{2}\right)\hbar\omega$$

The state of lowest energy $E_0 = \hbar\omega/2$ is the ground state. The energies E_n of the higher states differ from the ground-state energy by the energy of n quanta, each having the energy $\hbar\omega$ of Planck's quantum (see Chapter 1).

The eigenfunctions, normalized in ξ, can be represented in the form

$$\phi_n(\xi) = (\sqrt{\pi}\, 2^n n!)^{-1/2} H_n(\xi) e^{-\xi^2/2}, \qquad n = 0, 1, 2, \ldots$$

where the $H_n(\xi)$ are the *Hermite polynomials*. They are given by

$$H_0(\xi) = 1, \qquad H_1(\xi) = 2\xi$$

and for higher values of n by the recurrence relation

$$H_n(\xi) = 2\xi H_{n-1}(\xi) - 2(n-1)H_{n-2}(\xi), \qquad n = 2, 3, \ldots$$

Figure 6.3 shows the Hermite polynomials $H_n(\xi)$ and the normalized eigenfunctions $\phi_n(\xi)$ for low values of n.

The eigenfunctions $\varphi_n(x)$, normalized in x, are

$$\varphi_n(x) = \left(\sigma_0 \sqrt{\pi}\, 2^n n!\right)^{-1/2} H_n\left(\frac{x}{\sigma_0}\right) \exp\left(-\frac{x^2}{2\sigma_0^2}\right)$$

They are plotted in Figure 6.4a together with the potential energy $V(x)$. The dashed lines indicate the energy eigenvalues in relation to the bottom of the potential energy. They serve as zero lines for the corresponding φ_n. On the right-hand side the energy spectrum is shown. The exponential factor $\exp(-\xi^2/2)$ in the formula for φ_n ensures that

$$\varphi_n(x) \to 0 \qquad \text{for } |x| \to \infty$$

rendering these wave functions normalizable.

Figure 6.4b gives the probability densities $|\varphi_n(x)|^2$, showing that, even in regions where E is smaller than V, there is a certain probability of observing the particle. The absolute square of the wave function of the ground state formulated in terms of the position variable $x = \sigma_0\xi$ has the form

$$|\varphi_0|^2 = \frac{1}{\sqrt{\pi}\,\sigma_0} \exp\left(-\frac{x^2}{2\sigma_0^2/2}\right)$$

The exponent in this equation shows that the width of the probability density of the harmonic oscillator's ground state is $\sigma_0/\sqrt{2}$.

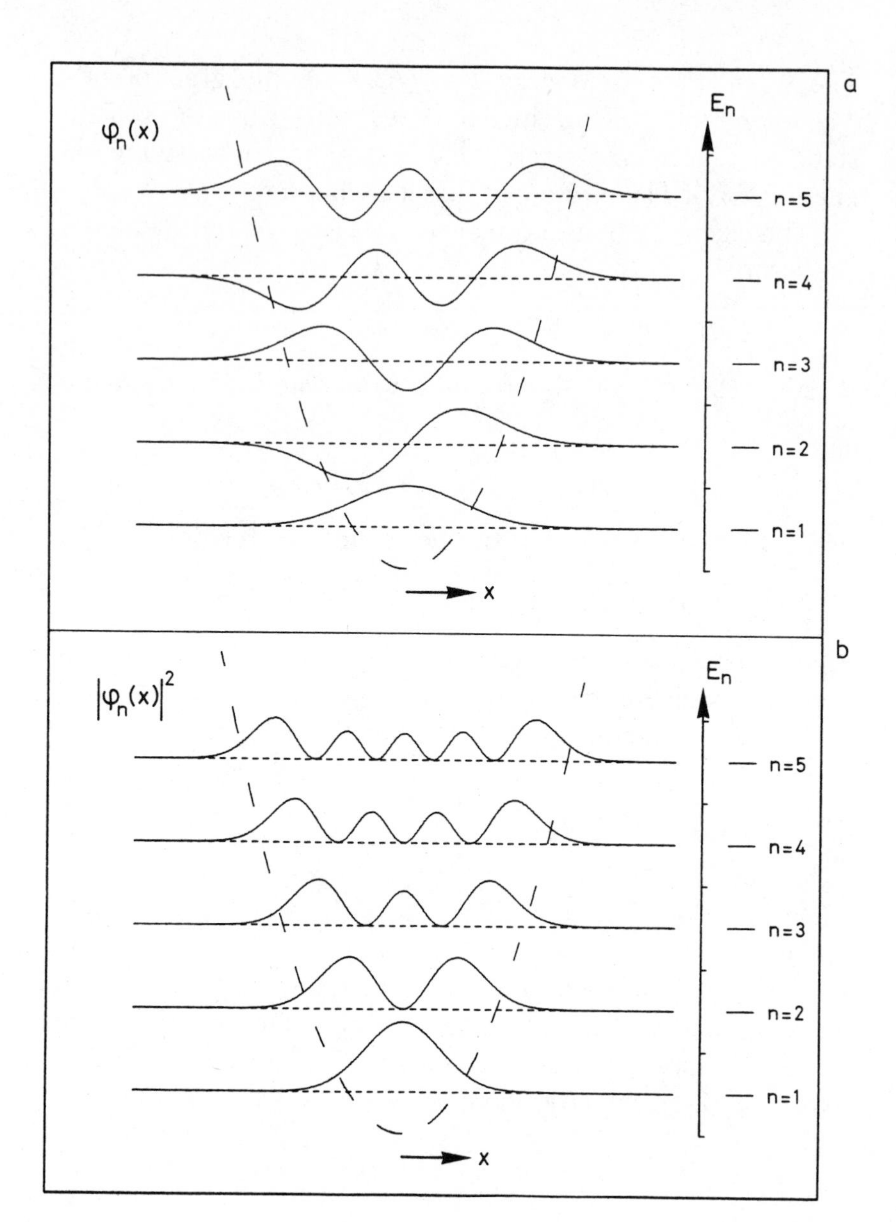

Figure 6.4 **Bound states in a harmonic oscillator potential. The potential is drawn as a long-dash line, a parabola. The eigenvalue spectrum of bound states is indicated by the horizontal lines on the right side. Repeated on the left as short-dash lines, they serve as zero lines for (a) the wave functions and (b) the probability densities of the bound states.**

6.4 Harmonic Particle Motion

We now consider the quantum-mechanical description of a particle moving under the influence of harmonic force. The particle at initial time $t = 0$ is at rest when placed in a position $x = x_0 \neq 0$, which is not the equilibrium position of the oscillator. In terms of a wave function, the initial state consists of a Gaussian wave packet of width σ with zero average momentum and an expectation value at position $x = x_0$ of the corresponding classical particle. This wave packet can be decomposed into a sum over eigenfunctions

$\varphi_n(x)$ of the harmonic oscillator,

$$\varphi(x) = \sum_{n=0}^{\infty} a_n \varphi_n(x)$$

The time-dependent solution of the Schrödinger equation with $\varphi(x)$ as initial wave function at $t = 0$ is then simply

$$\psi(x,t) = \sum_{n=0}^{\infty} a_n \varphi_n(x) \exp\left(-\frac{i}{\hbar} E_n t\right)$$

where $E_n = (n + \frac{1}{2})\hbar\omega$.

The infinite sum can be added up explicitly. For brevity we give here only the result for the absolute square of the wave function,

$$|\psi(x,t)|^2 = \frac{1}{\sqrt{2\pi}} \frac{2\sigma}{\sqrt{\sigma_0^4 s^2 + 4\sigma^4 c^2}} \times \exp\left[-\frac{2\sigma^2}{\sigma_0^4 s^2 + 4\sigma^4 c^2}(x - cx_0)^2\right]$$

where c and s represent $\cos \omega t$ and $\sin \omega t$, respectively, and where $\sigma_0/\sqrt{2}$ is the width of the probability distribution of the harmonic oscillator's ground state, as introduced in Section 6.3. This equation represents a Gaussian distribution with oscillating expectation value $x_0(t) = x_0 \cos \omega t$ and oscillating width $\sigma(t) = \sqrt{\sigma_0^4 \sin^2 \omega t + 4\sigma^4 \cos^2 \omega t}/(2\sigma)$. Of course, for the initial time $t = 0$ the time-dependent width $\sigma(t)$ reduces to the initial width σ.

Figure 6.5a shows a time development of a wave packet in the harmonic oscillator with initial width $\sigma < \sigma_0/\sqrt{2}$. As expected, the time dependence of the average position performs the same oscillation as the corresponding classical particle. The width oscillates with twice the frequency of the oscillator, starting with σ and increasing for the first quarter period $T/4 = \pi/2\omega$ to its maximum value $\sigma(T/4) = \sigma_0^2/(2\sigma)$. In Figure 6.5b the initial width is $\sigma > \sigma_0/\sqrt{2}$. Here the wave packet is wide initially and becomes narrower in the first quarter period, decreasing to the minimum value $\sigma_0^2/(2\sigma)$. The case $\sigma = \sigma_0/\sqrt{2}$ (Figure 6.5c), in which the width of the packet remains constant in time, represents the border line between the two situations. The particular value $\sigma_0/\sqrt{2}$ is exactly the width of the absolute square $|\varphi_0|^2$ of the ground-state wave function (shown in Figure 6.4b). The factor $\sqrt{2}$ appears since σ_0 was defined conventionally as the width of

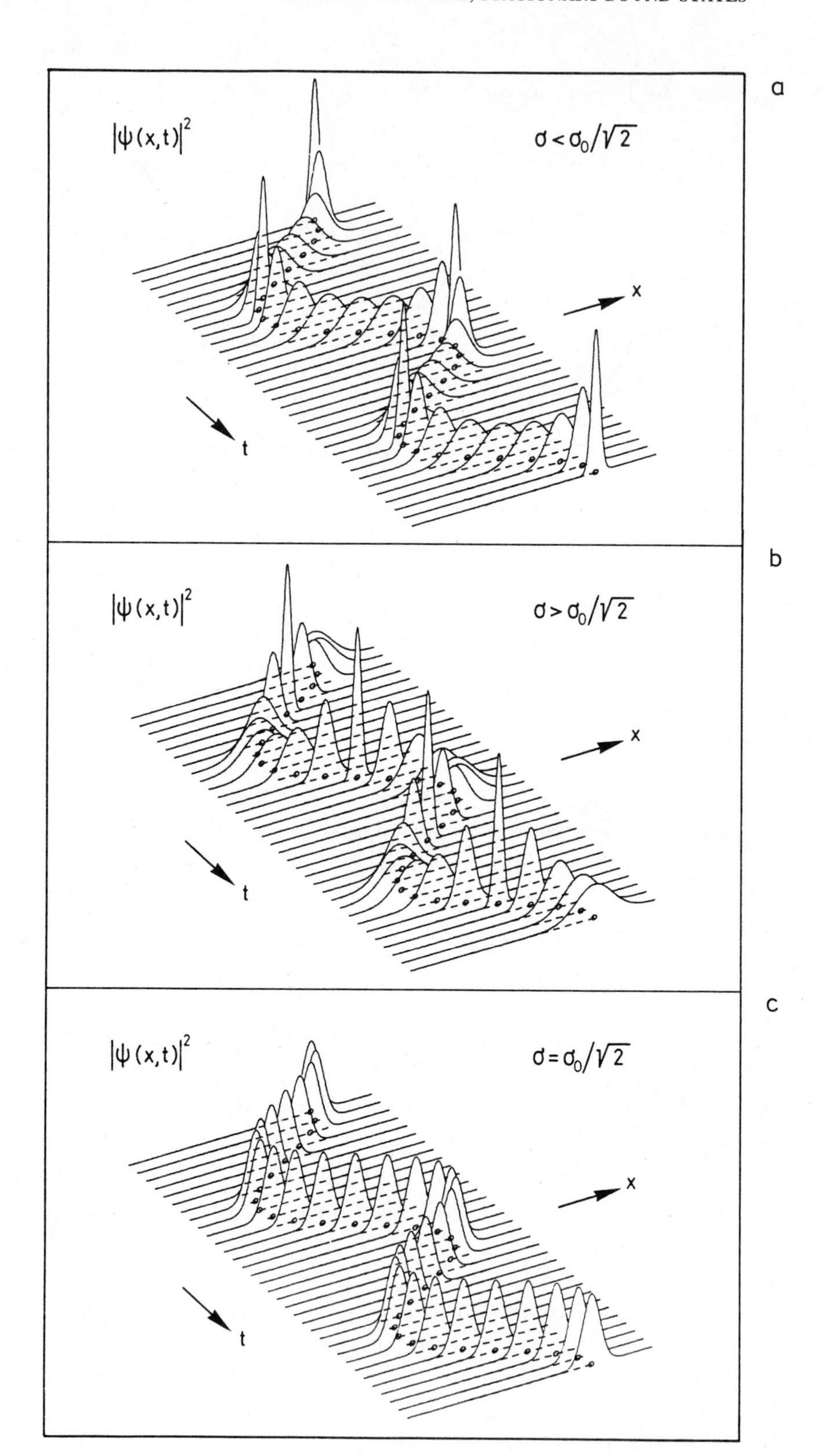

Figure 6.5 **Time development of a Gaussian wave packet, represented by its probability density, under the influence of a harmonic force. The circles show the motion of the corresponding classical particle. The broken lines extend between its turning points. The wave packet is initially at rest at an off-center position. (a) The initial width of the wave packet is smaller than that of the oscillator's ground state. (b) The initial width of the wave packet is greater. (c) Both widths are equal.**

the wave function φ_0 itself. In all three situations the behavior of the position expectation value is identical and equal to that of the classical particle.

We now look at half a period of the oscillation in more detail. In Figure 6.6a, which depicts this time interval, the time development of the probability density is plotted again for a wave packet with initial width smaller than the ground-state width $\sigma_0/\sqrt{2}$. The real and imaginary parts of the wave function are shown in Figures 6.6b and c. At the turning points, $t = 0$ and $t = T/2$, the wave function is purely Gaussian and either real or imaginary. For other moments in time, the wiggly structure originates from the superposition of the eigenfunctions of the harmonic oscillator. As is true of the eigenfunctions themselves, the distance between two nodes increases in the vicinity of the turning points. For a free harmonic wave the distance between two nodes is half the wavelength; a large wavelength signifies low momentum. We can therefore interpret the increasing distance between nodes in the vicinity of the turning points as the slowing down of the particle.

Finally, we look at the particular situation of a particle "at rest" in the center of the oscillator (Figure 6.7). Initially, the particle is sharply localized compared to the ground-state width, that is, $\sigma < \sigma_0/\sqrt{2}$. The expectation value in space remains at $x = 0$, just as the classical particle does. The width of the wave packet, however, oscillates with twice the oscillator frequency between its initial value σ and its maximum value $\sigma_0^2/(2\sigma)$. Only for initial width $\sigma = \sigma_0/\sqrt{2}$ does the absolute square of the wave packet keep its position as well as its shape.

The wave packet of Figure 6.5c is called a *coherent state* of the oscillator. While oscillating the wave packet keeps its width equal to the ground-state width of the oscillator. At all times it is a *minimum-uncertainty state*, that is to say, it fulfills Heisenberg's uncertainty principle as an equation $\Delta x\,\Delta p = \hbar/2$.

The ground state of the harmonic oscillator is a particular coherent state because it is also an eigenstate of the Hamilton operator. The other coherent states are not among the eigenstates but are particular superpositions of eigenstates of the harmonic oscillator. Since the various eigenstates differ in energy, a coherent state, except for the ground state, is a superposition of states with different numbers of energy quanta $\hbar\omega$. The weights $p(n)$, with which these states of different

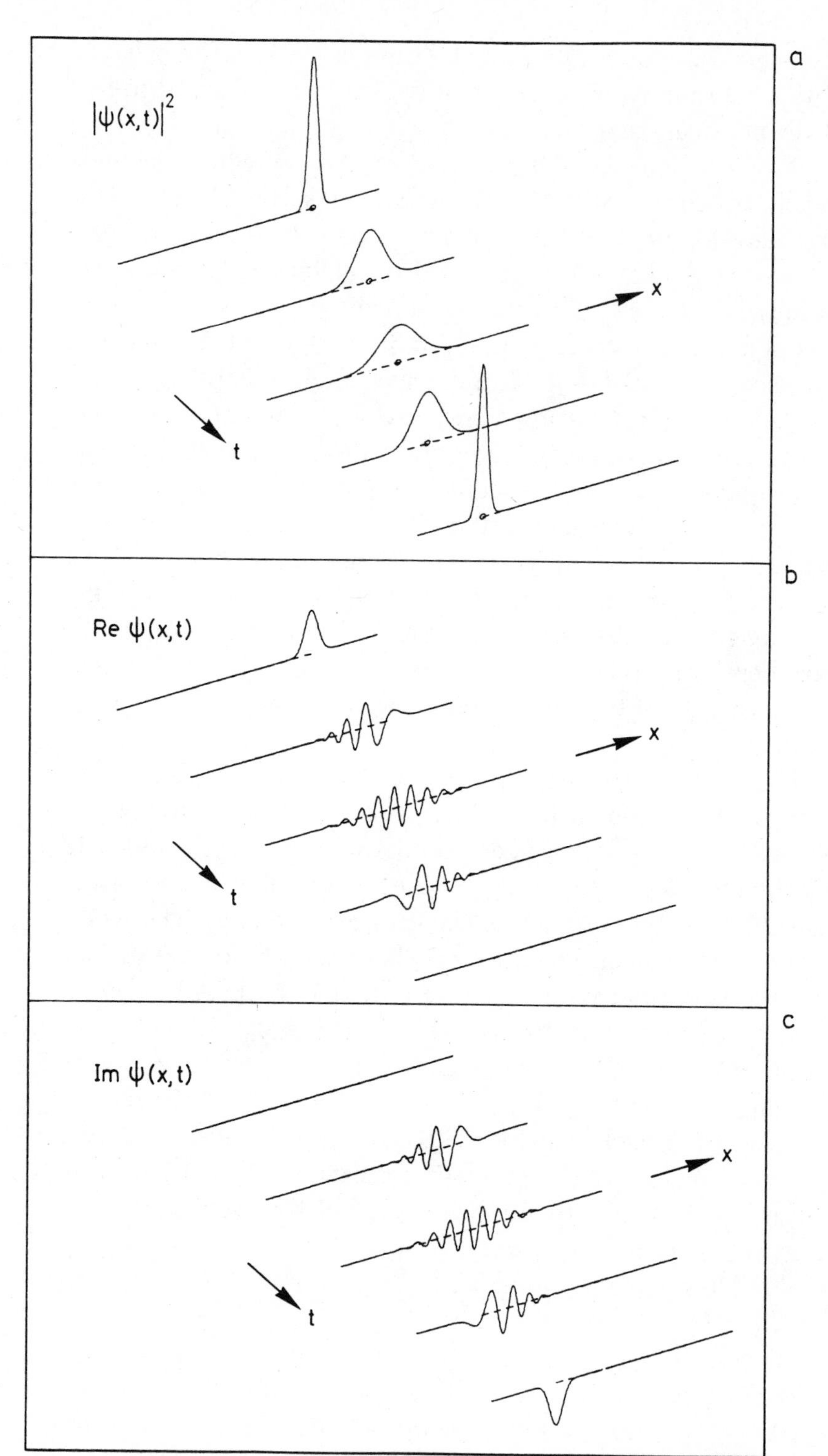

Figure 6.6 **Time development of a Gaussian wave packet under the influence of a harmonic force, observed over half an oscillation period. (a) Probability density. (b) Real part of the wave function. (c) Imaginary part of the wave function.**

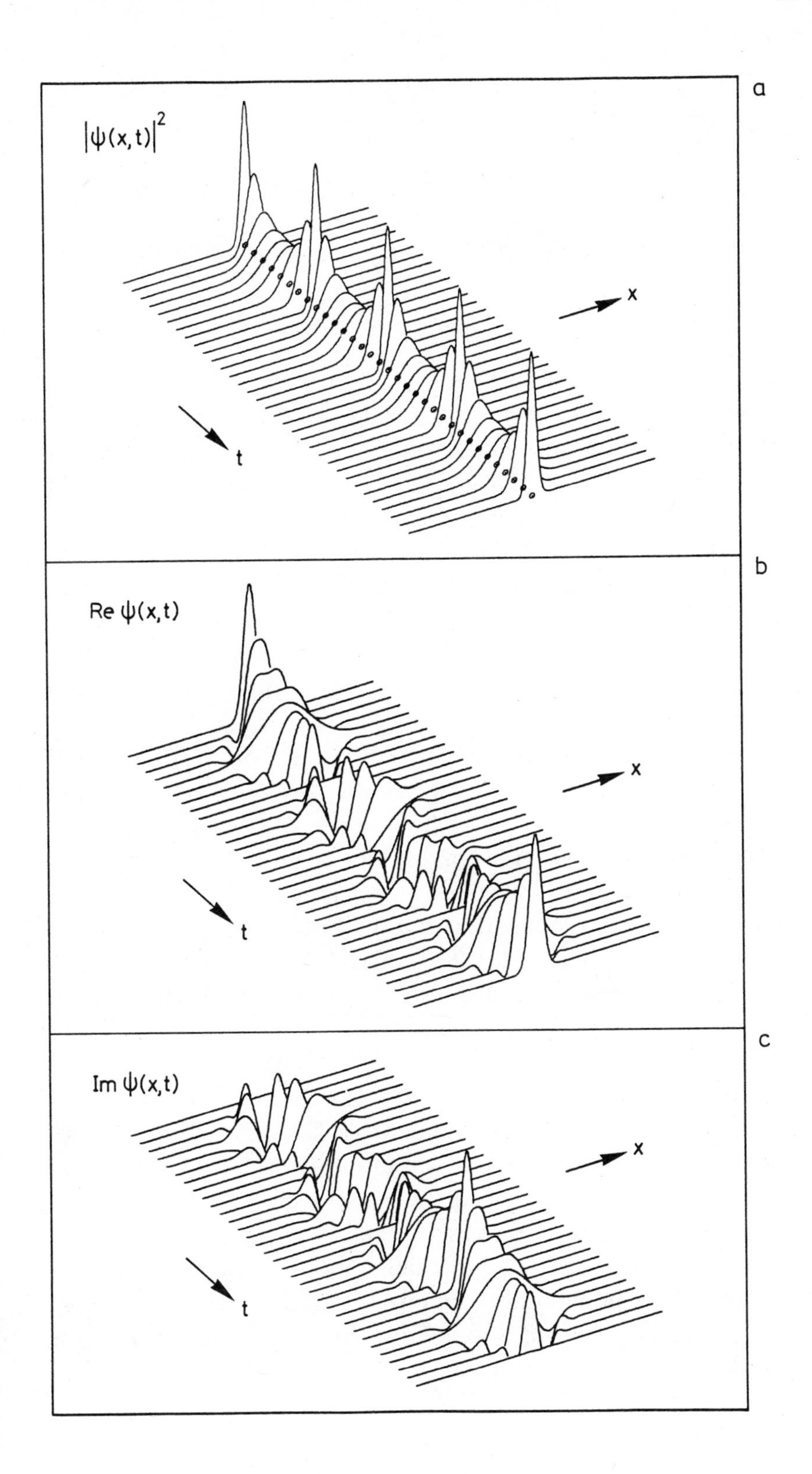

Figure 6.7 **Time development of a wave packet at rest in the center of a harmonic oscillator. The packet is represented by (a) its probability density and (b) the real part and (c) the imaginary part of its wave function. Since its initial width is different from that of the oscillator's ground state, the width of the packet oscillates in time with twice the oscillator frequency. Except for the initial position, all parameters are identical to those of Figure 6.5a.**

numbers n of energy quanta $\hbar\omega$ contribute to the coherent state, follow a Poisson distribution

$$p(n) = \frac{\langle n \rangle^n}{n!} e^{-\langle n \rangle}$$

Here $\langle n \rangle$ is the expectation value of the number of quanta given by

$$\langle n \rangle \hbar\omega = \langle E \rangle$$

where $\langle E \rangle$ is the energy expectation value of the coherent state. It therefore has a nonvanishing variance of the number of energy quanta and of the energy. If an external force acts upon a harmonic oscillator in its ground state, the oscillator responds with a transition into another coherent state. If the action of the external force is terminated at some time t_0, the state of the oscillator behaves as the coherent state of Figure 6.5c. It performs a harmonic oscillation along a classical trajectory with the frequency ω of the classical oscillator. Coherent states play an important role in quantum optics and quantum electronics.

The initial packets shown in Figures 6.5a and b are not coherent states. Their initial widths are different from the ground-state width $\sigma_0\sqrt{2}$. They are called *squeezed states*. Squeezed states are not minimum-uncertainty states at all moments of time. Twice during one period of oscillation, however, they develop into minimum-uncertainty states. As we have seen in Figures 6.5a and b, wave packets representing squeezed states also oscillate so that their expectation values follow the classical trajectories. Their widths, however, vary with time. They oscillate forth and back between a minimum and a maximum value. The distribution of the numbers of the energy quanta contributing to a squeezed state deviates from a Poisson distribution. Not being minimum-uncertainty states, squeezed states allow one observable quantity of an oscillator to be less uncertain than it is in the ground state, at the cost of the other observables occurring in Heisenberg's uncertainty principle. For this reason squeezed states are of great interest in the theory of measurement of weak signals.

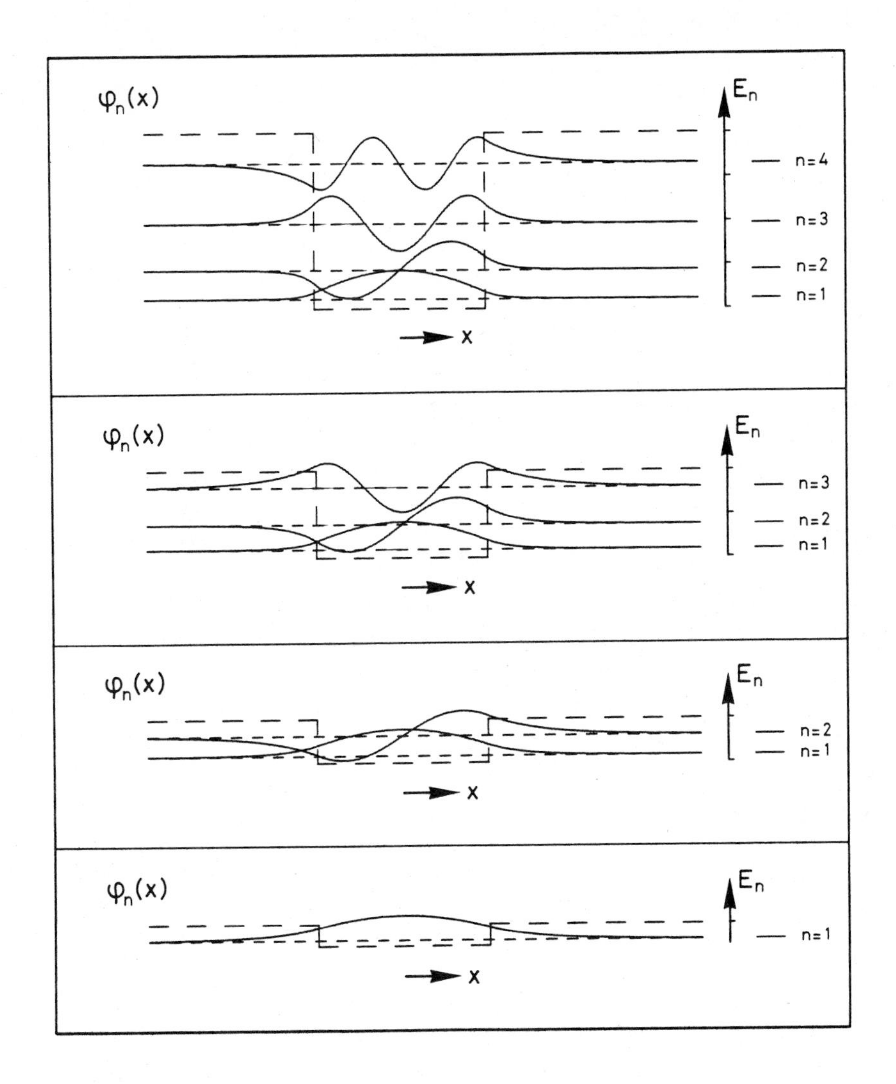

Figure 6.8 **Bound-state wave functions and energy spectra for square-well potentials of different finite depths but identical widths. The number of bound states increases with the depth of the potential.**

6.5 Spectra of Square-Well Potentials of Finite Depths

In Section 4.3 we studied the stationary bound states in a square-well potential. We found that these states exist only for discrete negative energy eigenvalues, which form the discrete spectrum of bound-state energies. The probability densities of these states are concentrated for the most part in the square well. We now discuss the bound-state spectra for different shapes of the square well.

Figure 6.8 shows the wave functions and the energy spectra for several square-well potentials of equal widths but different depths. For a well of finite depth, there is only a finite number of bound states. Their number increases with depth. In contrast to the wave functions of an infinitely deep well, the wave functions of a finite square well are different from zero outside the well but drop there exponentially to

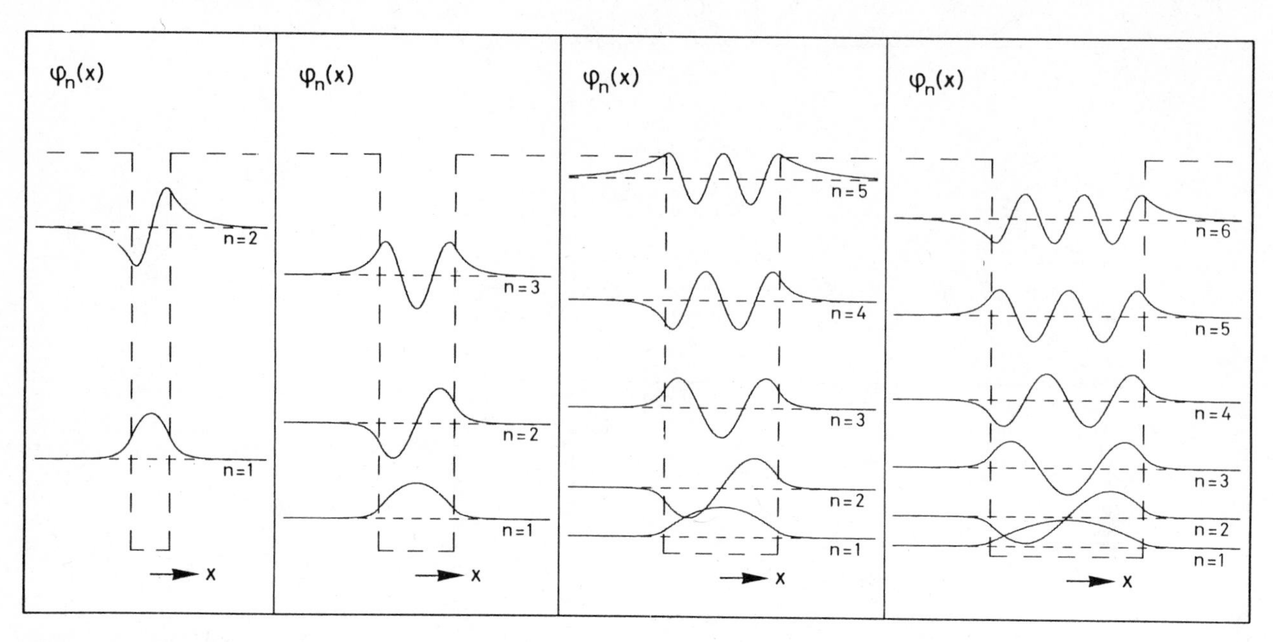

Figure 6.9 **Bound-state wave functions for square-well potentials of identical depth but different widths. The number of bound states increases with the width of the well.**

zero. The exponential falloff is fastest for the ground state. Figure 6.9 indicates that for a fixed depth the number of bound states increases as the well becomes wider.

6.6 Periodic Potentials, Band Spectra

As a first step in discussing periodic potentials as they occur in *crystals*, let us look at two potential wells more or less distant from each other. Figure 6.10 shows such potentials as well as the spectra of eigenvalues and eigenfunctions. When the two wells have some distance between them, we observe pairs of energy eigenvalues group closely together. Of the eigenfunctions belonging to each pair, one is always symmetric, the other antisymmetric. Comparing the eigenfunctions of two single wells with those of a single well, we observe that in corresponding regions they strongly resemble one another. The symmetric wave function of the double well is a smooth symmetric match with the two wave functions of the two single wells. The antisymmetric wave function of the double well is an antisymmetric match. In the limiting case when the distance between the two wells becomes zero, that is, when the wall vanishes, the eigenfunctions and the spectra become those of a single well of double width.

We now need to study the structure of the pairs of wave functions in two wells in more detail. The relation of their

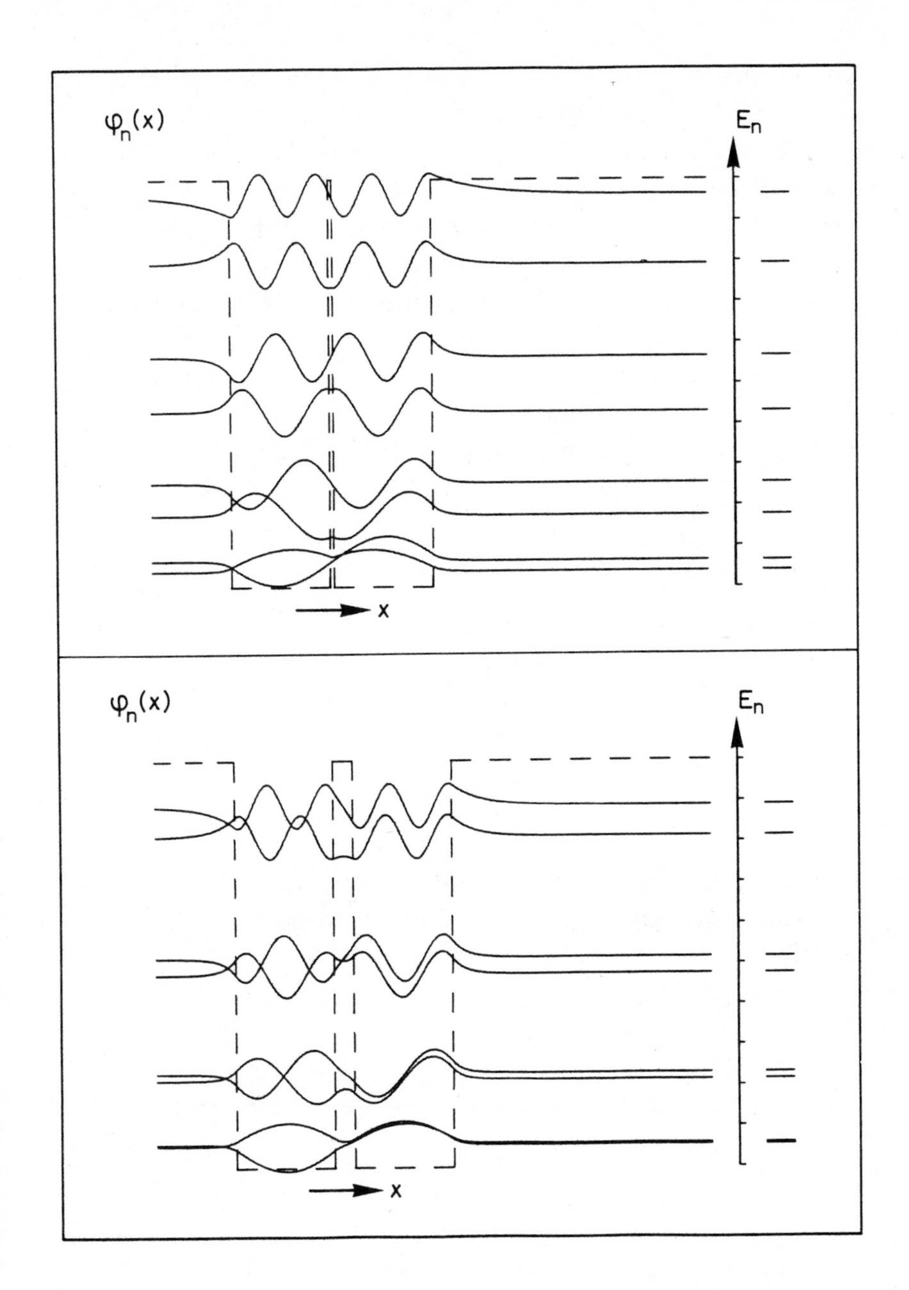

Figure 6.10 **Bound-state wave functions and energy spectra for systems of two square wells. In one system the wells are very close together, in the other some distance apart.**

structure to that of the wave functions in a single well is easily explained, using the same reasoning given in Section 4.3. To this end we divide the x-axis into five regions,

I	$-\infty < x < -d_2,$	$V(x) = 0$
II	$-d_2 < x < -d_1,$	$V(x) = -V_0$
III	$-d_1 < x < d_1,$	$V(x) = 0$
IV	$d_1 < x < d_2,$	$V(x) = -V_0$
V	$d_2 < x < \infty,$	$V(x) = 0$

where the potential has a constant value. Notice that the potential is completely symmetric with respect to point $x = 0$. That is, it does not change if x is replaced by $-x$. In regions I and V the wave function must show exponential falloff for large values of $|x|$. In regions II and IV it oscillates as a superposition of two complex exponentials.

The behavior of the wave function is determined in particular by its structure in region III, which encompasses the origin. In this domain the wave function is a linear combination of real exponentials which, because of the symmetry of the problem, are either symmetric (s) or antisymmetric (a):

$$\varphi^s_{\mathrm{III}} = A_s \tfrac{1}{2}(e^{\varkappa_s x} + e^{-\varkappa_s x}) = A_s \cosh(\varkappa_s x)$$

and

$$\varphi^a_{\mathrm{III}} = A_a \tfrac{1}{2}(e^{\varkappa_a x} - e^{-\varkappa_a x}) = A_a \sinh(\varkappa_a x)$$

The parameters $\varkappa_a$, $\varkappa_s$, are given by

$$\varkappa_s = -\frac{i}{\hbar} p'_s = \frac{1}{\hbar}\sqrt{-2mE_s}$$

$$\varkappa_a = -\frac{i}{\hbar} p'_a = \frac{1}{\hbar}\sqrt{-2mE_a}$$

where E_s and E_a are the negative bound-state energies of the symmetric and antisymmetric solutions, respectively. The wave function in region III connects the wave functions of regions II and IV. It therefore determines the overall symmetry. The total wave function is symmetric if in region III it is of the symmetric type, $\varphi^s_{\mathrm{III}} = A_s \cosh(\varkappa_s x)$. Since the antisymmetric solution has the larger average curvature, it possesses the greater kinetic energy

$$E_{\mathrm{kin}} = -\int \varphi(x) \frac{\hbar^2}{2m} \frac{d^2}{dx^2} \varphi(x)\, dx$$

compared to the symmetric solution. This explains why the splitting of the two energy eigenvalues of the bound states increases when the two wells approach each other. When the separating wall in region III has disappeared, the symmetric solution no longer has a dent in the middle.

It is now plausible that for a potential consisting of a periodic repetition of N neighboring wells, each single-well eigenvalue reflects itself in a set of N bound states of the periodic system. The spacing of the energy eigenvalues of these states may be very narrow. They are said to form an *energy band*. A crystal consists of a large number ($N \approx 10^{23}$) of regularly spaced atoms. They form a periodic electric

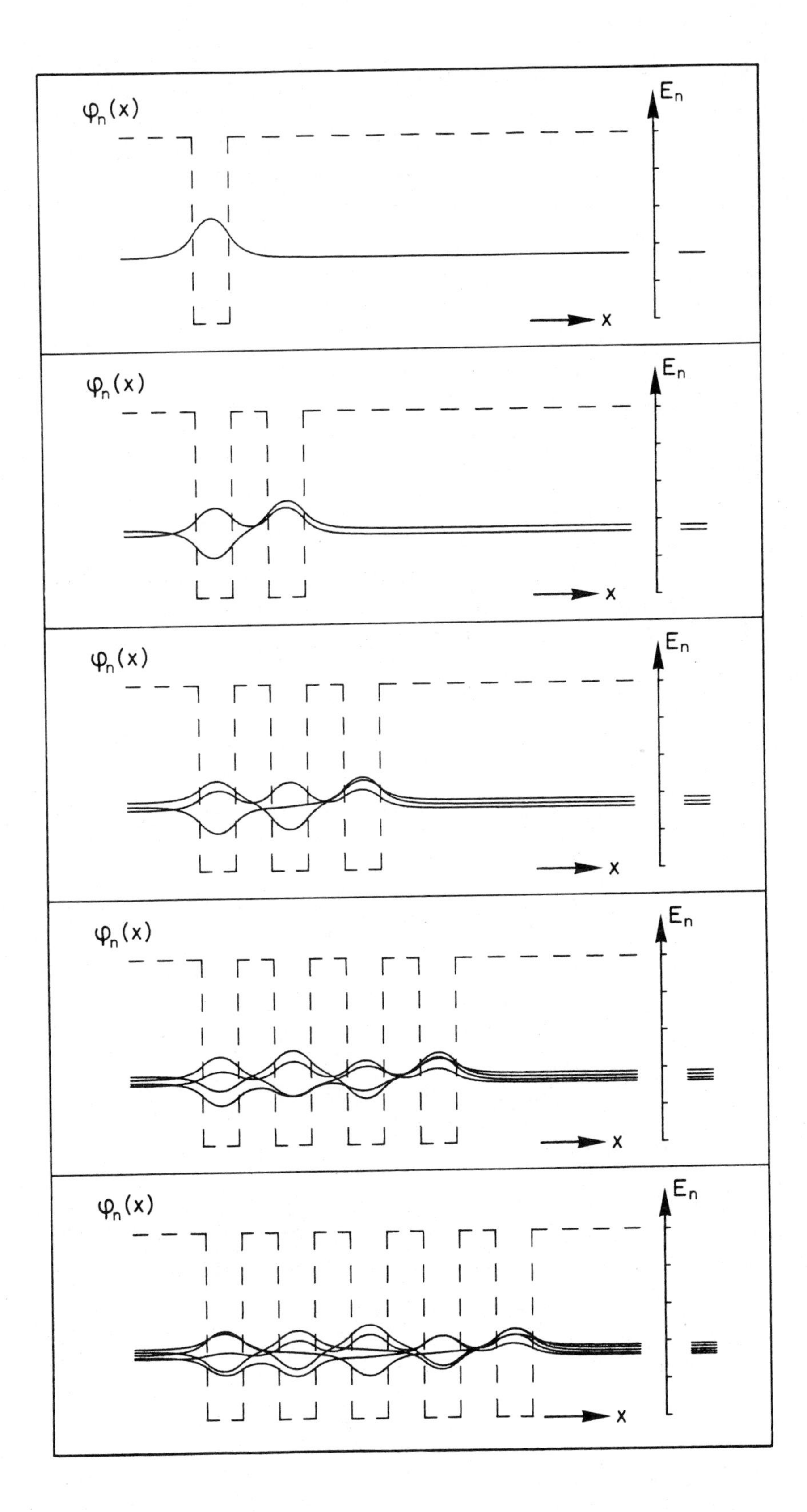

Figure 6.11 **Bound-state wave functions and energy spectra for a potential well and for potentials consisting of two, three, four, and five neighboring wells. The states have very similar energies.**

potential pattern in three dimensions giving rise to analogous band structures.

Figure 6.11 shows how the band structure, starting with the ground state of a single square well, takes form when two, three, four, and five potential wells are placed at equal distances next to it and to one another. The number of states forming the band is equal to the number of potential wells. Their spacing in energy becomes narrower as the number of wells increases. Certainly, for large numbers of potential wells forming a periodic structure, each individual band contains a large number of states represented by periodic wave functions. The wave functions of a single band can be linearly combined to form wave packets describing localized particles. If the time dependence of the eigenstates is included in the superposition (see Section 6.2), the wave packets describe particles moving freely in the periodic potential structure. In this way the free motion of electrons in the *conduction band* of the lattice of a metal or a semiconductor can be explained.

Problems

6.1 Calculate the integrals over the products of the eigenfunctions $\varphi_n(x)$ as given in Section 6.1 for the bound states of the deep square well,

$$\int_{-d/2}^{d/2} \varphi_n(x)\varphi_m(x)\,dx$$

6.2 What determines the frequency of the oscillation in the deep square well shown in Figure 6.2? What determines the wavelength of the interference wiggles in Figure 6.2?

6.3 Show that for $n = 0, 1$ the functions $\phi_n(\xi)$, $\xi = x/\sigma_0$, given in Section 6.3, are solutions of the stationary Schrödinger equation for the harmonic oscillator.

6.4 In terms of the momentum operator $\hat{p} = (\hbar/i)(d/dx)$, the operator of total energy of a harmonic oscillator is

$$H = \frac{1}{2m}\hat{p}^2 + \frac{m}{2}\omega^2 x^2$$

In a bound state of the harmonic oscillator, the expectation values $\langle p\rangle$ and $\langle x\rangle$ of momentum and position vanish. Thus the expectation values

$$\langle p^2 \rangle = (\Delta p)^2 + \langle p \rangle^2 = (\Delta p)^2$$

$$\langle x^2 \rangle = (\Delta x)^2 + \langle x \rangle^2 = (\Delta x)^2$$

are equal to the squares of the uncertainties of momentum and position. Use the uncertainty principle, from Section 3.3, to calculate the minimum energy of a bound state in a harmonic oscillator potential.

6.5 In Figures 6.5a, b, and c the behavior in time of a harmonic oscillator is plotted. The initial average values of position and momentum are identical in the three figures. The initial width of the wave packet differs, however. Explain the time developments of the widths in pictures 6.5a and b within the framework of a classical assembly of particles with corresponding classical initial conditions. Take advantage of Heisenberg's uncertainty principle in discussing the initial conditions.

6.6 Give an argument why the real and imaginary parts of the wave functions of Figures 6.6b and c have long wavelengths to the left or right when they are close to their left or right classical turning points, but not when the wave packet is in the center of the oscillator potential.

6.7 Compare the ratio $R = E_2/E_1$ of the energies E_2, E_1 of the two lowest levels in the different parts of Figure 6.9 with the corresponding ratio in the infinitely deep potential well; they are given in Section 6.1. Explain your result.

6.8 A rough approximation of the wave functions of the multiple square-well potentials in Figure 6.11 is given by

$$\varphi_n(x) = \sqrt{\frac{B_N}{N}}\, \varphi_n(x, B_N) \sum_{l=1}^{N} \varphi_1(x - x_l, d)$$

Here $\varphi_1(x - x_l, d)$ is the ground-state wave function of a single potential well of width d and depth V_0 symmetric around $x = x_l$. With B_N the width of the whole arrangement of all N square-well potentials, including their $N - 1$ separating walls, $\varphi_n(x, B_N)$ is the eigenfunction of wave number n of the square-well potential with depth V_0 and width B_N.

Using Figure 6.9 for $\varphi_n(x, B_N)$ and Figure 6.11 (top) for $\varphi_1(x - x_l, d)$, sketch the wave functions $\varphi_n(x)$ for $n = 1, 2, \ldots, N$ and $N = 2, 3, 4, 5$. Compare their appearance with that of the wave functions in Figure 6.11. Discuss their symmetry properties.

6.9 What is the parity of the ground state with respect to reflection about the symmetry point of the potential for all examples given in this chapter? Explain the result, using the square well and the harmonic oscillator potential as examples.

7. Coupled Harmonic Oscillators: Distinguishable Particles

So far we have always studied the motion of a single particle under the influence of an external potential. This potential is, however, often caused by another particle. The hydrogen atom, for example, consists of a nucleus, the proton, carrying a positive electric charge, and a negatively charged electron. The electric force between proton and electron is described by the Coulomb potential. The proton exerts a force on the electron, and—according to Newton's third law—the electron exerts a force on the proton. The proton has a mass about 2000 times the electron mass. Therefore the motion of the proton relative to the center of mass of the atom can usually be ignored. In this approximation the electron can be regarded as moving under the influence of an external potential. Generally, however, we have to describe the motion of both particles in a two-particle system. For simplicity we shall consider one-dimensional motion only; that is, both particles move only in x-direction.

7.1 The Two-Particle Wave Function

We have seen that the basic entity of quantum mechanics is the wave function describing a system, and we have discussed its interpretation as a probability amplitude. A system consisting of two particles is described by a complex wave function

$\psi = \psi(x_1, x_2, t)$ depending on time t and on two spatial coordinates x_1 and x_2. Its absolute square $|\psi(x_1, x_2, t)|^2$ is the *joint probability density* for finding at time t the two particles at locations x_1 and x_2. Of course, the wave function is assumed to be normalized, since the probability $\int_{-\infty}^{\infty}\int_{-\infty}^{\infty}|\psi(x_1, x_2, t)|^2 \, dx_1 \, dx_2$ of observing the particles anywhere in space has to be one. If the two particles differ in kind, such as a proton and an electron forming the hydrogen atom, they are said to be *distinguishable*. Two particles of the same kind, having the same masses, charges, and so on, as two electrons do, are said to be *indistinguishable*. For distinguishable particles the absolute square $|\psi(x_1, x_2, t)|^2 \, dx_1 \, dx_2$ describes the probability of finding at time t particle 1 in an interval dx_1 around position x_1 and simultaneously particle 2 in an interval dx_2 around x_2.

Figure 7.1a illustrates the joint probability density

$$\rho(x_1, x_2, t) = |\psi(x_1, x_2, t)|^2$$

for a fixed time t. Here a Cartesian coordinate system is spanned by the position variables x_1, x_2, and ρ is plotted in the direction perpendicular to the x_1, x_2-plane. In this way the function $\rho(x_1, x_2)$ appears as a surface. On two margins of the coordinate plane, functions of only one variable, x_1, or the other, x_2, are also shown. They are defined by

$$\rho_1(x_1, t) = \int_{-\infty}^{+\infty} \rho(x_1, x_2, t) \, dx_2$$

and

$$\rho_2(x_2, t) = \int_{-\infty}^{+\infty} \rho(x_1, x_2, t) \, dx_1$$

These *marginal distributions* describe the probability of observing one particle at a certain location, irrespective of the position of the second particle.

The black dot under the hump over the x_1, x_2-plane marks the expectation values $\langle x_1 \rangle$ and $\langle x_2 \rangle$ of the positions of particles 1 and 2, respectively. From the shape of the surface as well as from the marginal distributions, it is clear that in our example particle 2 is localized more sharply than particle 1.

The function shown in the three parts of Figure 7.1 is a Gaussian distribution of the two variables x_1, x_2. It has the

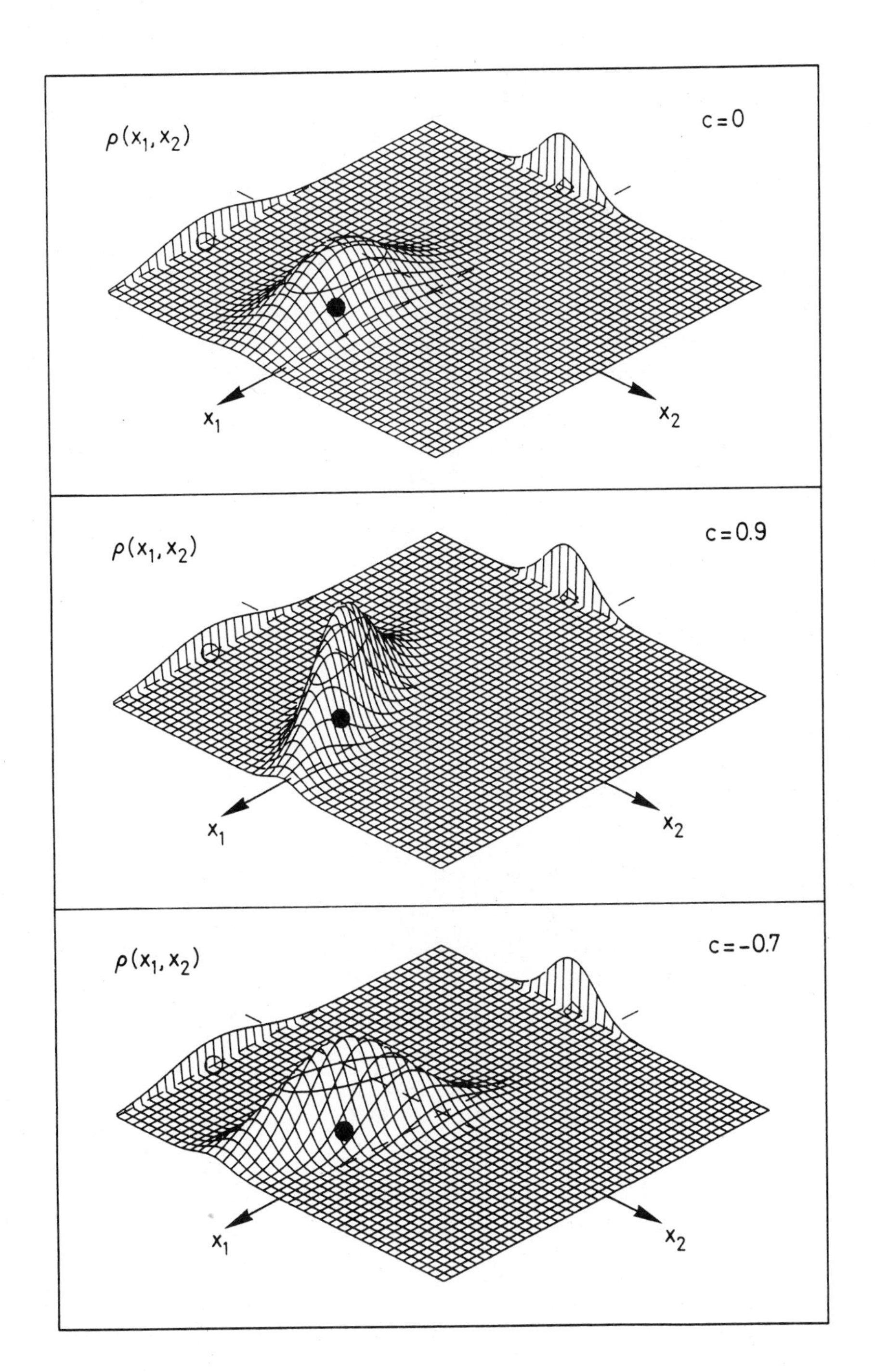

Figure 7.1 **Joint probability density $\rho(x_1, x_2)$ for a system of two particles. It forms a surface over the x_1, x_2-plane. The marginal distributions $\rho_1(x_1)$ and $\rho_2(x_2)$ are plotted as curves over the margins parallel to the x_1-axis and the x_2-axis, respectively. In each plot classical position x_{10}, x_{20} is indicated by a black dot in the x_1, x_2-plane as well as by its projections on the margins. Also shown is the covariance ellipse. The three plots apply to the cases of (a) uncorrelated variables and (b) positive and (c) negative correlation between x_1 and x_2.**

general form

$$\rho(x_1, x_2) = A \exp\left\{ -\frac{1}{2(1-c^2)} \left[\frac{(x_1 - \langle x_1 \rangle)^2}{\sigma_1^2} - 2c \frac{(x_1 - \langle x_1 \rangle)}{\sigma_1} \frac{(x_2 - \langle x_2 \rangle)}{\sigma_2} + \frac{(x_2 - \langle x_2 \rangle)^2}{\sigma_2^2} \right] \right\}$$

where the normalization constant is given by

$$A = \frac{1}{(2\pi)\sigma_1\sigma_2\sqrt{1 - c^2}}$$

This *bivariate Gaussian* depends on five parameters. They are the expectation values $\langle x_1 \rangle$ and $\langle x_2 \rangle$, the widths σ_1 and σ_2, and the *correlation coefficient* c. The expectation values correspond to the classical positions of the two particles.

The marginal distributions of the bivariate Gaussian are Gaussians of a single variable:

$$\rho_1(x_1) = \frac{1}{\sqrt{2\pi}\,\sigma_1} \exp\left[-\frac{(x - \langle x_1 \rangle)^2}{2\sigma_1^2} \right]$$

$$\rho_2(x_2) = \frac{1}{\sqrt{2\pi}\,\sigma_2} \exp\left[-\frac{(x - \langle x_2 \rangle)^2}{2\sigma_2^2} \right]$$

Each marginal distribution depends on two parameters only, the expectation value and the width of its variable.

Lines of constant probability density in x_1, x_2 are the lines of intersection between the surface $\rho(x_1, x_2)$ and a plane $\rho = a =$ const. These lines are ellipses. Their size depends on the constant a. One of these ellipses, if projected on the x_1-axis and the x_2-axis, yields lines of lengths $2\sigma_1$ and $2\sigma_2$, respectively. It is called the *covariance ellipse* of the bivariate Gaussian. The three plots in Figure 7.1 differ only by the value c of the covariance. The covariance ellipses are shown as lines of constant probability on the surfaces $\rho(x_1, x_2)$. For $c = 0$ the principal axes of the covariance ellipse are parallel to the coordinate axes. In this situation variables x_1 and x_2 are *uncorrelated*, that is, knowledge that $x_1 > \langle x_1 \rangle$ holds true does not tell us which is more probable, observing $x_2 > \langle x_2 \rangle$ or $x_2 < \langle x_2 \rangle$. For uncorrelated variables the relation between the joint probability density and the marginal distribution is simple, $\rho(x_1, x_2) = \rho_1(x_1)\rho_2(x_2)$.

The situation is different for correlated variables, that is, for $c \neq 0$. For a positive correlation, $c > 0$, the major axis of the ellipse lies along a direction between those of the x_1-axis and the x_2-axis. If we know that $x_1 > \langle x_1 \rangle$ is valid it is more probable to have $x_2 > \langle x_2 \rangle$ than to have $x_2 < \langle x_2 \rangle$. If, on the other hand, the correlation is negative, $c < 0$, the major axis has a direction between those of the x_1-axis and the negative x_2-axis. In this situation, once it is known that $x_1 > \langle x_1 \rangle$ is valid, $x_2 < \langle x_2 \rangle$ is more probable than $x_2 > \langle x_2 \rangle$.

The amount of correlation is measured by the numerical value of c, which can vary in the range $-1 < c < 1$. In the limiting case of total correlation, $c = \pm 1$, the covariance ellipse degenerates to a line, the principal axis. The joint probability density is completely concentrated along this line. That is, knowing the value x_1 of one variable, we also know the value x_2 of the other.

7.2 Coupled Harmonic Oscillators

As a particularly simple and instructive dynamical system, let us investigate the motion of two distinguishable particles of equal mass in external oscillator potentials. Both particles are coupled by another harmonic force. The external potentials are assumed to have the same form,

$$V(x_1) = \frac{k}{2}x_1^2, \quad V(x_2) = \frac{k}{2}x_2^2, \qquad k > 0$$

The potential energy of the coupling is

$$V_c(x_1, x_2) = \frac{\varkappa}{2}(x_1 - x_2)^2, \qquad \varkappa > 0$$

The Schrödinger equation for the wave function $\psi(x_1, x_2, t)$ is then

$$i\hbar \frac{\partial}{\partial t}\psi(x_1, x_2, t) = H\psi(x_1, x_2, t)$$

where H is the Hamilton operator of the form

$$H = -\frac{\hbar^2}{2m}\frac{\partial^2}{\partial x_1^2} + V(x_1) - \frac{\hbar^2}{2m}\frac{\partial^2}{\partial x_2^2} + V(x_2) + V_c(x_1, x_2)$$

This equation is written, in analogy to the single-particle equation, so that its right-hand side is the sum of the kinetic and potential energies of the two particles.

The Schrödinger equation is solved with an initial condition that places the expectation values of the two particles into positions $x_{10} = \langle x_1(t_0)\rangle$ and $x_{20} = \langle x_2(t_0)\rangle$ at time $t = t_0$. We consider the particular situation in which the expectation values of the initial momenta of the two particles are zero. In quantum mechanics there is an infinite variety of wave functions with the expectation values $\langle x_1(t_0)\rangle = x_{10}$, $\langle p_1(t_0)\rangle = 0$, and $\langle x_2(t_0)\rangle = x_{20}$, $\langle p_2(t_0)\rangle = 0$ at initial time t_0 describing the particles. Even if we restrict ourselves to the bell-shaped form of a Gaussian wave packet at t_0, we still have to specify

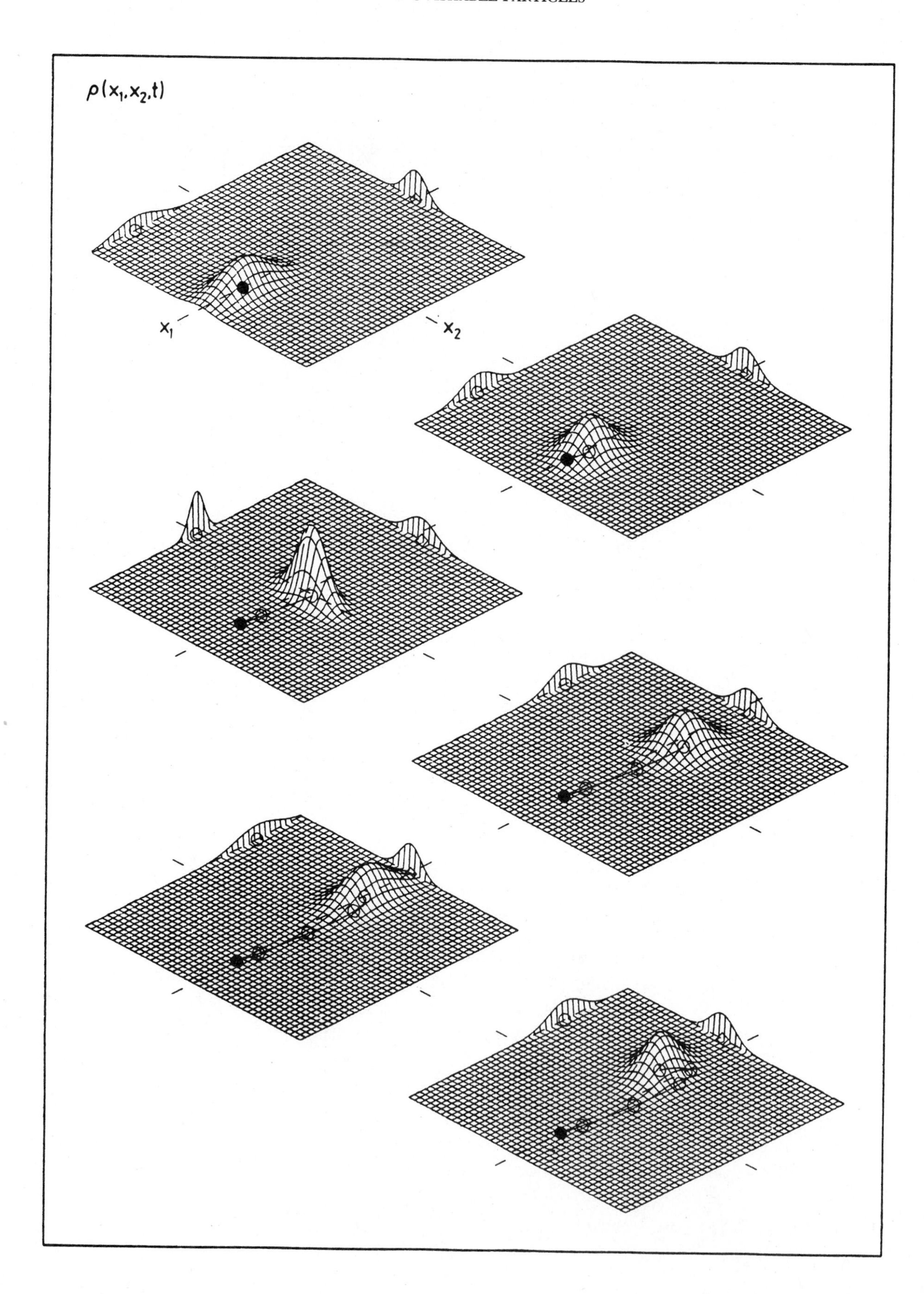

$\rho(x_1, x_2, t)$
x_1
x_2

its widths and correlation. For later moments of time $t > t_0$, the time-dependent solution evolving out of the initial wave packet according to the Schrödinger equation for two coupled harmonic oscillators maintains the Gaussian form. Its parameters, however, become time-dependent.

In Figure 7.2 the joint probability distribution $\rho(x_1, x_2, t)$ is shown for several times $t = t_0, t_1, \ldots, t_N$ together with its marginal distributions $\rho_1(x_1, t)$ and $\rho_2(x_2, t)$. We observe rather complex behavior. The hump where the probability density is large moves in the x_1, x_2-plane and at the same time changes its form; that is, the widths σ_1, σ_2 as well as the correlation coefficient c are time-dependent. The motion of the position expectation values $\langle x_1 \rangle, \langle x_2 \rangle$ is shown as a trajectory in the x_1, x_2-plane, and the initial positions x_{10}, x_{20} at $t = t_0$ are marked as a black dot at the beginning of the trajectory. The last dot on the trajectory corresponds to the time for which the probability density is plotted. A crude survey can be made by looking only at the marginal distributions.

Figure 7.3a shows the time developments of the marginal distributions of the system in Figure 7.2. The left-hand part contains the marginal distribution $\rho_1(x_1, t)$, the right-hand part $\rho_2(x_2, t)$. The symbols on the x_1- and x_2-axes indicate position expectation values of the particles that are identical to the classical positions. The initial momenta were chosen so that the particles are, classically speaking, initially at rest. Particle 1 is initially in an off-center position, particle 2 in the center. It is obvious from Figure 7.3a that the position expectation values have the well-known energy exchange pattern of coupled oscillators. The oscillation amplitude of particle 1 decreases with time, whereas that of particle 2 increases until it has reached the initial amplitude of particle 1. At this moment the two particles have interchanged their roles, and the energy is now transferred from particle 2 to particle 1. The time developments of the widths in Figure 7.3a are much less clear.

***Figure* 7.2 Joint probability density $\rho(x_1, x_2, t)$ and marginal distributions $\rho_1(x_1, t)$, $\rho_2(x_2, t)$ for two distinguishable particles forming a system of coupled harmonic oscillators. The different plots apply to various times $t_j = t_0, t_1, \ldots, t_N$. The classical position of the two particles at the various moments in time is marked by a dot in the x_1, x_2-plane and by two dots on the margins. The initial dot for $t_j = t_0$ is black. The classical motion between t_0 and t_j is represented by the trajectory drawn in the x_1, x_2-plane.**

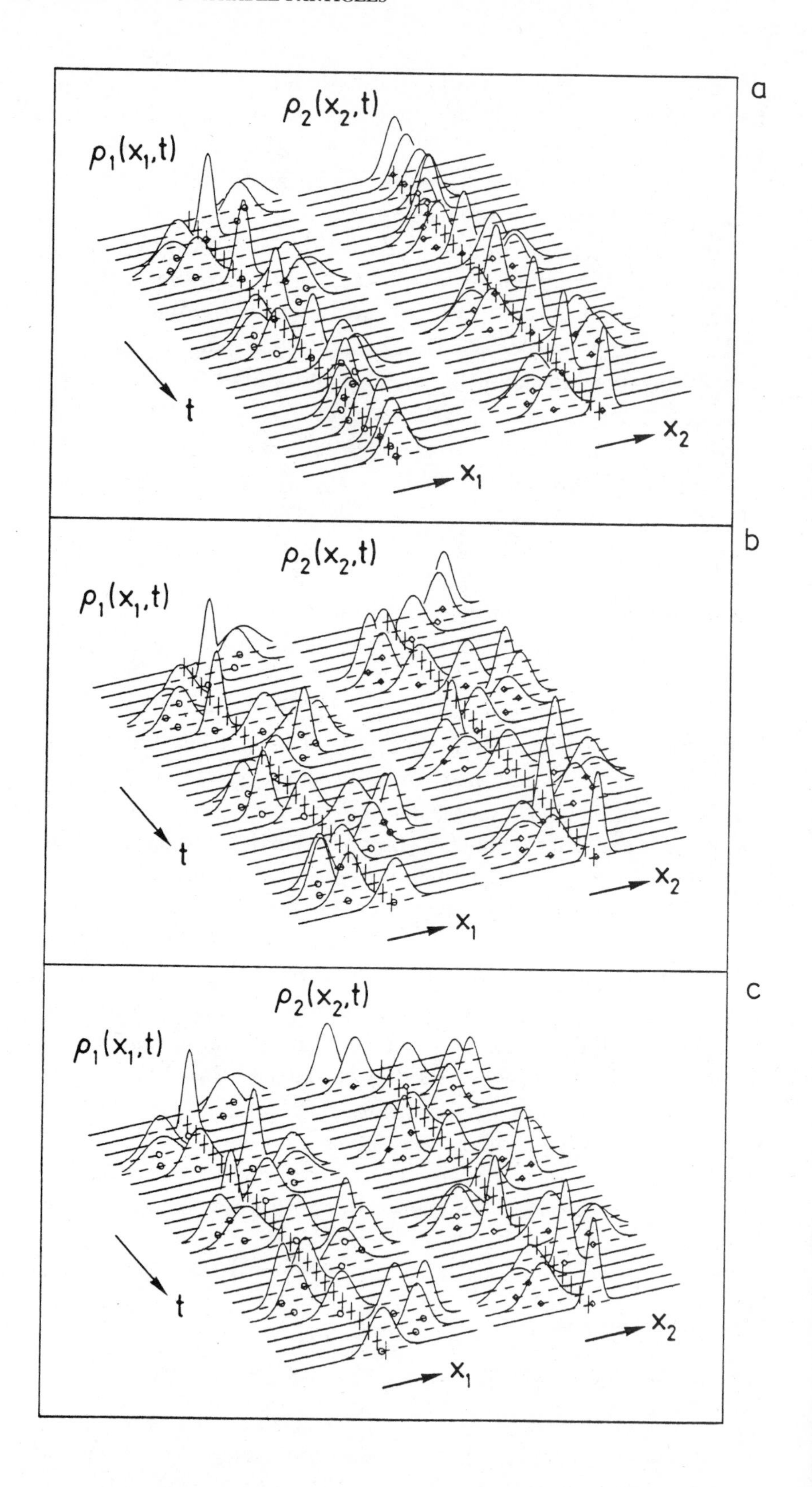
a
ρ₁(x₁,t)
ρ₂(x₂,t)
t
x₁
x₂
b
ρ₁(x₁,t)
ρ₂(x₂,t)
t
x₁
x₂
c
ρ₁(x₁,t)
ρ₂(x₂,t)
t
x₁
x₂

For a systematic study of coupled harmonic oscillators, it is important to note the following.

1. The time dependence of the expectation values $\langle x_1(t)\rangle, \langle x_2(t)\rangle$ is determined by their initial values and it is identical to that of classical particles. It is independent of the initial values σ_{10}, σ_{20} and c_0 of the widths and of the correlation coefficient.
2. The time dependence of the widths $\sigma_1(t), \sigma_2(t)$ and of the correlation coefficient $c(t)$ is given by the initial values of these quantities. It does not depend on the initial positions x_{10}, x_{20}.

The classical system of two coupled harmonic oscillators has two characteristic *normal oscillations*. They can be excited by choosing particular initial conditions. For one of the normal oscillations the center of mass remains at rest. This situation can be realized by choosing initial positions opposite to each other, $x_{10} = -x_{20}$, so that the center of gravity is initially at the origin. Since the sum of forces on the two masses in this position vanishes, the center of mass stays at rest. The oscillation occurs only in the relative coordinate $r = x_2 - x_1$. Its angular frequency is

$$\omega_r = \sqrt{(k + 2\varkappa)/m}$$

The second normal oscillation is brought about by initial conditions that make the force between the two masses vanish. That is, the two particles have the same initial position $x_{10} = x_{20} = R_0$, which is therefore also the initial position R of the center of mass. Since no force acts between the two particles, they stay together at all times, $x_1(t) = x_2(t)$. Now, however, because the sum of forces does not vanish, the center of mass moves under the influence of a linear force. Thus it performs a harmonic oscillation with the angular frequency

$$\omega_R = \sqrt{k/m}$$

Figure 7.3 Time development of the marginal distribution $\rho_1(x_1,t)$ on the left and marginal distribution $\rho_2(x_2,t)$ on the right for a system of coupled oscillators. The classical positions of the two distinguishable particles are plotted on the two axes with circles for particle 1 and diamonds for particle 2. They coincide with the expectation values computed with marginal distributions. (a) The initial position expectation value of particle 2 is zero. (b) The particles are excited in a normal oscillation in which the center of mass oscillates and there is no relative motion. (c) The particles are excited in a normal oscillation in which there is relative motion and the center of mass is at the rest. In all three cases the initial momentum expectation values of the two particles are zero.

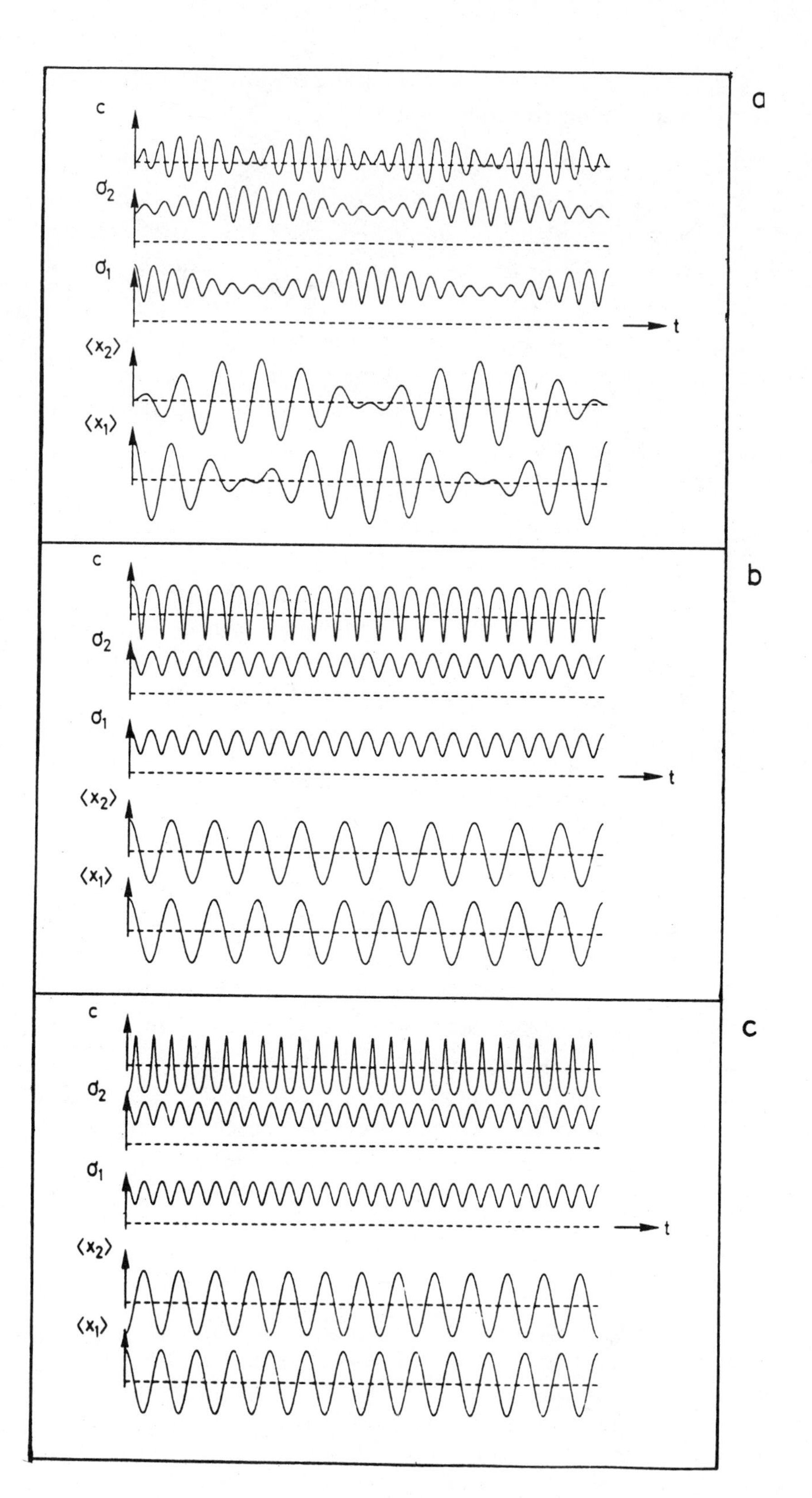

Figure 7.4 **Time dependences of the expectation values $\langle x_1(t)\rangle$, $\langle x_2(t)\rangle$, the widths $\sigma_1(t)$, $\sigma_2(t)$, and the correlation $c(t)$ for a system of coupled harmonic oscillators. (a) Rather general initial conditions were chosen. (b) The oscillation of the expectation values corresponds to an oscillation of the center of mass with frequency ω_R. The initial values $\sigma_1(t_0)$, $\sigma_2(t_0)$, and $c(t_0)$ were chosen so that the two widths and the correlation oscillate with frequency $2\omega_R$. (c) The oscillation of the expectation values corresponds to an oscillation in the relative motion with frequency ω_r; the widths and the correlation oscillate with frequency $2\omega_r$.**

Oscillations with arbitrary initial conditions can be described as superpositions of the two normal oscillations, causing such phenomena as the transfer of energy from one mass to the other. Normal oscillations can also be produced in the quantum-mechanical coupled oscillators by exactly the same prescription. Examples are given in Figures 7.3b and c.

Figure 7.4a presents the oscillations of the expectation values $\langle x_1(t)\rangle, \langle x_2(t)\rangle$, the widths $\sigma_1(t), \sigma_2(t)$, and the correlation $c(t)$ for a rather general set of initial conditions. All these quantities have beats. We already know that the beats in the time dependence of the expectation values come from superposition of the two normal oscillations.

As we know from the example of the single harmonic oscillator (Section 6.4), the width of the probability distribution oscillates with twice the frequency of the oscillator. We may therefore stipulate that the widths $\sigma_1(t), \sigma_2(t)$ and the correlation coefficient $c(t)$ will show periodicity with twice the normal frequencies if their initial values σ_{10}, σ_{20}, and c_0 are appropriately chosen.

Figure 7.4b shows such a particular situation. Here the dependences of the expectation values $\langle x_1(t)\rangle, \langle x_2(t)\rangle$ and of the widths and the correlation coefficient are plotted. The initial-position expectation values were chosen so that the oscillators have the normal frequency ω_R. The initial widths and correlation coefficient were selected so that the frequency of these quantities is $2\omega_R$. As stated earlier, the time dependence of σ_1, σ_2, and c is totally independent of the initial positions. In our example the positions were chosen to oscillate with frequency ω_R to allow for a simple comparison between the frequency ω_R of the positions and $2\omega_R$ of the widths.

Figure 7.4c gives the analogous plots for the other normal frequency ω_r. It is interesting to note that preparing the normal modes in the widths requires an initial condition $\sigma_{10} = \sigma_{20}$, a relation which then holds for all moments in time. The variation in time of σ_1 and σ_2 is actually a periodic oscillation of frequency $2\omega_R$ or $2\omega_r$ added to a constant. Furthermore, it should be remarked that the initial value c_0 of the correlation coefficient is different from zero in both cases.

For one particular set of initial values σ_1, σ_2, and c, these quantities remain constant independent of time, as shown in Figure 7.5. In this situation the correlation coefficient is always positive, which is easily understood if we remember the attractive force between the two oscillators. If the coordinate of one particle is known, the other one is probably in its

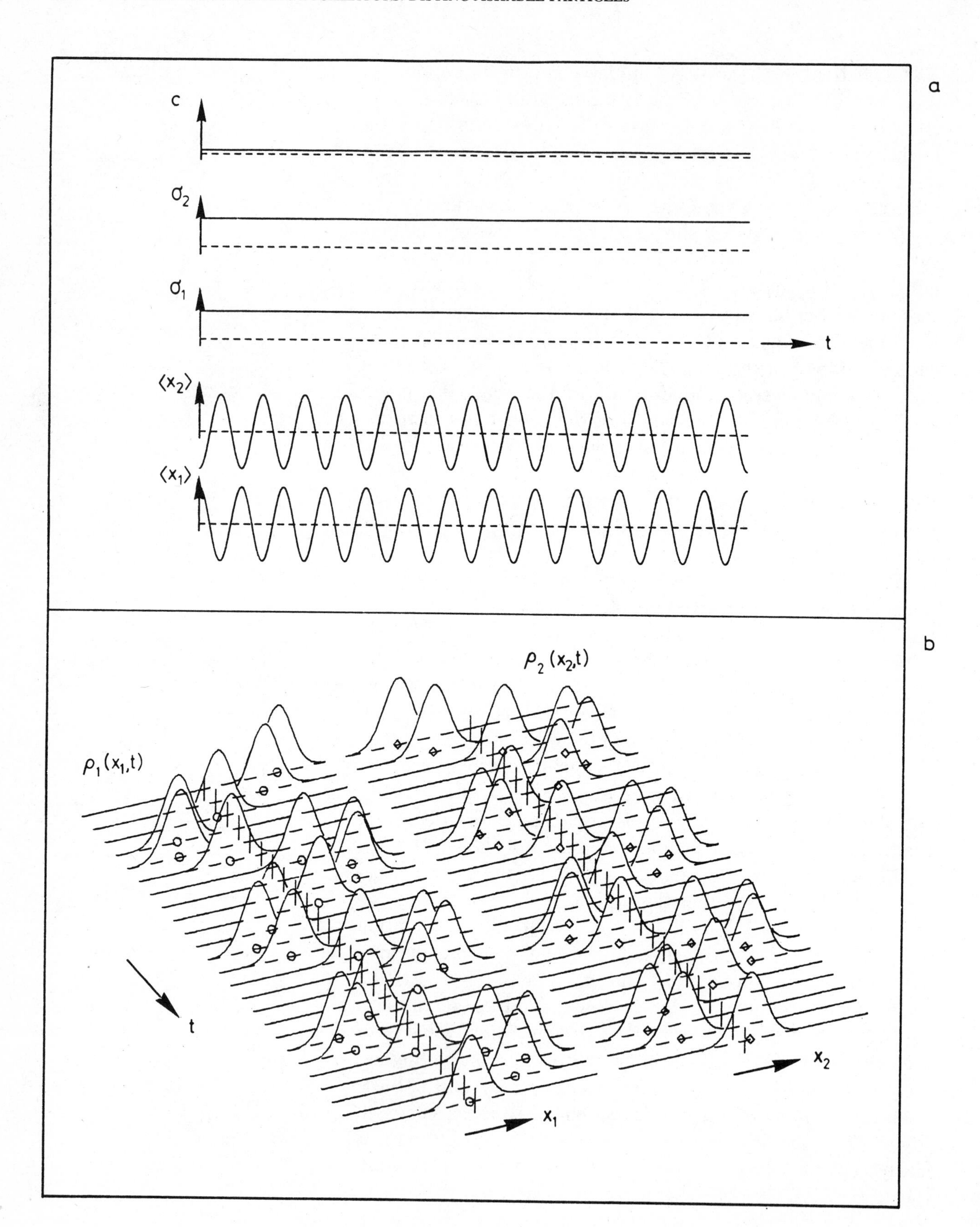
a
c
σ2
σ1
t
⟨x2⟩
⟨x1⟩
b
ρ2 (x2,t)
ρ1 (x1,t)
t
x2
x1

neighborhood rather than elsewhere. This probability constitutes the positive correlation between the variables x_1 and x_2.

7.3 Stationary States

The stationary wave functions φ_E are solutions of the time-independent Schrödinger equation

$$H\varphi_E(x_1, x_2) = E\varphi_E(x_1, x_2)$$

The Hamiltonian is that given at the beginning of Section 7.2. As in classical mechanics, the Hamiltonian can be separated into two terms,

$$H = H_R + H_r$$

where

$$H_R = -\frac{\hbar^2}{2M}\frac{d^2}{dR^2} + kR^2$$

governs the motion of the center of mass

$$R = \tfrac{1}{2}(x_1 + x_2)$$

and

$$H_r = -\frac{\hbar^2}{2\mu}\frac{d^2}{dr^2} + \frac{1}{2}\left(\frac{k}{2} + \varkappa\right)r^2$$

determines the dynamics of the relative motion in the relative coordinate

$$r = x_2 - x_1$$

Here $M = 2m$ denotes the total mass, $\mu = m/2$ the "reduced mass" of the system.

The separation of the Hamiltonian permits a factored *ansatz* for the stationary wave functions,

$$\varphi_E(x_1, x_2) = U_N(R)\,u_n(r)$$

with the factors fulfilling the equations

$$H_R U_N(R) = (N + \tfrac{1}{2})\hbar\omega_R U_N(R)$$

$$H_r u_n(r) = (n + \tfrac{1}{2})\hbar\omega_r u_n(r)$$

***Figure* 7.5 Coupled harmonic oscillators. The initial conditions $\langle x_1(t_0)\rangle$, $\langle x_2(t_0)\rangle$ are the same as in Figure 7.4c, corresponding to an oscillation in the relative motion. The parameters $\sigma_1(t_0)$, $\sigma_2(t_0)$, and $c(t_0)$, however, were chosen so that the widths and the correlation coefficient remain constant independent of time. (a) Time dependences of the quantities $\langle x_1(t)\rangle$, $\langle x_2(t)\rangle$, $\sigma_1(t)$, $\sigma_2(t)$, $c(t)$. (b) Time developments of the marginal distributions.**

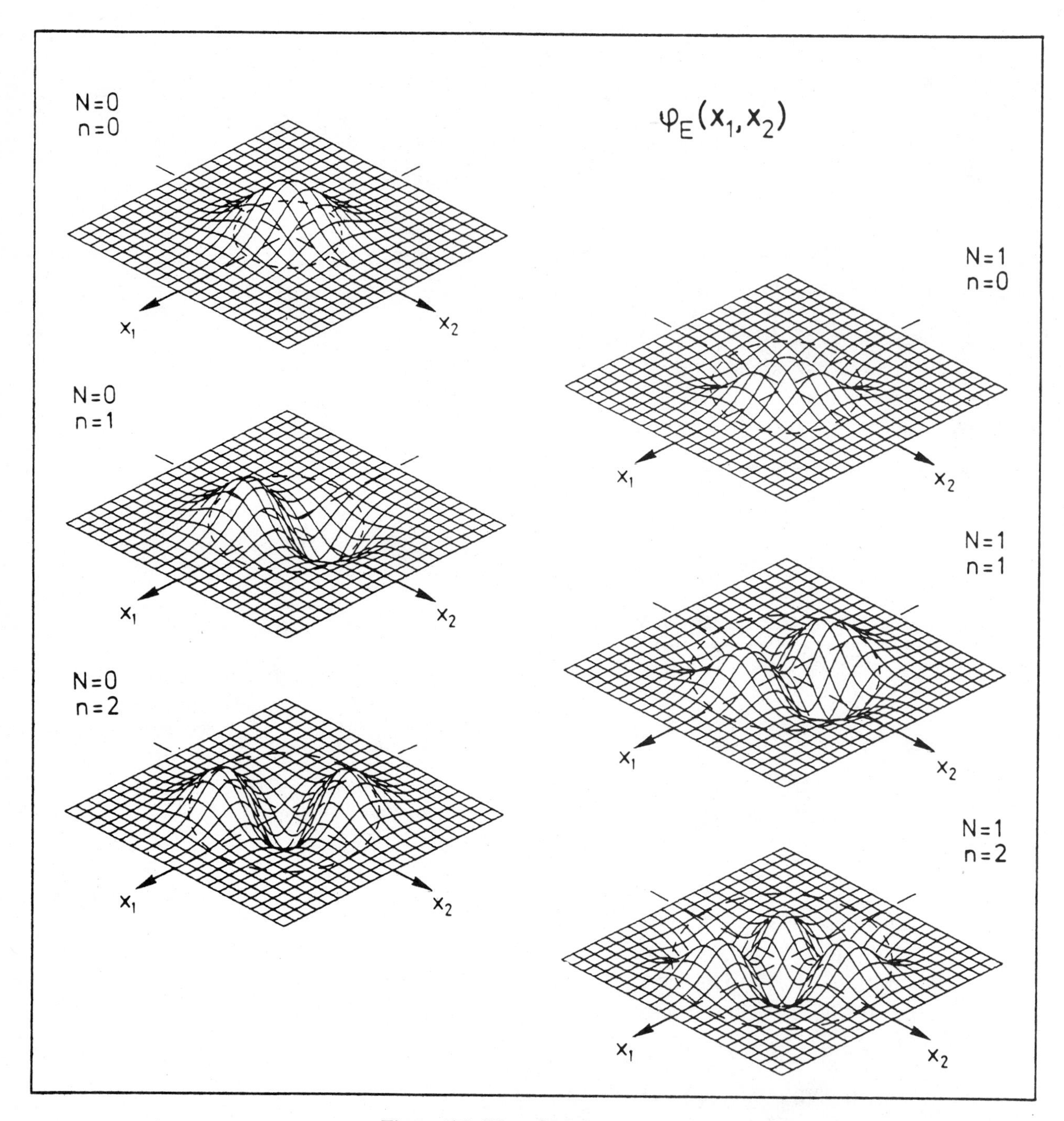

Figure 7.6 Wave function $\varphi_E(x_1, x_2)$ for stationary states of a system of two coupled harmonic oscillators for low values of the quantum numbers N and n. Note that $\varphi_E(x_1, x_2)$ is symmetric with respect to the permutation $(x_1, x_2) \rightarrow (x_2, x_1)$ for n even and antisymmetric for n odd. The dashed ellipse in the x_1, x_2-plane corresponds to the energetically allowed region for classical particles.

for the center-of-mass and relative motions, respectively. The functions $U_N(R)$ and $u_n(r)$ are thus the eigenfunctions for harmonic oscillators of single particles, as discussed in Section 6.3. The total energy E is simply the sum of the center-of-mass and relative energies:

$$E = (N + \tfrac{1}{2})\hbar\omega_R + (n + \tfrac{1}{2})\hbar\omega_r$$

The energy spectrum now has two independent quantum numbers, N for center-of-mass excitations and n for relative excitations. Figure 7.6 shows the stationary states $\varphi_E(x_1, x_2)$ for the lowest values of quantum numbers N and n.

Problems

7.1 Determine the coordinate transformation and thus the new coordinates ξ_1, ξ_2 that transform the exponent of the Gaussian function for $\rho(x_1, x_2)$, as given in Section 7.1, to normal form so that we have

$$\rho(x_1, x_2) = A \exp\left\{-\left[\frac{(\xi_1 - \langle x_1\rangle)^2}{\sigma_1'^2} + \frac{(\xi_2 - \langle x_2\rangle)^2}{\sigma_2'^2}\right]\right\}$$

7.2 Give an argument why the shape of the wave packet $\rho(x_1, x_2, t)$ in Figure 7.2 changes with time as it does. It may help you to look carefully at Figure 6.5 for the single harmonic oscillator.

7.3 Derive the relations given in Section 7.2 for the two normal frequencies ω_r and ω_R for a classical system of coupled harmonic oscillators.

7.4 Verify that the Hamiltonian for a system of two coupled oscillators can be decomposed into the Hamiltonian H_R for the center-of-mass motion and the Hamiltonian H_r for relative motion, as given at the beginning of Section 7.3.

7.5 In Section 7.2 the oscillators decouple for $\varkappa = 0$. The stationary Schrödinger equation can be solved by a product *ansatz* in the variables x_1, x_2,

$$\varphi_E(x_1, x_2) = \varphi_{E_1}(x)\varphi_{E_2}(x_2), \qquad E = E_1 + E_2$$

Show that $\varphi_{E_1}(x_1), \varphi_{E_2}(x_2)$ are then solutions of the stationary Schrödinger equation for the one-dimensional harmonic oscillator.

8. Coupled Harmonic Oscillators: Indistinguishable Particles

8.1 The Two-Particle Wave Function for Indistinguishable Particles

The probability density $\rho(x_1, x_2, t) = |\psi(x_1, x_2, t)|^2$ used in the last chapter described the joint probability of observing particle 1 at position x_1 and particle 2 at x_2. There is no difficulty with this notion as long as particle 1 can be unambiguously attributed to position x_1 and particle 2 to x_2. To so attribute them, however, presupposes that particles 1 and 2 have different identities, that they can be distinguished by properties other than being at different locations or having different momenta. They must have different intrinsic properties, for instance, different masses or different electric charges. A system consisting of an electron and a proton is one in which the two particles have different intrinsic properties. A system consisting of two electrons is not. For such a system it is impossible in principle to distinguish the two particles if they are close to each other.

To be more precise, we call two particles close to each other if their position expectation values $\langle x_1 \rangle, \langle x_2 \rangle$ differ by no more than the uncertainty to which these positions are known. As usual, we denote the uncertainties in the two positions by σ_1 and σ_2. Then the two particles are close if

$$(\langle x_1 \rangle - \langle x_2 \rangle)^2 \leq \sigma_1^2 + \sigma_2^2$$

For a system of two indistinguishable particles close to each other, the two situations

1. Particle 1 is at x_1, particle 2 at x_2.
2. Particle 2 is at x_1, particle 1 at x_2.

cannot be distinguished, and we can only assert that one of the two particles is at x_1 and the other at x_2.

Thus, in general, the probability density for such a situation does not allow us to differentiate between the two particles. We therefore have to require that the probability density $|\psi(x_1, x_2, t)|^2$ remain unaltered if the two particles 1 and 2 are interchanged, that is, if their coordinates x_1 and x_2 are permuted in the argument of ψ,

$$|\psi(x_1, x_2, t)|^2 = |\psi(x_2, x_1, t)|^2$$

Nor can any of the measurable quantities distinguish the two particles. This means that the potential energy of the two particles must be a symmetric function in the two position variables

$$V(x_1, x_2) = V(x_2, x_1)$$

which, in turn, implies that the Hamiltonian of the two particles is also symmetric not only in the momenta $p_1 = -i\hbar\partial/\partial x_1$, $p_2 = -i\hbar\partial/\partial x_2$ but also in the two position variables x_1, x_2:

$$\begin{aligned} H(p_1, p_2, x_1, x_2) &= -\frac{\hbar^2}{2m}\frac{\partial^2}{\partial x_1^2} - \frac{\hbar^2}{2m}\frac{\partial^2}{\partial x_2^2} + V(x_1, x_2) \\ &= H(p_2, p_1, x_2, x_1) \end{aligned}$$

Therefore, together with the solution $\psi'(x_1, x_2, t)$ of the Schrödinger equation

$$i\hbar\frac{\partial}{\partial t}\psi'(x_1, x_2, t) = H\psi'(x_1, x_2, t)$$

the function $\psi'(x_2, x_1, t)$ obtained by exchanging the arguments (x_1, x_2) is also a solution of the Schrödinger equation. Thus any superposition

$$\psi(x_1, x_2, t) = a\psi'(x_1, x_2, t) + b\psi'(x_2, x_1, t)$$

where a and b are complex numbers, solves the Schrödinger equation

$$i\hbar\frac{\partial}{\partial t}\psi(x_1, x_2, t) = H\psi(x_1, x_2, t)$$

The symmetry of the probability density $|\psi(x_1, x_2, t)|^2$ under the permutation of x_1 and x_2 puts constraints on the coefficients a and b. We have

$$\begin{aligned}|\psi(x_1, x_2, t)|^2 = a^*a|\psi'(x_1, x_2, t)|^2 + b^*b|\psi'(x_2, x_1, t)|^2 \\ +a^*b\psi'^*(x_1, x_2, t)\psi'(x_2, x_1, t) \\ +ab^*\psi'^*(x_2, x_1, t)\psi'(x_1, x_2, t)\end{aligned}$$

Comparing this equation with the corresponding formula for $|\psi(x_2, x_1, t)|^2$, we conclude that the equations for the coefficients are

$$a^*a = b^*b, \; a^*b = b^*a$$

With factoring into absolute value and phase factor,

$$a = |a|e^{i\alpha}, \qquad b = |b|e^{i\beta}$$

we find

$$|a| = |b|, \qquad e^{2i\alpha} = e^{2i\beta}$$

The periodicity of the exponential function fixes phase 2β relative to 2α modulus 2π only, that is,

$$2\beta = 2\alpha + 2n\pi, \qquad n = 0, \pm 1, \pm 2, \ldots$$

Thus only two values for the phase factor $e^{i\beta}$ remain,

$$e^{i\beta} = e^{i(\alpha + n\pi)} = \pm e^{i\alpha},$$

and therefore

$$b = \pm a$$

For the superposition we find

$$\psi(x_1, x_2, t) = a[\psi'(x_1, x_2, t) \pm \psi'(x_2, x_1, t)]$$

The overall phase $e^{i\alpha}$ is arbitrary for any wave function, and the absolute value $|a|$ is fixed by the normalization condition for the wave function $\psi(x_1, x_2, t)$. Putting everything together, we conclude that the wave function for two indistinguishable particles is either symmetric

$$\psi(x_1, x_2, t) = \psi(x_2, x_1, t)$$

or antisymmetric

$$\psi(x_1, x_2, t) = -\psi(x_2, x_1, t)$$

under permutation of the two coordinates x_1 and x_2.

The behavior of these two types of wave function is characteristically different. The particles having a symmetric two-particle wave function are called Bose-Einstein particles or *bosons*, those with an antisymmetric two-particle wave function Fermi-Dirac particles or *fermions*. The distinction between bosons and fermions becomes clear if we look at the values of their wave functions for the particular locations $x_1 = x_2$. The symmetric wave function is not restricted for these locations, whereas the antisymmetric solution must vanish for them:

$$\psi(x, x, t) = 0$$

Thus, in particular, the probability density for two fermions at the same position vanishes. Furthermore, if the two-particle wave function $\psi(x_1, x_2, t)$ is a product of two identical single-particle wave functions, the antisymmetric two-particle wave function vanishes:

$$\psi(x_1, x_2, t) = \varphi(x_1, t)\varphi(x_2, t) - \varphi(x_2, t)\varphi(x_1, t) = 0$$

This result must be interpreted as saying that two fermions cannot populate the same state, or that fermions must always populate different states. This phenomenon was discovered in 1925 by Wolfgang Pauli when he was trying to explain the fact that N electrons always populate the N lowest-lying states in atomic shells. The postulate of antisymmetric wave functions for fermions is called the *Pauli exclusion principle*.

8.2 Stationary States

As a first example, we look at the wave functions $\varphi_E(x_1, x_2)$ for the stationary states of two bosons or two fermions. They are obtained from solutions of the time-dependent Schrödinger equation factored in time and space dependence in the form

$$\psi(x_1, x_2, t) = \exp\left(-\frac{i}{\hbar}Et\right)\varphi_E(x_1, x_2)$$

For the stationary wave function the result of the last section requires symmetry for bosons,

$$\varphi_E^B(x_1, x_2) = \varphi_E^B(x_2, x_1)$$

or antisymmetry for fermions,

$$\varphi_E^F(x_1, x_2) = -\varphi_E^F(x_2, x_1)$$

For the motion of two indistinguisable particles in a system of coupled harmonic oscillators, we start with the solutions obtained in Section 7.3 for distinguishable particles. The function $u_n(r)$, being a solution of the one-particle Schrödinger equation for harmonic motion in the relative coordinate, is itself either symmetric, $u_n(-r) = u_n(r)$ for even n, or antisymmetric, $u_n(-r) = -u_n(r)$ for odd n. Therefore the wave functions for two bosons are simply

$$\varphi_E^B(x_1, x_2) = U_N(R)u_n(r), \qquad n \text{ even}$$

and correspondingly the wave functions for two fermions are

$$\varphi_E^F(x_1, x_2) = U_N(R)u_n(r), \qquad n \text{ odd}$$

The two sets of wave functions together constitute the complete set that we found for distinguishable particles. The symmetry or antisymmetry is apparent in Figure 7.6. The spectrum of energy eigenvalues of coupled harmonic oscillators made up of distinguishable particles splits in two, one describing the bosons,

$$E = (N + \tfrac{1}{2})\hbar\omega_R + (n + \tfrac{1}{2})\hbar\omega_r, \qquad n \text{ even}$$

the other one the fermions,

$$E = (N + \tfrac{1}{2})\hbar\omega_R + (n + \tfrac{1}{2})\hbar\omega_r, \qquad n \text{ odd}$$

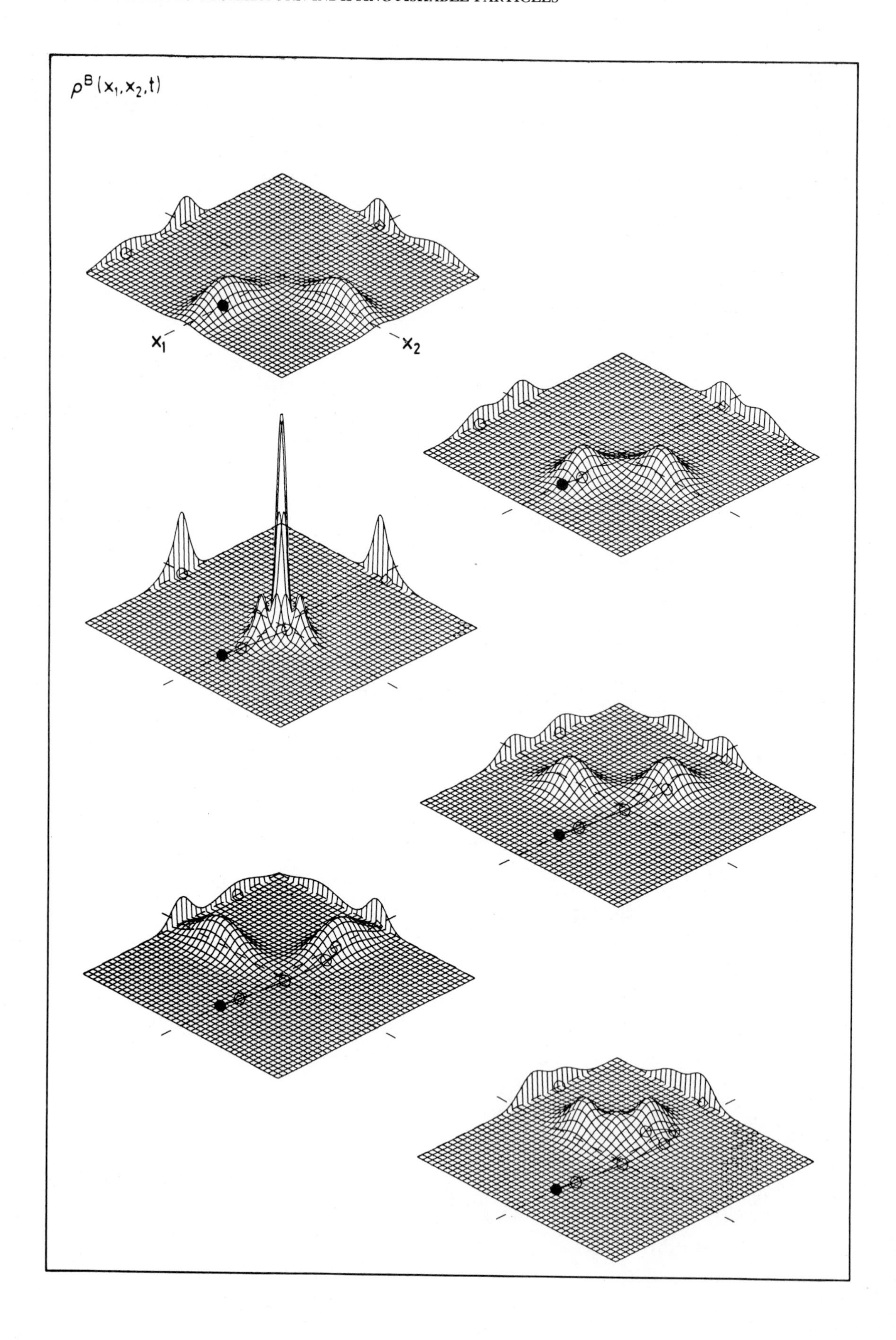

ρ^B(x1,x2,t)
x1
x2

8.3 Motion of Wave Packets

In order to describe motions in our system of coupled harmonic oscillators, we have to solve the time-dependent Schrödinger equation

$$i\hbar \frac{\partial \psi}{\partial t} = H\psi$$

If $\psi(x_1, x_2, t)$ is a solution with initial condition $\psi(x_1, x_2, 0)$, then $\psi(x_2, x_1, t)$ is also a solution corresponding to the initial condition $\psi(x_2, x_1, 0)$. This is guaranteed by the symmetry of the Hamiltonian in coordinates and momenta of indistinguishable particles, as discussed in Section 8.1.

Again, by symmetrization or antisymmetrization, we obtain still other solutions of the time-dependent Schrödinger equation. They are

$$\psi^B(x_1, x_2, t) = a_B[\psi(x_1, x_2, t) + \psi(x_2, x_1, t)]$$

$$\psi^F(x_1, x_2, t) = a_F[\psi(x_1, x_2, t) - \psi(x_2, x_1, t)]$$

and correspond, of course, to symmetric or antisymmetric initial conditions. The numerical factors a_B, a_F ensure normalization of the corresponding wave packets.

As a first example, let us consider the motion of two bosons forming a system of coupled harmonic oscillators. In Figure 8.1 the joint probability density

$$\rho^B(x_1, x_2, t) = |\psi^B(x_1, x_2, t)|^2$$

and the marginal distributions $\rho_1^B(x_1, t)$ and $\rho_2^B(x_2, t)$ are shown for several times $t = t_0, t_1, \ldots, t_N$. Except for the symmetrization of the wave function, all parameters are the same as those for distinguishable particles, whose motion was illustrated in Figure 7.2. In particular, the trajectory of the classical particles in the x_1, x_2-plane is identical in both figures. Since the position expectation values x_{10}, x_{20} at initial time

Figure 8.1 **Joint probability density and marginal distributions for two bosons forming a system of coupled harmonic oscillators. The joint probability density $\rho^B(x_1, x_2, t)$ is shown as a surface over the x_1, x_2-plane, the marginal distribution $\rho_1^B(x_1, t)$ as a curve over the margin parallel to the x_1-axis, and the marginal distribution $\rho_2^B(x_2, t)$ as a curve over the other margin. The distributions are shown for various times $t_j = t_0, t_1, \ldots, t_N$. The positions of the classical particles are indicated by dots in the plane and on the margins; their motion is represented by the trajectory in the x_1, x_2-plane.**

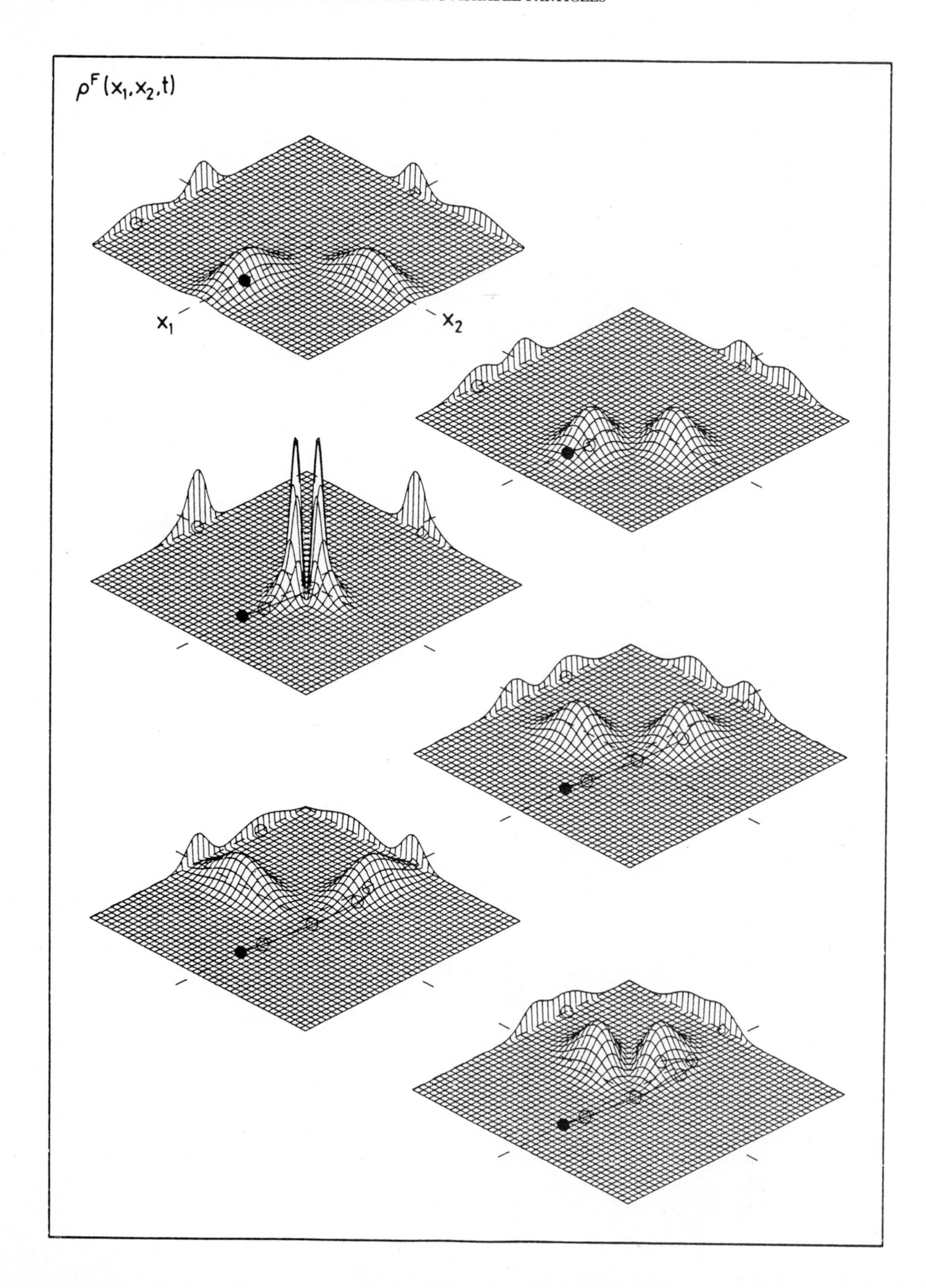
$\rho^F(x_1, x_2, t)$
x_1
x_2

$t = t_0$ are farther apart than the width of the unsymmetrized wave packet in Figure 7.2, we observe for $t = t_0$ two well-separated humps corresponding to points $x_1 = x_{10}$, $x_2 = x_{20}$, and $x_1 = x_{20}$, $x_2 = x_{10}$, respectively. The marginal distribution $\rho_1^B(x_1, 0)$, which describes the probability that one particle of the two will be observed at x_1, irrespective of the position of the other one, also has two humps. The two humps again reflect the fact that the two particles cannot be distinguished. Then, of course, the marginal distribution $\rho_2(x_2, t)$ has to be identical to the marginal distribution $\rho_1(x_1, t)$. In pursuit of their motion, the particles attain a distance smaller than the width of the unsymmetrized wave packet. In this situation the two humps are no longer separated but merge into one. For a later moment in time they are again separated, and so on.

Figure 8.2 shows the corresponding motion of two fermions. For $t = t_0$, when the two particles are well separated, the situation looks qualitatively similar, but it becomes strikingly different when the particles move close to each other. The hump splits along the direction $x_1 = x_2$, where the probability density is exactly zero as a consequence of the Pauli exclusion principle. In fact, for fermions the probability density vanishes for locations $x_1 = x_2$ at all moments in time. At no time can two fermions be at the same place.

Figures 8.3b and c show the time developments of the marginal distributions $\rho^{B,F}(x, t)$ for two bosons and for two fermions forming a system of coupled harmonic oscillators. The difference between the two is much less striking than that between the corresponding probability distributions of Figures 8.1 and 8.2. But still a trace of the Pauli exclusion principle is visible in the marginal distributions. Near the center of Figures 8.3b and c, where the particles are close to each other, the two humps are farther apart for fermions than for bosons. For purposes of comparison, the time developments of the two marginal distributions for the corresponding system of two distinguishable particles are given in Figure 8.3a.

***Figure* 8.2 Joint probability density and marginal distributions for two fermions. All initial conditions are the same as those for Figure 8.1.**

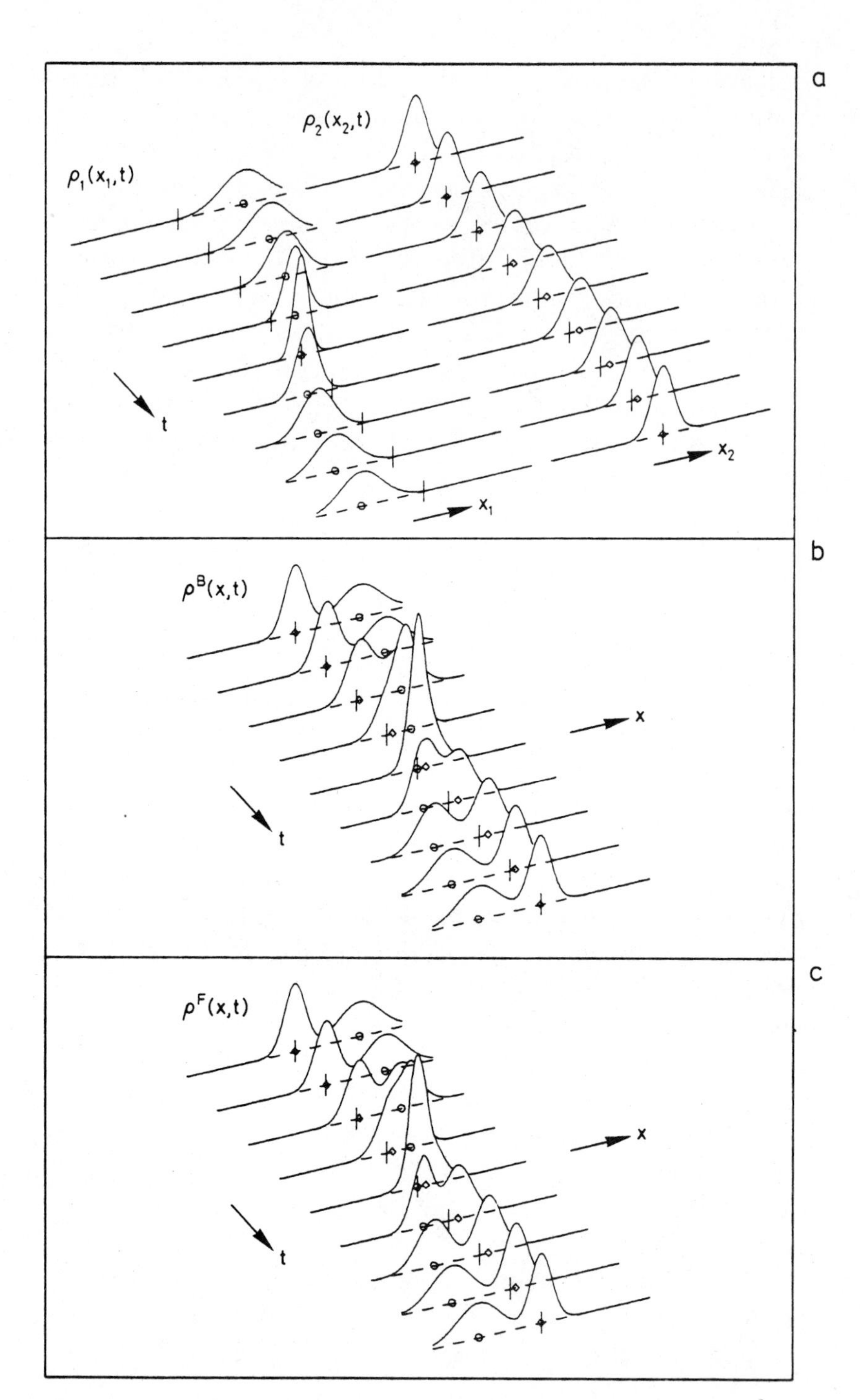

Figure 8.3 (a) Time developments of the two marginal distributions for two distinguishable particles forming a system of coupled harmonic oscillators. Time developments of the marginal distributions $\rho^{B,F}(x,t)$ for the corresponding systems of (b) two bosons and (c) two fermions.

8.4 Indistinguishable Particles from a Classical Point of View

The quantum-mechanical description of the motion of indistinguishable particles poses the question whether the classical concept of the trajectory of a particle can still be upheld or whether it has to be given up. Looking at the joint probability distributions for indistinguishable particles in Figures 8.1 and 8.2, we observe two distinct humps as long as the classical positions are far apart. The center of either of them moves along its classical trajectory with the initial positions

$$x_1 = x_{10}, \qquad x_2 = x_{20}$$

or

$$x_1 = x_{20}, \qquad x_2 = x_{10}$$

where x_{10}, x_{20} are the position expectation values of the probability distribution for distinguishable particles. This observation suggests that although the particles are indistinguishable in their intrinsic properties, they can under the given circumstances be distinguished by their position. Thus if we call the particle that at $t = t_0$ is in the neighborhood of x_{10} particle 1 and the particle that is close to x_{20} particle 2, it is perfectly consistent to say that particle 1 stays in the neighborhood of the trajectory $\langle x_1(t)\rangle$ and particle 2 in that of $\langle x_2(t)\rangle$, as long as the two humps are well separated. Here, $\langle x_1(t)\rangle$ is the expectation value of the coordinate x_1 for the wave packet of distinguishable particles and also the classical position of particle 1 at time t. As soon as the particles come closer to each other than the widths of the humps, there is no longer a clear correspondence between the classical trajectory and the structure of the probability density. Once the positions are separated again, a new correspondence can be established.

A look at the relevant formulas justifies this reasoning. The wave functions ψ^B, for bosons, and ψ^F, for fermions, were obtained from that for distinguishable particles, ψ, by symmetrization and antisymmetrization,

$$\psi^{B,F}(x_1, x_2, t) = a_{B,F}[\psi(x_1, x_2, t) \pm \psi(x_2, x_1, t)]$$

The probability density is found by taking the absolute square,

$$\begin{aligned}\rho^{B,F}(x_1, x_2, t) &= |\psi^{B,F}(x_1, x_2, t)|^2 \\ &= |a_{B,F}|^2[\rho(x_1, x_2, t) + \rho(x_2, x_1, t) \\ &\qquad \pm \tau(x_1, x_2, t)]\end{aligned}$$

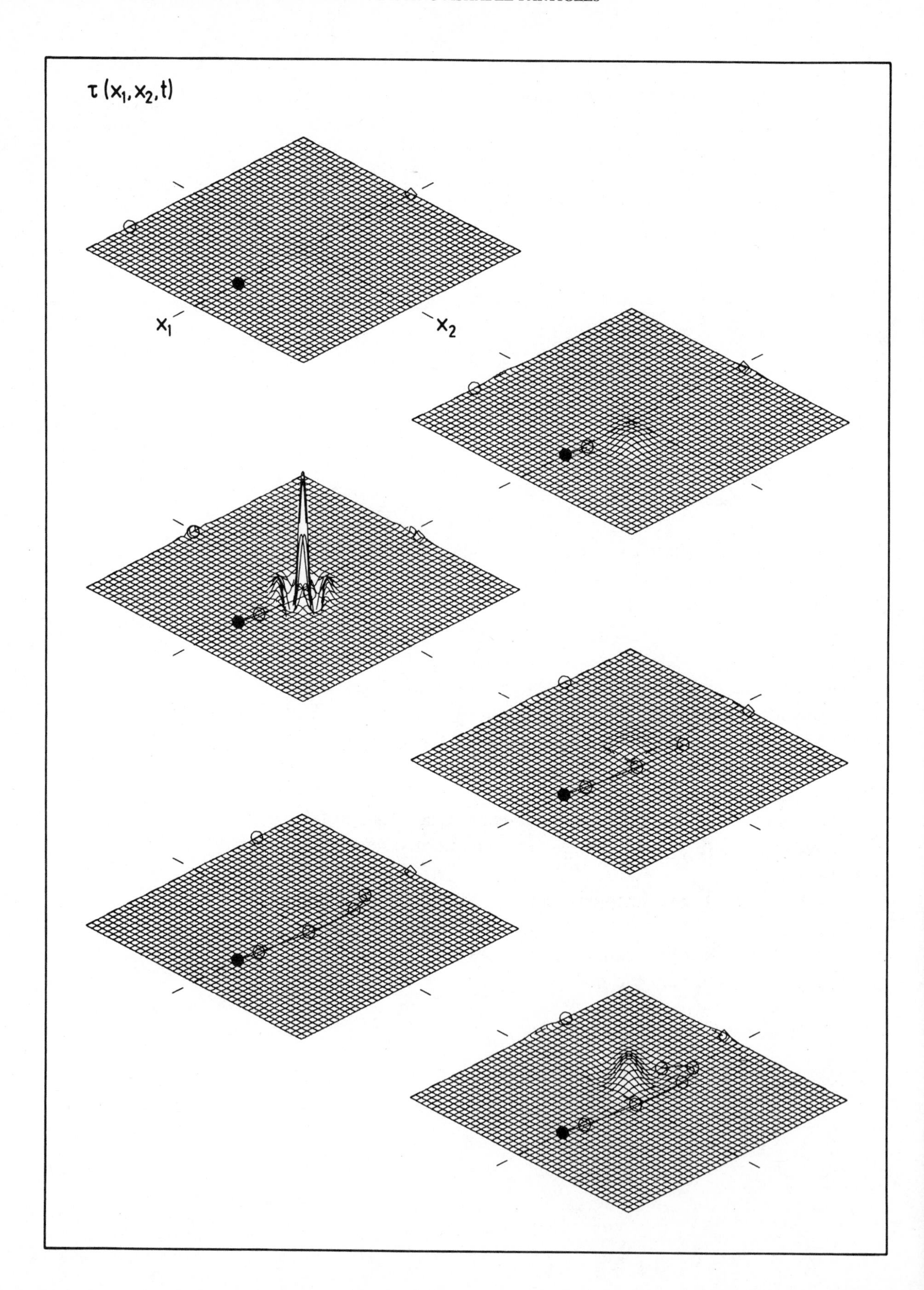

$\tau(x_1, x_2, t)$
x_1
x_2

Here

$$\rho(x_1, x_2, t) = |\psi(x_1, x_2, t)|^2$$

is the joint probability distribution for distinguishable particles with coordinate x_1 corresponding to particle 1 and coordinate x_2 to particle 2. The density

$$\rho(x_2, x_1, t) = |\psi(x_2, x_1, t)|^2$$

describes the situation in which particles 1 and 2 are interchanged.

The term

$$\tau(x_1, x_2, t) = \psi^*(x_1, x_2, t)\psi(x_2, x_1, t) + \psi^*(x_2, x_1, t)\psi(x_1, x_2, t)$$

is called the *interference term*. This term is practically zero unless the two particles are closer to each other than the width of the single hump. To show this, we consider the particular point $x_1 = x_{10}$, $x_2 = x_{20}$ in Figure 7.2. Clearly here $\psi(x_{10}, x_{20}, t)$ and its complex conjugate have large amplitudes, whereas $\psi(x_{20}, x_{10}, t)$ and its complex conjugate practically vanish. Figure 8.4, which shows the interference term $\tau(x_1, x_2, t)$ for various times $t = t_0, t_1, \ldots, t_N$, verifies the nature of the interference term. The figure corresponds in all conditions to those of Figures 8.1 and 8.2. In fact, these figures were obtained using the complex formula for $\rho^{B,F}(x_1, x_2, t)$ given earlier. In Figure 8.5 only the sum of the first two terms—the interference term is excluded—is plotted. We see that the interference is comparable to the sum of the

Figure 8.4 **The interference term for two indistinguishable particles forming a system of coupled harmonic oscillators. The distribution is shown for various times $t_j = t_0, t_1, \ldots, t_N$. All initial conditions are the same as those for Figure 8.1.**

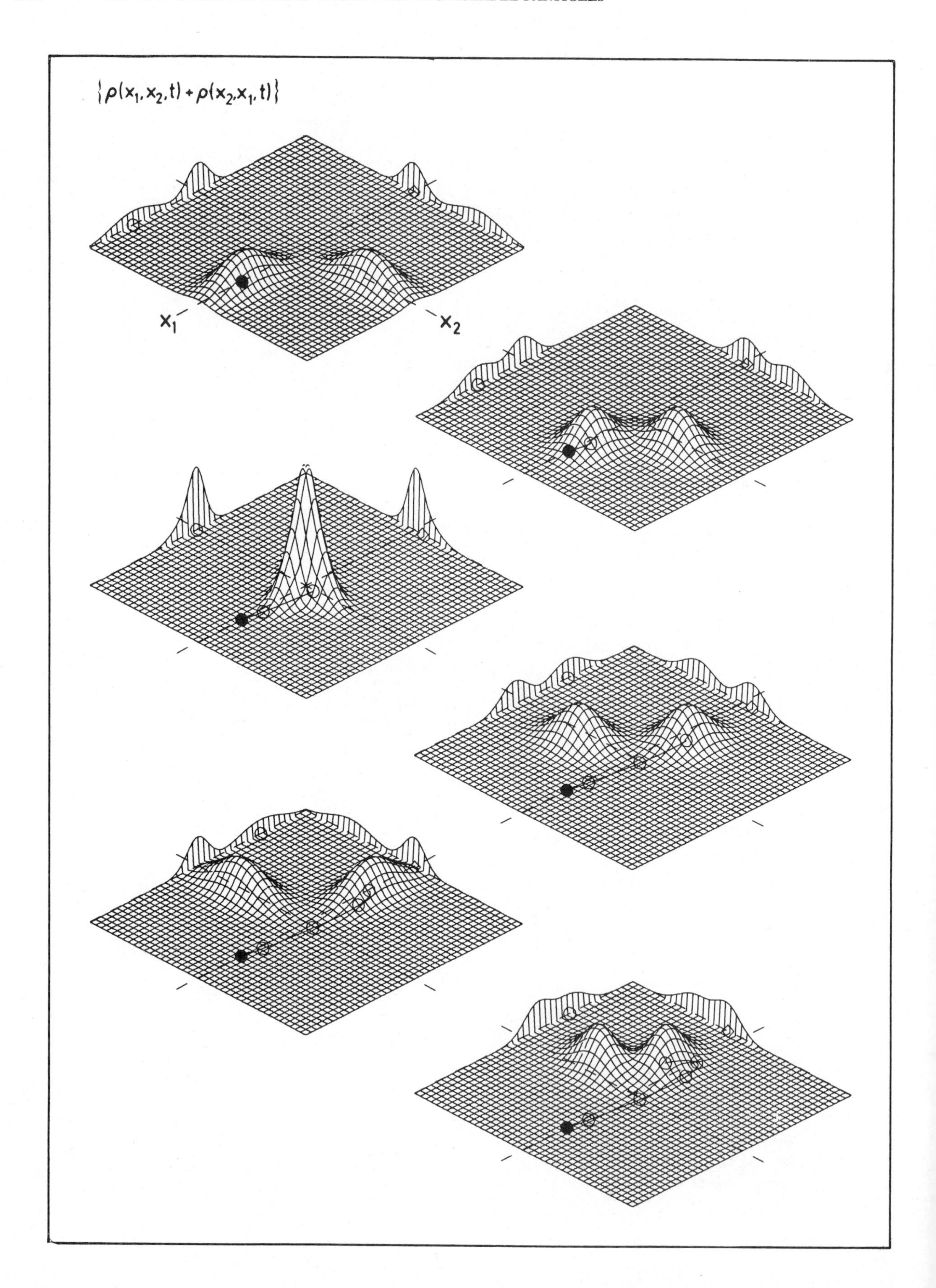
$\{\rho(x_1, x_2, t) + \rho(x_2, x_1, t)\}$
x_1
x_2

other two only when these overlap, that is, when the particles are close to each other.

The probability densities for bosons in Figure 8.1 and for fermions in Figure 8.2 are obtained from the symmetrized probability density for distinguishable particles given in Figure 8.5 and the interference term given in Figure 8.4. We summarize this discussion by emphasizing that the probability density for indistinguishable particles is obtained by symmetrizing the probability density for distinguishable particles and adding or subtracting the interference term. This term contributes only if the particles are sufficiently close to each other. Thus the concept of classical trajectories can be maintained as long as we are able to distinguish the particles by their initial positions and as long as we refrain from localizing them individually in the overlap region.

Finally, Figure 8.6a gives the marginal distribution for the symmetrized probability density for distinguishable particles, which, of course, is nothing but the sum of the two marginal distributions for distinguishable particles. Figure 8.6b shows the marginal distribution for the interference term. Again, the marginal distributions for bosons can be constructed by adding the distributions of Figures 8.6a and b, those for fermions by subtracting the distribution of Figure 8.6b from that of Figure 8.6a.

***Figure* 8.5 Symmetrized probability density for two distinguishable particles forming a system of coupled harmonic oscillators. All initial conditions are the same as those for Figure 8.1.**

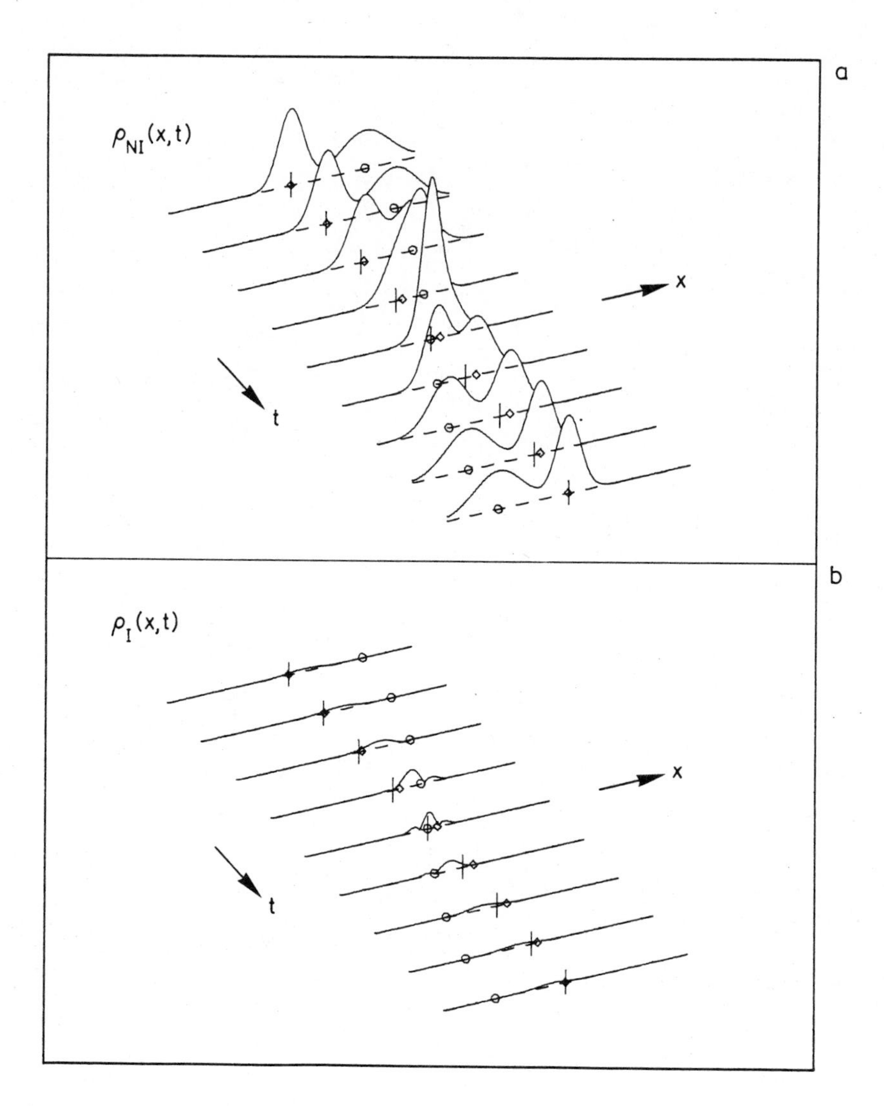

Figure 8.6 **Time developments of (a) the marginal distribution for the symmetrized probability density for two distinguishable particles and (b) the marginal distribution for the interference term for two indistinguishable particles. The particles form a system of coupled harmonic oscillators. All initial conditions are the same as those for Figure 8.3.**

Problems

8.1 Which eigenstates of the system of two coupled harmonic oscillators, as plotted in Figure 7.6, can be occupied by bosons, which by fermions?

8.2 Show that the eigenfunctions for the coupled harmonic oscillators must have the symmetry properties with respect to the permutation of x_1, x_2 observed in Figure 7.6.

8.3 Compare Figures 8.1 and 8.2 with Figures 8.4 and 8.5 and characterize the role of the interference term in distinguishing bosons and fermions.

8.4 Electrons are fermions. They possess intrinsic angular momentum which is called spin s and can assume the two projections $\pm\hbar/2$. The wave function for an electron in a one-dimensional potential is fully characterized by the spatial wave function $\varphi(x)$ and the spin projection. The Pauli exclusion principle then allows two electrons to occupy the same spatial state since they can assume two spin projections.

A number N of electrons is to be accommodated in a potential well of width d with infinitely high walls. What is the minimum total energy of all electrons? For the minimum total energy, what is the highest energy an electron assumes? Express it in terms of the ground-state energy! How does this compare to the situation in which the potential is occupied by N bosons?

8.5 Solve the preceding problem for the harmonic oscillator potential.

9. Wave Packet in Three Dimensions

9.1 Momentum

The position of the classical particle in three-dimensional space is described by the components x, y, z of the *position vector*:

$$\mathbf{r} = (x, y, z)$$

Similarly, the three components of momentum form the *momentum vector*:

$$\mathbf{p} = (p_x, p_y, p_z)$$

Following our one-dimensional description in Section 3.3, we now introduce operators for all three components of momentum:

$$\hat{p}_x = \frac{\hbar}{i}\frac{\partial}{\partial x}, \quad \hat{p}_y = \frac{\hbar}{i}\frac{\partial}{\partial y}, \quad \hat{p}_z = \frac{\hbar}{i}\frac{\partial}{\partial z}$$

The three operators form the *vector operator of momentum*,

$$\hat{\mathbf{p}} = (\hat{p}_x, \hat{p}_y, \hat{p}_z) = \frac{\hbar}{i}\left(\frac{\partial}{\partial x}, \frac{\partial}{\partial y}, \frac{\partial}{\partial z}\right) = \frac{\hbar}{i}\nabla$$

which is the differential operator ∇, called *nabla* or *del*, multiplied by $\hbar/i$.

The three-dimensional stationary plane wave

$$\begin{aligned}\varphi_{\mathbf{p}}(\mathbf{r}) &= \frac{1}{(2\pi\hbar)^{1/2}}\exp\left(\frac{i}{\hbar}p_x x\right) \\ &\quad \times \frac{1}{(2\pi\hbar)^{1/2}}\exp\left(\frac{i}{\hbar}p_y y\right)\frac{1}{(2\pi\hbar)^{1/2}}\exp\left(\frac{i}{\hbar}p_z z\right) \\ &= \frac{1}{(2\pi\hbar)^{3/2}}\exp\left(\frac{i}{\hbar}\mathbf{p}\cdot\mathbf{r}\right)\end{aligned}$$

with

$$\mathbf{p}\cdot\mathbf{r} = p_x x + p_y y + p_z z$$

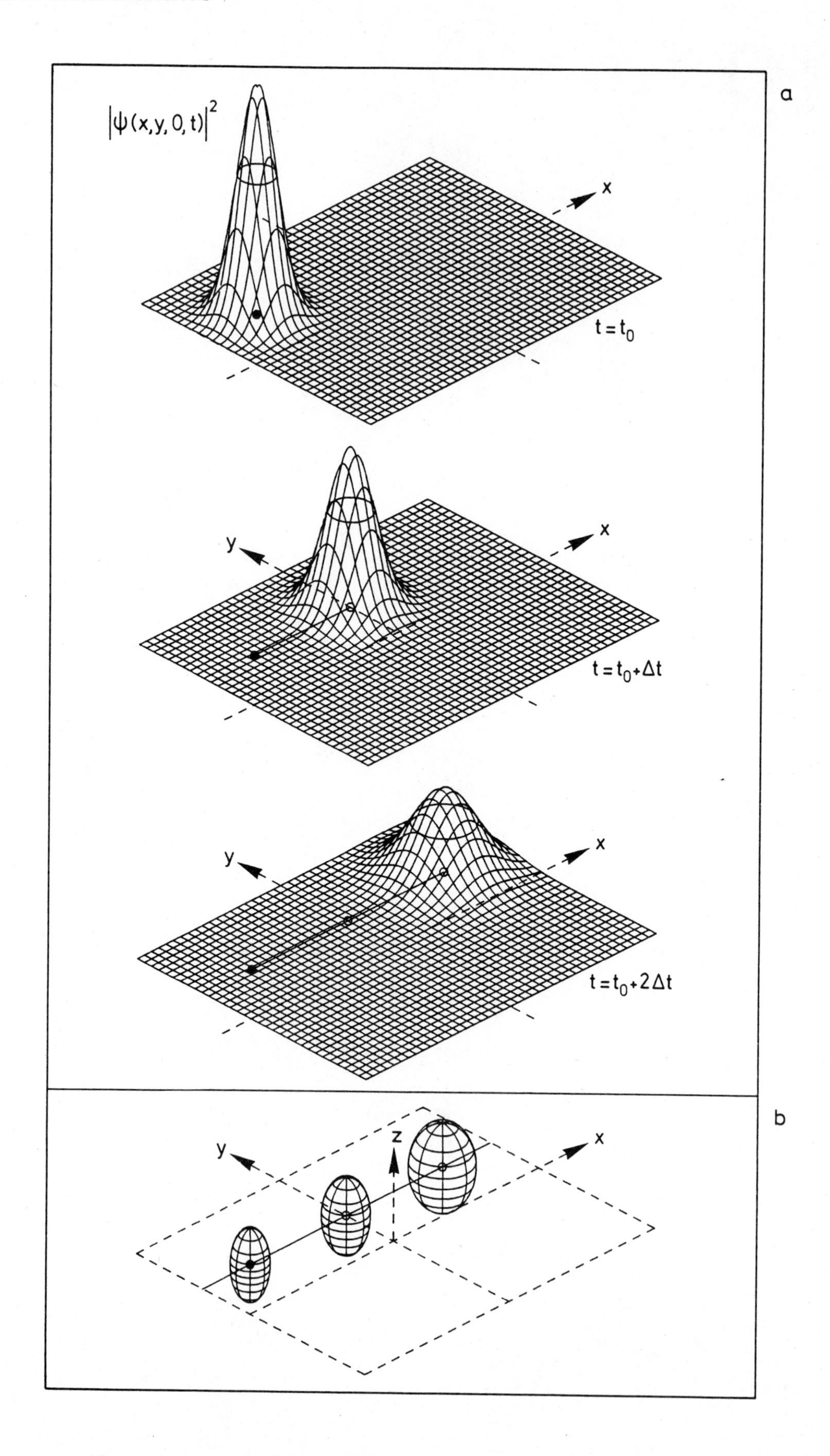
a
$|\psi(x,y,0,t)|^2$
x
$t=t_0$
y
x
$t=t_0+\Delta t$
y
x
$t=t_0+2\Delta t$
b
y
z
x

is simply the product of three one-dimensional stationary waves of the momentum components p_x, p_y, and p_z corresponding to the three directions x, y, and z in space. The surfaces of constant phase δ are given by

$$\frac{i}{\hbar}\mathbf{p}\cdot\mathbf{r} = \delta$$

They are planes perpendicular to the *wave vector*

$$\mathbf{k} = \frac{\mathbf{p}}{\hbar}$$

The wave vector is the three-dimensional generalization of wave number k in one dimension, as introduced in Section 2.1. It determines the wavelength through the relation

$$\lambda = \frac{2\pi}{|\mathbf{k}|}$$

The three-dimensional stationary plane wave is a simultaneous solution—also called a simultaneous eigenfunction—of the three equations

$$\hat{p}_x\varphi_{\mathbf{p}}(\mathbf{r}) = p_x\varphi_{\mathbf{p}}(\mathbf{r}), \quad \hat{p}_y\varphi_{\mathbf{p}}(\mathbf{r}) = p_y\varphi_{\mathbf{p}}(\mathbf{r}), \quad \hat{p}_z\varphi_{\mathbf{p}}(\mathbf{r}) = p_z\varphi_{\mathbf{p}}(\mathbf{r})$$

The three numbers p_x, p_y and p_z forming the vector $\mathbf{p}$ are called the momentum eigenvalues of the plane wave $\varphi_{\mathbf{p}}(\mathbf{r})$.

The three-dimensional time-dependent wave function, like the one-dimensional, is obtained by multiplying the stationary eigenfunction $\varphi_{\mathbf{p}}(\mathbf{r})$ by the energy-dependent phase factor,

$$\exp\left(-\frac{i}{\hbar}Et\right), \qquad E = \frac{\mathbf{p}^2}{2M} = \frac{1}{2M}\left(p_x^2 + p_y^2 + p_z^2\right)$$

that is,

$$\begin{aligned}\psi_{\mathbf{p}}(\mathbf{r}, t) &= \frac{1}{(2\pi\hbar)^{3/2}}\exp\left(-\frac{i}{\hbar}Et\right)\exp\left(\frac{i}{\hbar}\mathbf{p}\cdot\mathbf{r}\right)\\ &= \psi_{p_x}(x, t)\psi_{p_y}(y, t)\psi_{p_z}(z, t)\end{aligned}$$

Figure 9.1 A three-dimensional Gaussian wave packet moves freely in space. Its position expectation value moves on a straight line in the x, y-plane. (a) The first three illustrations show for three equidistant moments in time the probability density in the x, y-plane as a bell-shaped surface, the expectation value as a dot on the plane, and the trajectory of the corresponding classical particle as a straight line in the plane. The covariance ellipse encircling the surface comprises a fixed fraction of the total probability. It contains the complete probability density information for the x, y-plane. (b) The complete information for the three-dimensional probability distribution is given by the probability ellipsoid. It is centered around the position expectation value and shown for the three moments in time that are depicted separately in part a. The classical trajectory in the x, y-plane is also shown.

This time-dependent expression for the three-dimensional harmonic wave also factors into exponentials corresponding to the three dimensions.

The three-dimensional free, unaccelerated motion of a particle is again described by a superposition of these plane waves with a spectral function,

$$f(\mathbf{p}) = f_x(p_x) f_y(p_y) f_z(p_z)$$

$$f_a(p_a) = -\frac{1}{(2\pi)^{1/4}\sqrt{\sigma_{p_a}}} \exp\left[-\frac{(p_a - p_{a0})^2}{4\sigma_{p_a}^2}\right], \qquad a = x, y, z$$

which is the product of three Gaussian spectral functions centered around the expectation values $(p_{x0}, p_{y0}, p_{z0}) = \mathbf{p}_0$ with the widths $\sigma_{p_x}, \sigma_{p_y}, \sigma_{p_z}$ as introduced in Section 3.2. The superposition of the functions $\psi_{\mathbf{p}}(\mathbf{r} - \mathbf{r}_0, t)$ with the spectral function $f(\mathbf{p})$ is given by

$$\psi(t, \mathbf{r}) = \int f(\mathbf{p}) \psi_{\mathbf{p}}(\mathbf{r} - \mathbf{r}_0, t) \, d^3\mathbf{p}$$

It represents the moving wave packet that starts at $t = 0$ around point $\mathbf{r}_0$ with the average momentum $\mathbf{p}_0$. Because of the product forms of $f(\mathbf{p})$ and $\psi_{\mathbf{p}}(\mathbf{r} - \mathbf{r}_0, t)$, the equation can also be written in product form,

$$\psi(\mathbf{r}, t) = M_x(x, t) e^{i\phi_x(x,t)} M_y(y, t) e^{i\phi_y(y,t)} M_z(z, t) e^{i\phi_z(z,t)}$$

where the meaning of the symbols can easily be inferred from the one-dimensional wave packet of Section 3.2.

Figure 9.1a shows the probability distribution $|\psi(x, y, 0, t)|^2$ in the x, y-plane of the moving wave packet for the initial moment in time, t_0 and two later ones. The straight line in the x, y-plane marks the classical trajectory that has been chosen to lie in this plane. The dots indicate the positions of the corresponding classical particle at the three moments in time. The probability distribution shown is a two-dimensional Gaussian dispersing in time. The ellipse encircling the bell-shaped bump comprises a certain fraction of the total probability. It is the covariance ellipse, which, in a different context, has already been discussed in Section 7.1. As the wave packet disperses, this ellipse grows in size. For a Gaussian wave packet this ellipse completely characterizes the position and the degree of localization of the particle in the x, y-plane. The complete three-dimensional Gaussian wave packet is then characterized by a *covariance ellipsoid*. Figure 9.1b shows the ellipsoids that correspond to the three situations of Figure 9.1a.

9.2 Angular Momentum, Spherical Harmonics

Three-dimensional motion is further characterized by *angular momentum*. For a classical particle it is simply the vector product of the position vector and the momentum vector,

$$\mathbf{L} = \mathbf{r} \times \mathbf{p}$$

or in components,

$$L_x = yp_z - zp_y, \quad L_y = zp_x - xp_z, \quad L_z = xp_y - yp_x$$

The quantum-mechanical analog is obtained by inserting the operator of momentum $\hat{\mathbf{p}} = (\hbar/i)\nabla$ into the classical expression for $\mathbf{L}$. This yields the *vector operator of angular momentum*,

$$\hat{\mathbf{L}} = \mathbf{r} \times \hat{\mathbf{p}} = \frac{\hbar}{i}\mathbf{r} \times \nabla$$

or in components,

$$\hat{L}_x = \frac{\hbar}{i}\left(y\frac{\partial}{\partial z} - z\frac{\partial}{\partial y}\right), \quad \hat{L}_y = \frac{\hbar}{i}\left(z\frac{\partial}{\partial x} - x\frac{\partial}{\partial z}\right),$$

$$\hat{L}_z = \frac{\hbar}{i}\left(x\frac{\partial}{\partial y} - y\frac{\partial}{\partial x}\right)$$

Whereas the components of momentum commute with each other, that is, $[\hat{p}_x, \hat{p}_y] = \hat{p}_x\hat{p}_y - \hat{p}_y\hat{p}_x = 0$, and so on, the components of angular momentum do not. In fact, the *commutation relations* are

$$[\hat{L}_x, \hat{L}_y] = i\hbar\hat{L}_z, \quad [\hat{L}_y, \hat{L}_z] = i\hbar\hat{L}_x, \quad [\hat{L}_z, \hat{L}_x] = i\hbar\hat{L}_y$$

Because the commutators do not vanish, an eigenfunction of $\hat{L}_z$ cannot be an eigenfunction of $\hat{L}_y$ as well. If, in addition to the eigenvalue equation

$$\hat{L}_z Y = l_z Y$$

the relation

$$\hat{L}_y Y = l_y Y$$

would also hold, we would in general have a contradiction to the commutator relation $[\hat{L}_y, \hat{L}_z] = i\hbar\hat{L}_x$ when applied to the eigenfunction Y:

$$(\hat{L}_y\hat{L}_z - \hat{L}_z\hat{L}_y)Y = (l_y l_z - l_z l_y)Y = 0 \neq i\hbar\hat{L}_x Y$$

This observation is tantamount to the statement that noncommuting operators do *not* have simultaneous eigenfunctions, except for trivial ones.

There is, however, another operator, the square of the vector operator of angular momentum,

$$\hat{\mathbf{L}}^2 = \hat{L}_x^2 + \hat{L}_y^2 + \hat{L}_z^2$$

which does commute with any of the components:

$$[\hat{\mathbf{L}}^2, \hat{L}_a] = 0, \qquad a = x, y, z$$

This relation is easily verified with the help of the commutation relations, for example,

$$\begin{aligned}
\left[\hat{L}_x^2 + \hat{L}_y^2 + \hat{L}_z^2, \hat{L}_z\right] &= [\hat{L}_x^2, \hat{L}_z] + [\hat{L}_y^2, \hat{L}_z] \\
&= \hat{L}_x[\hat{L}_x, \hat{L}_z] + [\hat{L}_x, \hat{L}_z]\hat{L}_x + \hat{L}_y[\hat{L}_y, \hat{L}_z] + [\hat{L}_y, \hat{L}_z]\hat{L}_y \\
&= \hat{L}_x(-i\hbar\hat{L}_y) - i\hbar\hat{L}_y\hat{L}_x + \hat{L}_y(i\hbar)\hat{L}_x + i\hbar\hat{L}_x\hat{L}_y = 0
\end{aligned}$$

Thus simultaneous eigenfunctions for $\hat{\mathbf{L}}^2$ and any of the components, for example, $\hat{L}_z$, can be found. For the following discussion it is convenient to use *polar coordinates* r, ϑ, and ϕ rather than Cartesian coordinates x, y, and z. In a polar coordinate system a point is given by its distance r from the origin, its *polar angle* ϑ, and its *azimuth* ϕ. The angles are illustrated in the top right corner of Figure 9.5. The relations between the coordinates of the two systems are

$$\begin{aligned}
x &= r \sin\vartheta \cos\phi \\
y &= r \sin\vartheta \sin\phi \\
z &= r \cos\vartheta
\end{aligned}$$

In polar coordinates the operators of angular momentum are

$$\begin{aligned}
\hat{L}_x &= i\hbar\left(\sin\phi\frac{\partial}{\partial\vartheta} + \operatorname{cotan}\vartheta\cos\phi\frac{\partial}{\partial\phi}\right) \\
\hat{L}_y &= -i\hbar\left(\cos\phi\frac{\partial}{\partial\vartheta} - \operatorname{cotan}\vartheta\sin\phi\frac{\partial}{\partial\phi}\right) \\
\hat{L}_z &= -i\hbar\frac{\partial}{\partial\phi} \\
\hat{\mathbf{L}}^2 &= -\hbar^2\left[\frac{1}{\sin\vartheta}\frac{\partial}{\partial\vartheta}\left(\sin\vartheta\frac{\partial}{\partial\vartheta}\right) + \frac{1}{\sin^2\vartheta}\frac{\partial^2}{\partial\phi^2}\right]
\end{aligned}$$

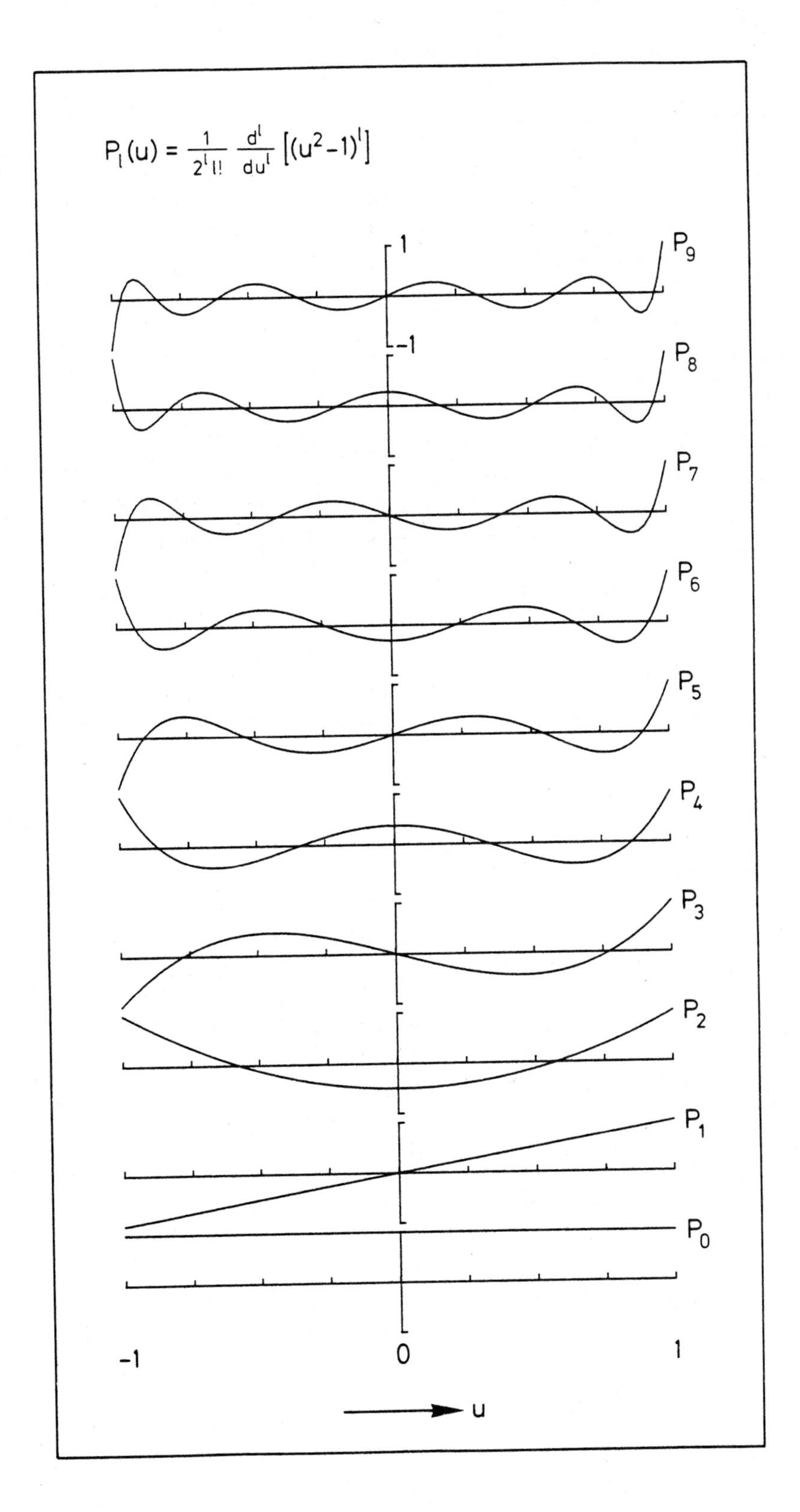

***Figure* 9.2 Graphs of the first ten Legendre polynomials $P_l(u)$.**

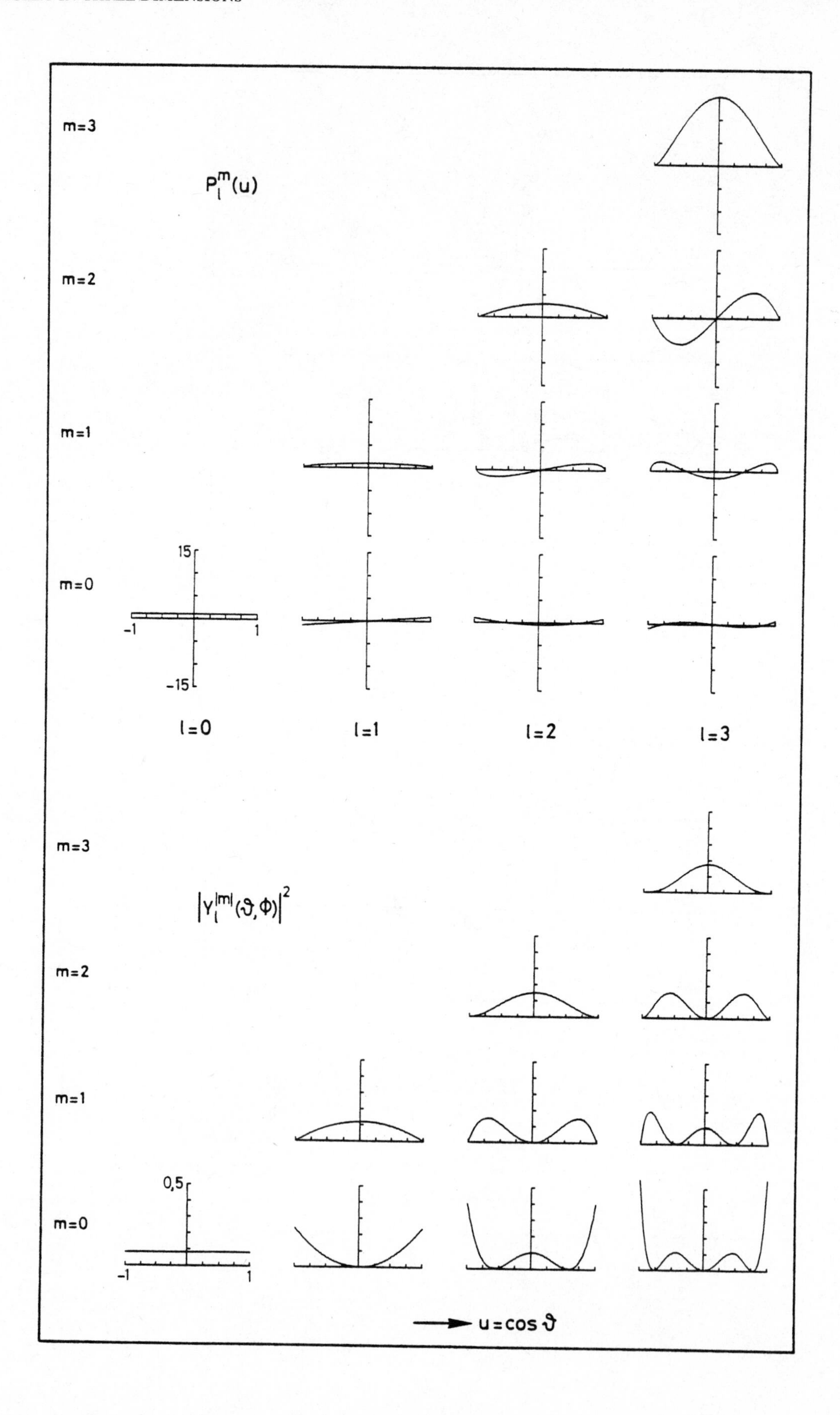

m=3
$P_l^m(u)$
m=2
m=1
m=0
15
-15
-1
1
l=0
l=1
l=2
l=3
m=3
$|Y_l^{|m|}(\vartheta,\phi)|^2$
m=2
m=1
m=0
0,5
-1
1
u=cos ϑ

We can write eigenvalue equations for the two operators $\hat{\mathbf{L}}$ and $\hat{L}_z$:

$$\hat{\mathbf{L}}Y_{lm} = l(l+1)\hbar^2 Y_{lm}$$

$$\hat{L}_z Y_{lm} = m\hbar Y_{lm}$$

Both operators have as eigenfunctions the *spherical harmonics* $Y_{lm}(\vartheta, \phi)$, which are discussed in the next paragraphs. The eigenvalues of the square of angular momentum are $l(l+1)\hbar^2$. This *quantum number of angular momentum* l can take on only integer values $l = 0, 1, 2, \ldots$. Thus, in contrast to classical mechanics, the square of angular momentum can take only discrete values that are integer multiples of $\hbar^2$. Correspondingly, the eigenvalues of the *z-component* L_z *of angular momentum* are $m\hbar$. The quantum number m can vary only in the range $-l \le m \le l$. In fact, m takes on only integer numbers in this range. For historical reasons quantum number m is sometimes called *magnetic quantum number.*

The spherical harmonics $Y_{lm}(\vartheta, \phi)$ have an explicit representation which is commonly based on the *Legendre polynomials*

$$P_l(u) = \frac{1}{2^l l!} \frac{d^l}{du^l}\left[(u^2 - 1)^l\right]$$

Figure 9.2 shows the plots of these polynomials for $l = 0, 1, 2, \ldots, 9$ and the domain $-1 \le u \le 1$.

The Legendre polynomials are special cases of the *associated Legendre functions* P_l^m, which are defined by

$$P_l^m(u) = (1 - u^2)^{m/2} \frac{d^m}{du^m} P_l(u), \qquad m = 0, 1, 2, \ldots, l$$

Figure 9.3 **Graphs of the associated Legendre functions $P_l^m(u)$, top, and of the absolute squares of the spherical harmonics $Y_{lm}(\vartheta, \phi)$, bottom. Except for a normalization factor, the absolute squares of the spherical harmonics are the squares of the associated Legendre functions.**

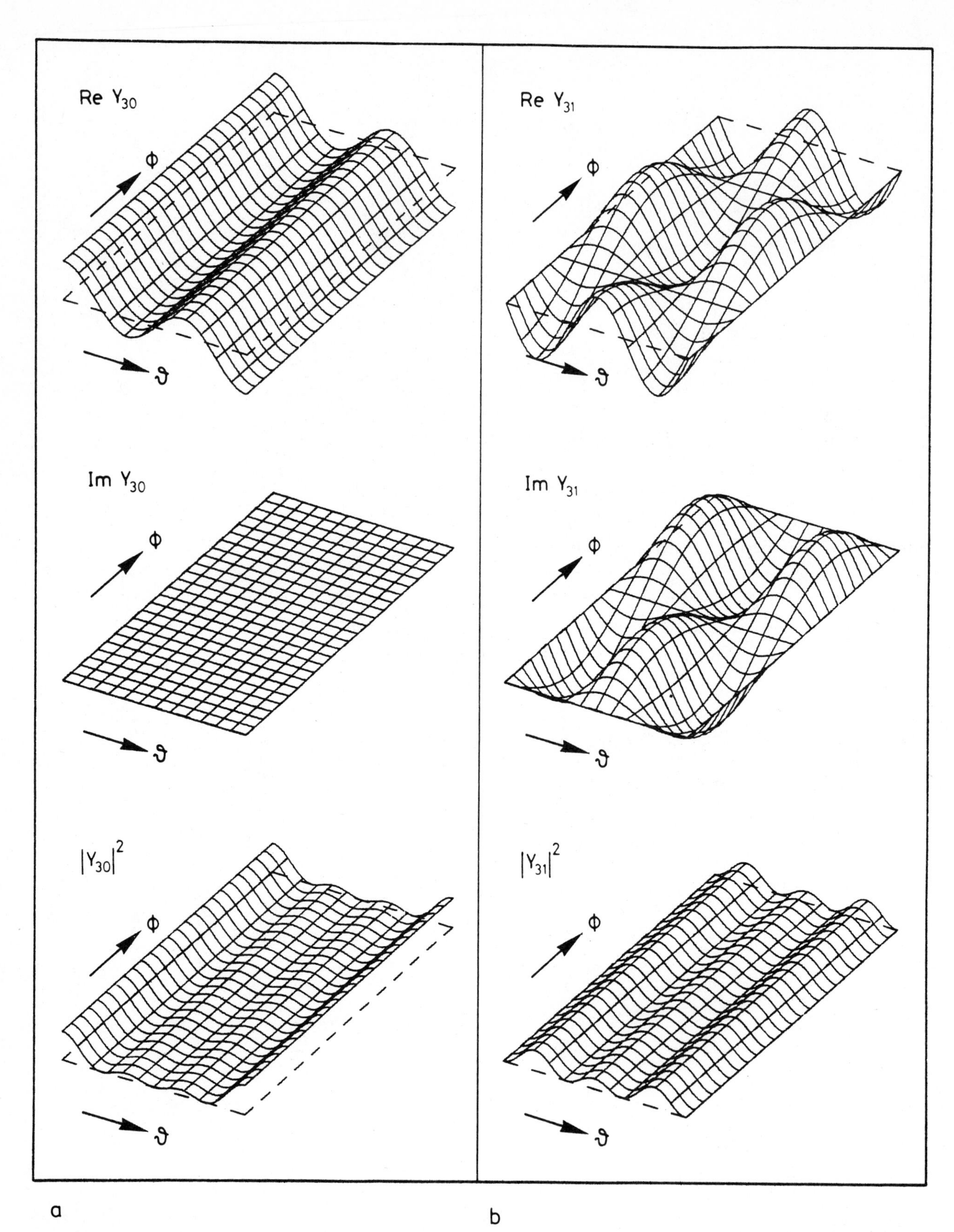

Figure 9.4 **The spherical harmonics Y_{lm} are complex functions of the polar angle ϑ, with $0 \leq \vartheta \leq \pi$, and the azimuth ϕ, with $0 \leq \phi \leq 2\pi$. They can be visualized by showing their real and imaginary parts and their absolute square over the ϑ, ϕ-plane. Such graphs are shown here for $l = 3$ and $m = 0, 1, 2, 3$.**

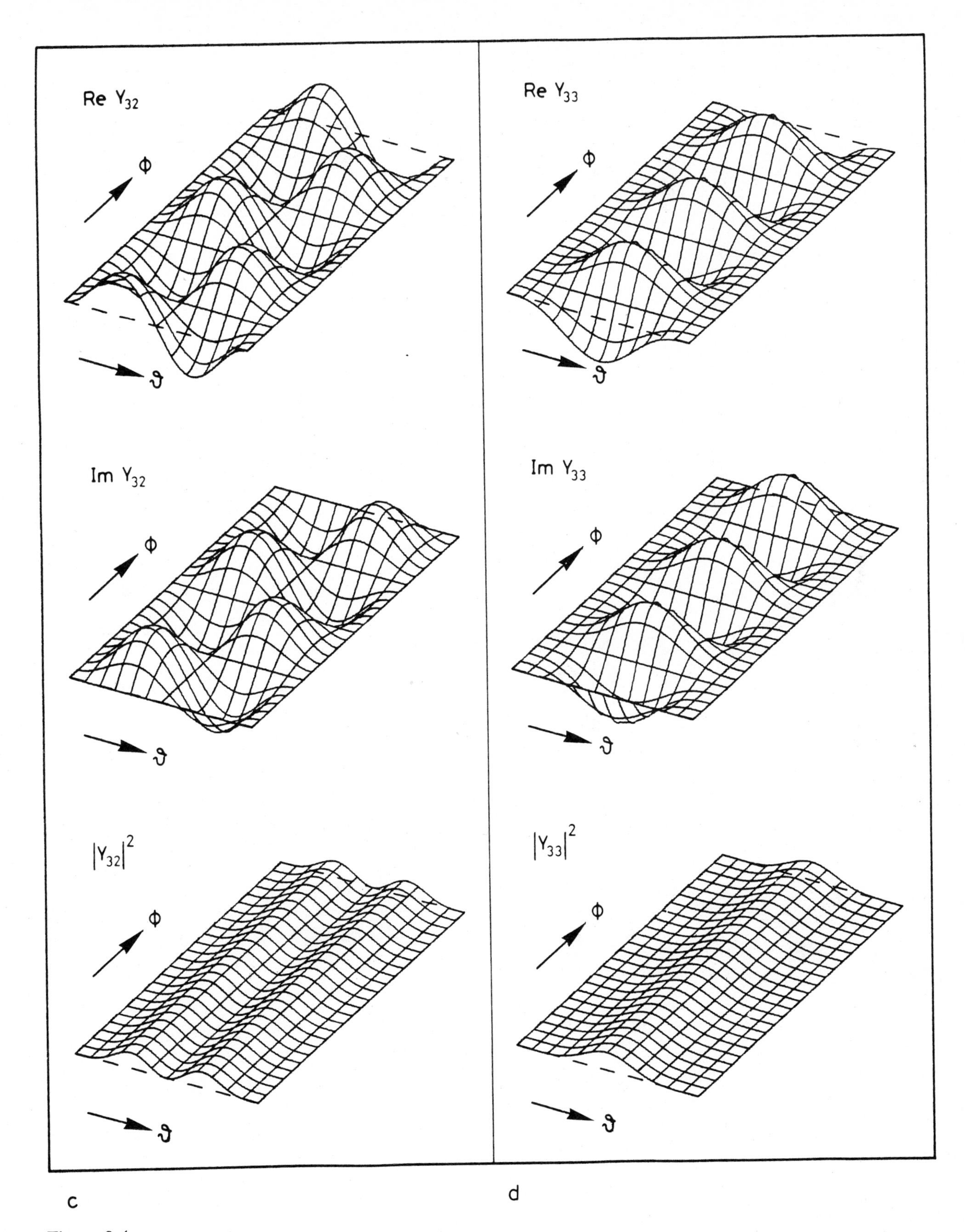

Figure 9.4

The top part of Figure 9.3 gives their graphs for $l = 0, 1, 2, 3$.

Finally, for $m \geq 0$, the spherical harmonics Y_{lm} have the representation

$$Y_{lm}(\vartheta, \phi) = (-1)^m \sqrt{\frac{2l+1}{4\pi} \cdot \frac{(l-m)!}{(l+m)!}}\, P_l^m(\cos\vartheta) e^{im\phi}$$

For negative $m = -1, -2, \ldots, -l$ the spherical harmonics are

$$Y_{l,-m}(\vartheta, \phi) = (-1)^m Y_{lm}^*(\vartheta, \phi)$$

Whereas the Legendre polynomials $P_l(u)$ and the associated Legendre functions $P_l^m(u)$ are real functions of the argument u, the spherical harmonics Y_{lm} are complex functions of their arguments. As an example, Figure 9.4 shows the real and imaginary parts as well as the absolute square of $Y_{3m}(\vartheta, \phi)$. As the definition and the plots indicate, $|Y_{lm}|^2$ depends only on ϑ. In fact, except for the normalization factor, it is equal to $[P_l^m(\cos\vartheta)]^2$. For comparison, the bottom of Figure 9.3 plots $|Y_{lm}|^2$ below P_l^m for $l = 0, 1, 2, 3$.

Since the variables of the spherical harmonics are the polar angle ϑ and the azimuth ϕ of a spherical coordinate system, it is advantageous to represent $|Y_{lm}|^2$ in such a coordinate system. This is done in Figure 9.5 where $|Y_{lm}(\vartheta, \phi)|^2$ is the length of the radius subtended under the angles ϑ and ϕ from the origin to the surface. In this way $|Y_{00}|^2 = 1/(4\pi)$ turns out to be a sphere. For all possible values l and m the functions $|Y_{lm}|^2$ are rotationally symmetric around the z-axis. They can vanish for certain values of ϑ. These are called ϑ-nodes if they occur for values of ϑ other than zero or $\pi/2$. It should be noted that $|Y_{ll}|^2$ does not have nodes, whereas $|Y_{lm}|^2$ possesses $l - |m|$ nodes.

Figure 9.5 **Polar diagrams of the absolute squares of the spherical harmonics. The distance from the origin of the coordinate system to a point on the surface seen under the angles ϑ and ϕ is equal to $|Y_{lm}(\vartheta, \phi)|^2$.**

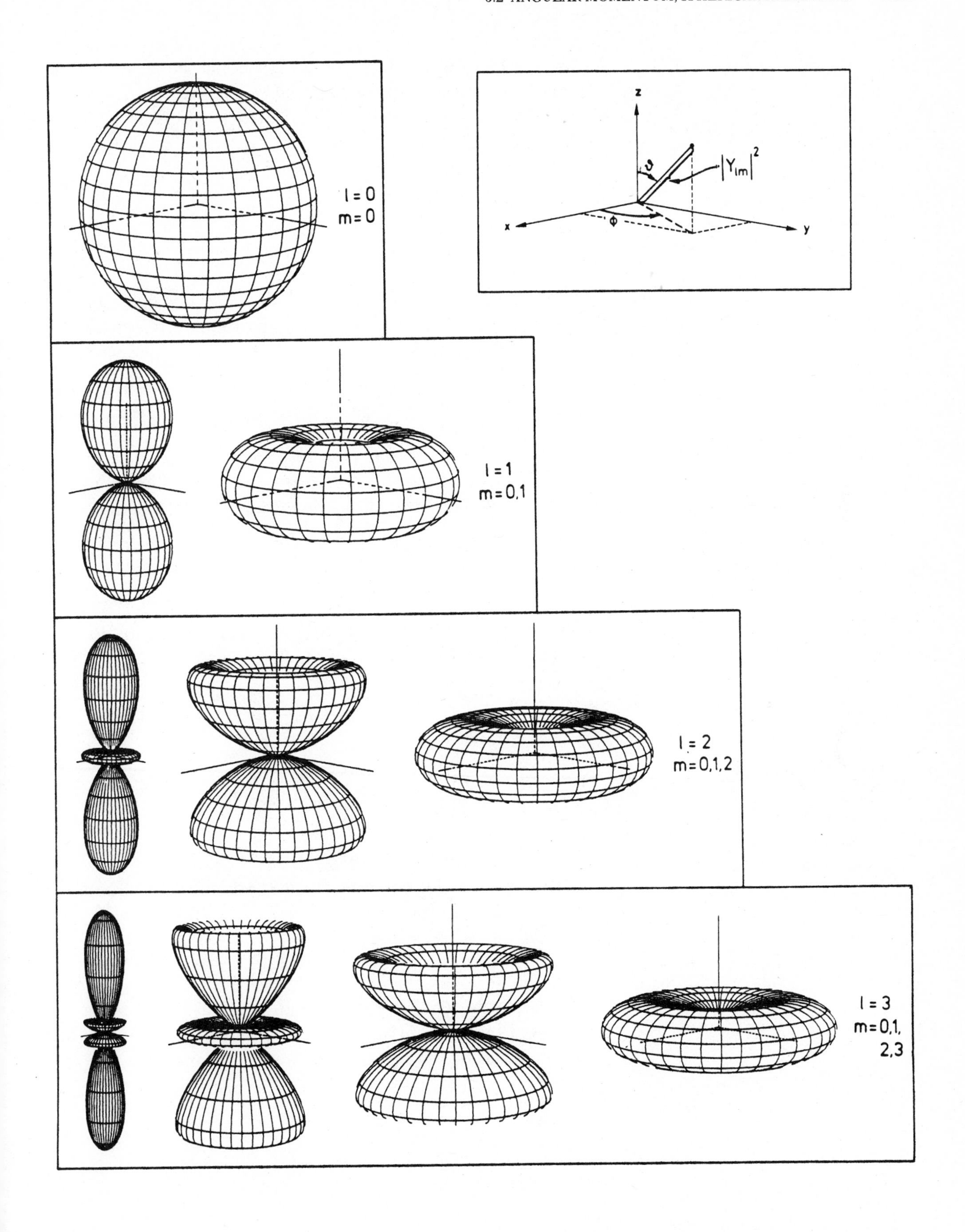
z
$|Y_{lm}|^2$
ϑ
φ
x
y
l = 0
m = 0
l = 1
m = 0,1
l = 2
m = 0,1,2
l = 3
m = 0,1,
2,3

9.3 Schrödinger Equation

As we did for the one-dimensional harmonic wave in Section 3.2, let us compare time and spatial derivatives of the three-dimensional harmonic wave $\psi_{\mathbf{p}}(\mathbf{r}, t)$, which was introduced in Section 9.1. They are

$$i\hbar \frac{\partial}{\partial t} \psi_{\mathbf{p}}(\mathbf{r}, t) = E \psi_{\mathbf{p}}(\mathbf{r}, t)$$

$$-\frac{\hbar^2}{2M} \nabla^2 \psi_{\mathbf{p}}(\mathbf{r}, t) = \frac{\mathbf{p}^2}{2M} \psi_{\mathbf{p}}(\mathbf{r}, t)$$

Here M is the mass of the particle.[1] The *Laplace operator* ∇^2 is simply the sum of the three second-order derivatives with respect to the coordinates:

$$\nabla^2 = \frac{\partial^2}{\partial x^2} + \frac{\partial^2}{\partial y^2} + \frac{\partial^2}{\partial z^2}$$

Making use of the relation $E = \mathbf{p}^2/(2M)$ between energy E, momentum $\mathbf{p}$, and mass M of a free particle, we obtain the Schrödinger equation for three-dimensional unaccelerated motion,

$$i\hbar \frac{\partial}{\partial t} \psi_{\mathbf{p}}(\mathbf{r}, t) = -\frac{\hbar^2}{2M} \nabla^2 \psi_{\mathbf{p}}(\mathbf{r}, t)$$

We may consider the operator on the right-hand side of this equation as the operator of kinetic energy,

$$\begin{aligned} T = \frac{\hat{\mathbf{p}}^2}{2M} &= \frac{1}{2M}\left(\hat{p}_x^2 + \hat{p}_y^2 + \hat{p}_z^2\right) \\ &= \frac{1}{2M}\left(-\hbar^2 \frac{\partial^2}{\partial x^2} - \hbar^2 \frac{\partial^2}{\partial y^2} - \hbar^2 \frac{\partial^2}{\partial z^2}\right) \\ &= -\frac{\hbar^2}{2M} \nabla^2 \end{aligned}$$

Thus the Schrödinger equation for three-dimensional free motion has the simple form

$$i\hbar \frac{\partial}{\partial t} \psi_{\mathbf{p}}(\mathbf{r}, t) = T \psi_{\mathbf{p}}(\mathbf{r}, t)$$

The equation can be extended to motion in a force field represented by a potential energy $V(\mathbf{r})$ by substituting for the operator of kinetic energy T the Hamiltonian operator of total energy,

$$H = T + V$$

[1] From here on we denote the mass of a particle by the capital letter M. This is done to avoid confusion with magnetic quantum number m.

The Schrödinger equation for motion under the influence of a force therefore reads

$$i\hbar\frac{\partial}{\partial t}\psi_{\mathbf{p}}(\mathbf{r},t) = H\psi_{\mathbf{p}}(\mathbf{r},t) = \left[-\frac{\hbar^2}{2M}\nabla^2 + V(\mathbf{r})\right]\psi_{\mathbf{p}}(\mathbf{r},t)$$

With the *ansatz*

$$\psi_{\mathbf{p}}(\mathbf{r},t) = \exp\left[-\frac{i}{\hbar}Et\right]\varphi_E(\mathbf{r})$$

which factors the wave function $\psi_{\mathbf{p}}(\mathbf{r},t)$ into a time-dependent exponential and the time-independent, stationary wave function $\varphi_E(\mathbf{r})$, we obtain the stationary Schrödinger equation

$$\left[-\frac{\hbar^2}{2M}\nabla^2 + V(\mathbf{r})\right]\varphi_E(\mathbf{r}) = E\varphi_E(\mathbf{r})$$

9.4 Solution of the Schrödinger Equation of Free Motion

Besides the solutions $\psi_{\mathbf{p}}(\mathbf{r},t)$ of the free Schrödinger equation, which represent harmonic plane waves with momentum $\mathbf{p}$, there are equivalent solutions which are determined by the quantum numbers l and m of angular momentum and energy E. To find these solutions, we express the Laplace operator in polar coordinates r, ϑ, and ϕ:

$$\nabla^2\varphi = \frac{1}{r}\frac{\partial^2}{\partial r^2}r\varphi - \frac{1}{r^2}\frac{1}{\hbar^2}\hat{\mathbf{L}}^2\varphi$$

Since the operator $\hat{\mathbf{L}}^2$ of the square of angular momentum, as discussed in Section 9.2, depends only on ϑ and ϕ, we now solve the Schrödinger equation using an *ansatz*

$$\varphi_{Elm}(\mathbf{r}) = R(r)Y_{lm}(\vartheta,\phi)$$

which is a product of two functions. The first function $R(r)$ depends only on the radial coordinate. The second function is the spherical harmonic $Y_{lm}(\vartheta,\phi)$, which was recognized in Section 9.2 as the eigenfunction for $\hat{\mathbf{L}}^2$. We obtain

$$\begin{aligned}
&-\frac{\hbar^2}{2M}\nabla^2\varphi_{Elm}(\mathbf{r}) \\
&= -\frac{\hbar^2}{2M}\left[\frac{1}{r}\frac{\partial^2}{\partial r^2}rR(r) - \frac{l(l+1)}{r^2}R(r)\right]Y_{lm}(\vartheta,\phi) \\
&= ER(r)Y_{lm}(\vartheta,\phi)
\end{aligned}$$

and conclude that

$$-\frac{\hbar^2}{2M}\left[\frac{1}{r}\frac{\partial^2}{\partial r^2}r - \frac{l(l+1)}{r^2}\right]R_{El}(r) = ER_{El}(r)$$

is the eigenvalue equation for the *radial wave function* $R_{El}(r)$ for positive values of r. Here we explicitly indicate the dependence of the radial wave function on energy E and the total angular momentum l. We call $\varphi_{Elm}(\mathbf{r}) = R_{El}(r)Y_{lm}(\vartheta,\phi)$ a *partial wave* of angular momentum l and z-component m. The solutions of this "free radial Schrödinger equation" are discussed in some detail in the next section.

9.5 Spherical Bessel Functions

Let us consider the solutions of the linear differential equation that depends on the integer parameter l,

$$\left[\frac{1}{\rho}\frac{d^2}{d\rho^2}\rho - \frac{l(l+1)}{\rho^2} + 1\right]f_l(\rho) = 0$$

For $\rho = kr$, $k = (1/\hbar)\sqrt{2ME}$, it is equivalent to the free radial Schrödinger equation.

The complex solutions of this linear differential equation are the *spherical Hankel functions* of the first $(+)$ and second $(-)$ kind,

$$h_l^{(\pm)}(\rho) = C_l^{\pm}\frac{e^{\pm i\rho}}{\rho}$$

where the complex coefficients C_l are polynomials of ρ^{-1} of the form

$$C_l^{\pm} = (\mp i)^l \sum_{s=0}^{l} \frac{1}{2^s s!}\frac{(l+s)!}{(l-s)!}(\mp i\rho)^{-s}$$

The first few of the Hankel functions are

$$h_0^{(\pm)} = \frac{e^{\pm i\rho}}{\rho}, \qquad h_1^{(\pm)} = \left(\mp i + \frac{1}{\rho}\right)\frac{e^{\pm i\rho}}{\rho}$$

An equivalent set of solutions are the *spherical Bessel functions*, which are simply the linear combinations

$$j_l(\rho) = \frac{1}{2i}\left[h_l^{(+)}(\rho) - h_l^{(-)}(\rho)\right]$$

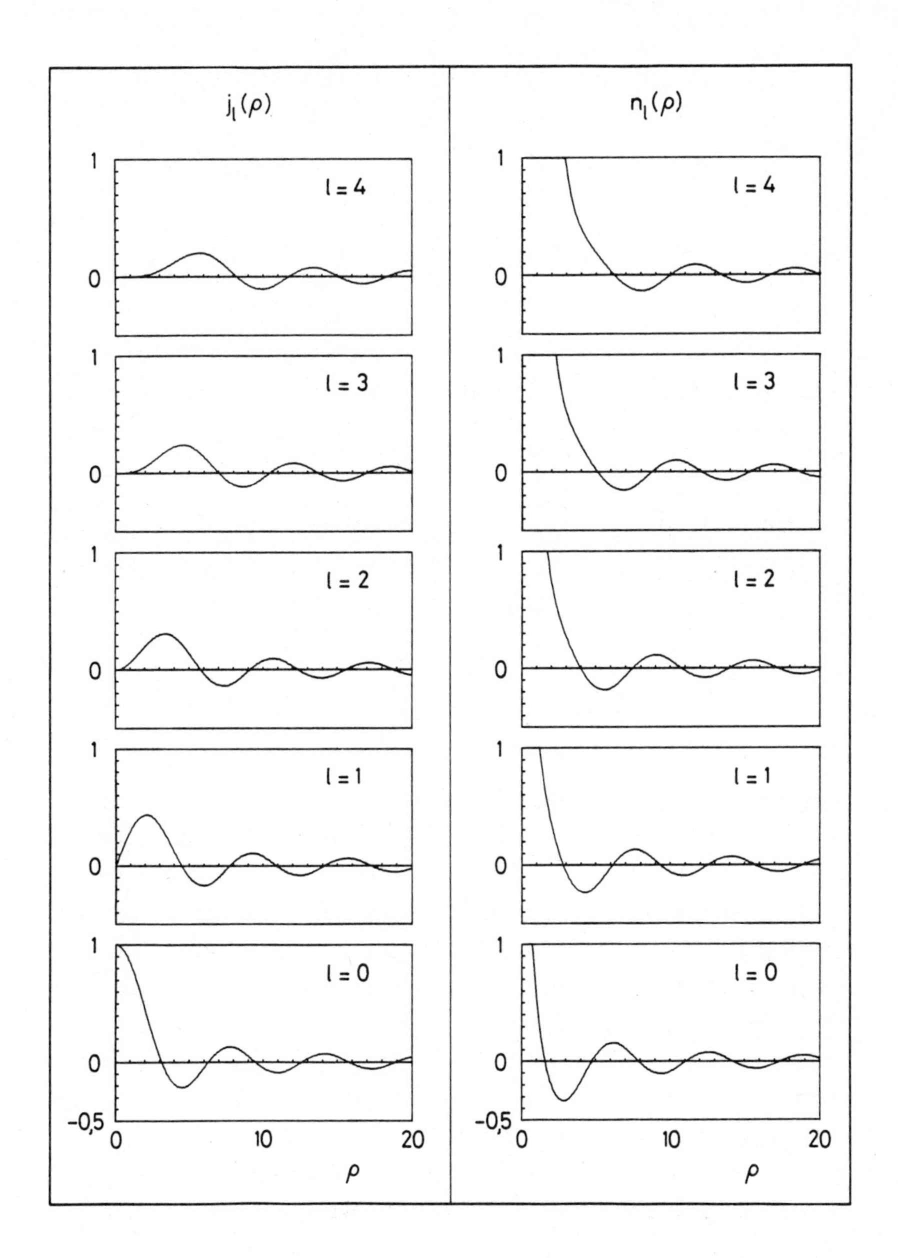

Figure 9.6 **Spherical Bessel functions $j_l(\rho)$ and spherical Neumann functions $n_l(\rho)$ for $l = 0, 1, \ldots, 4$.**

The *spherical Neumann functions*

$$n_l(\rho) = \tfrac{1}{2}\left[h_l^{(+)}(\rho) + h_l^{(-)}(\rho)\right]$$

are also solutions of the linear differential equation. In terms of the spherical Bessel and Neumann functions, the spherical Hankel functions can be expressed as

$$h_l^{(\pm)}(\rho) = n_l(\rho) \pm i j_l(\rho)$$

The first few spherical Bessel and Neumann functions are

$$j_0(\rho) = \frac{\sin\rho}{\rho}, \qquad j_1(\rho) = \frac{\sin\rho}{\rho^2} - \frac{\cos\rho}{\rho}$$

$$n_0(\rho) = \frac{\cos\rho}{\rho}, \qquad n_1(\rho) = \frac{\cos\rho}{\rho^2} + \frac{\sin\rho}{\rho}$$

The behavior of the spherical Bessel and Neumann functions for small values of the argument is

$$j_l(\rho) \sim \rho^l, \qquad n_l(\rho) \sim \rho^{-(l+1)}$$

and for large ρ

$$j_l(\rho) \underset{\rho\to\infty}{\longrightarrow} \frac{1}{\rho}\sin(\rho - \tfrac{1}{2}l\pi)$$

$$n_l(\rho) \underset{\rho\to\infty}{\longrightarrow} \frac{1}{\rho}\cos(\rho - \tfrac{1}{2}l\pi)$$

Since the spherical Neumann functions $n_l(\rho)$ diverge at the origin, only the spherical Bessel functions $j_l(\rho)$ are physical solutions of the free radial Schrödinger equation. The $n_l(\rho)$ as well as the spherical Hankel functions $h_l^{(\pm)}(\rho)$, however, are needed for the discussion of the radial Schrödinger equation for a square-well potential. Figure 9.6 plots the $j_l(\rho)$ and the $n_l(\rho)$ for $l = 0, \ldots, 4$.

In connection with the wave functions for a square-well potential, we encounter negative energies E_i that make values of wave number $k_i = \sqrt{2mE_i}/\hbar$ imaginary. Therefore the functions j_l, n_l, and $h_l^{(+)}$ are needed for imaginary arguments $\rho = i\eta$. Using the original definition, we can write

$$h_l^{(\pm)}(i\eta) = (\mp i)^{l\pm 1} \sum_{s=0}^{l} \frac{1}{2^s s!} \frac{(l+s)!}{(l-s)!} (\pm\eta)^{-s} \frac{e^{\mp\eta}}{\eta}$$

The $j_l(i\eta)$ and $n_l(i\eta)$ are again given by the linear combinations of the $h_l^{(+)}(i\eta)$ and the $h_l^{(-)}(i\eta)$. The values of these functions for such arguments are either real or purely imaginary. Figure 9.7 presents the functions

$$(-i)^l j_l(i\eta), \quad i^{l+1} n_l(i\eta), \quad i^{l+1} h_l^{(+)}(i\eta)$$

for $l > 0, 1, \ldots, 4$. The powers of i in front of $j_l(i\eta)$, $n_l(i\eta)$

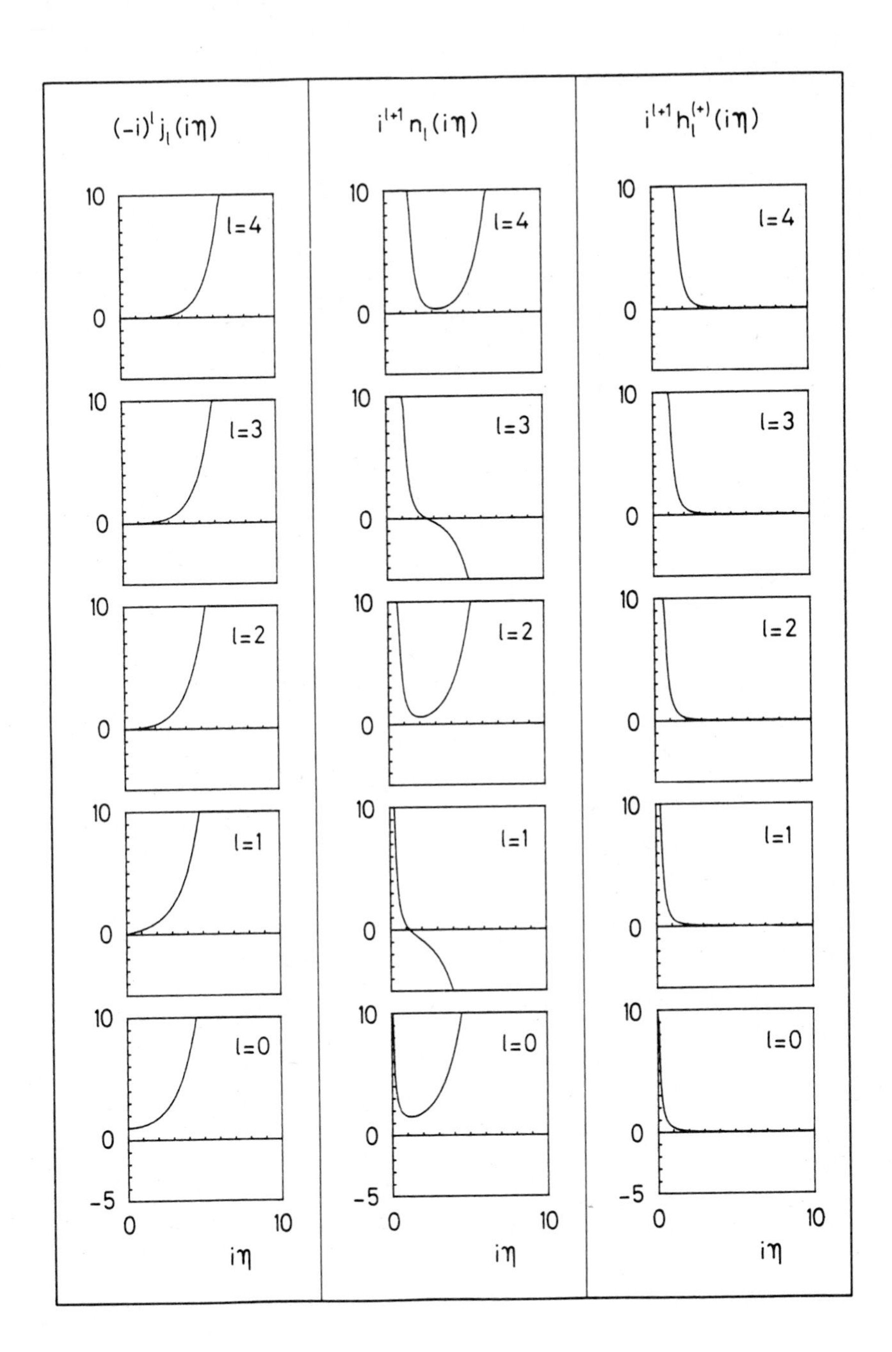

***Figure* 9.7** **For purely imaginary arguments $i\eta$, η real, the spherical Bessel functions j_l, the spherical Neumann functions n_l, and the spherical Hankel functions $h_l^{(+)}$ are either purely real or purely imaginary. The functions shown, that is, $(-i)^l j_l(i\eta)$, $i^{l+1} n_l(i\eta)$, and $i^{l+1} h_l^{(+)}(i\eta)$, are purely real.**

and $h_l^{(+)}(i\eta)$ ensure that the functions plotted in Figure 9.7 are real.

The $h_l^{(+)}(i\eta)$ play a role in describing bound states outside the potential well. Their asymptotic behavior for large η is

$$i^{l+1} h_l^{(+)}(i\eta) \sim \frac{e^{-\eta}}{\eta}, \qquad \eta \to \infty$$

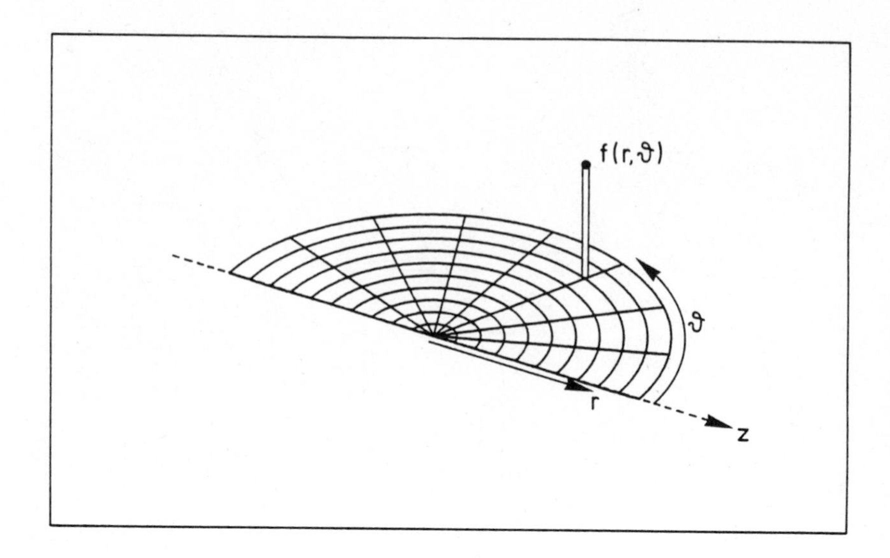

Figure 9.8 **The polar coordinate system used throughout the book for functions of the type $f = f(r, \vartheta)$. The admissible range of variables, $0 \le r < \infty, 0 \le \vartheta \le \pi$, corresponds to a half-plane. Here a half circle around the origin, $r = 0$, is viewed perspectively from a point outside the half-plane. The polar angle ϑ is measured against the z-axis, which points to the lower right. Lines of constant ϑ are straight lines beginning at the origin. Lines of constant r are half circles. Using the direction perpendicular to the half-plane to define an f-coordinate, we can represent a function $f(r, \vartheta)$ as a surface in r, ϑ, f-space. Figures 9.9 and 9.10 show lines of constant r and constant ϑ on this surface.**

9.6 Harmonic Plane Wave in Angular Momentum Representation

The spherical waves $j_l(kr)Y_{lm}(\vartheta, \phi)$, like the harmonic plane waves, form a complete set of functions which can also be used for constructing wave packets by superposition or for decomposing stationary solutions into spherical waves. In particular, we decompose the stationary harmonic plane wave into partial waves,

$$e^{i\mathbf{k}\cdot\mathbf{r}} = e^{ikz} = e^{ikr\cos\vartheta} = \sum_{l=0}^{\infty} (2l + 1)i^l j_l(kr)P_l(\cos\vartheta)$$

where the z-axis was chosen to be parallel to $\mathbf{k}$ and ϑ is therefore the angle between $\mathbf{k}$ and $\mathbf{r}$. Since the left-hand side of this relation does not depend on the azimuth ϕ, only the spherical harmonic function $Y_{l0} = \sqrt{(2l + 1)/4\pi}\, P_l$ occurs in the sum.

Figures 9.9 and 9.10 illustrate this decomposition. Polar coordinates r and ϑ are used to plot functions over the r, ϑ-half-plane. The polar coordinate system used throughout the book for functions of the type $f = f(r, \vartheta)$ is explained in Figure 9.8.

In the top right corner of Figure 9.9, the function $\cos(kz) = \mathrm{Re}\{e^{ikz}\}$ is presented. The left column contains the functions

$$(2l + 1)i^l j_l(kr)P_l(\cos\vartheta), \qquad l = 0, 2, \ldots, 8$$

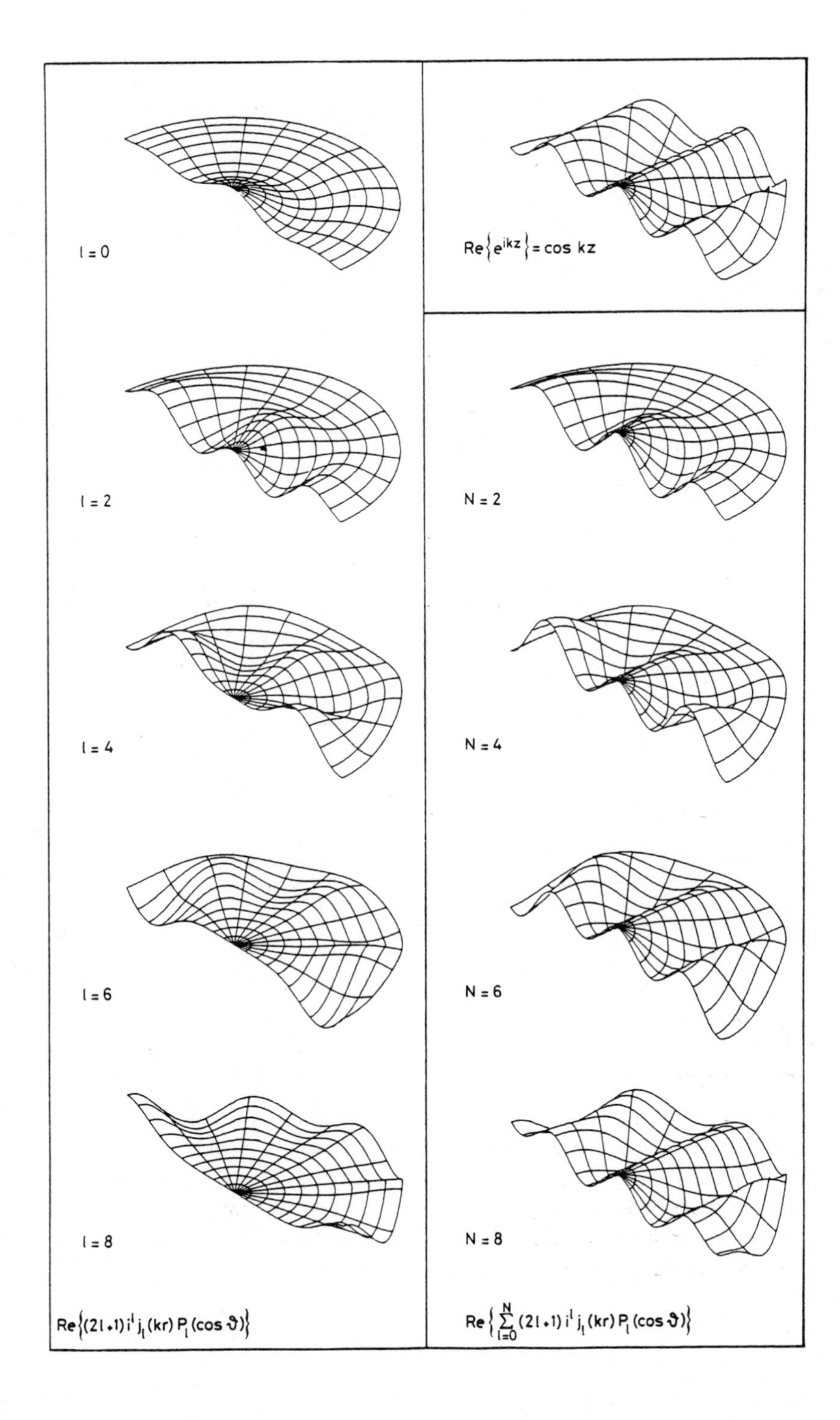

Figure 9.9 **Decomposition of a plane wave into spherical waves. The real part $\mathrm{Re}\{e^{ikz}\} = \cos(kz)$ of a plane wave is shown in the top right corner. The left column contains the terms of the decomposition that are purely real. The right column contains the sums of the first two terms ($N = 2$), three terms ($N = 4$), and so on of the left column.**

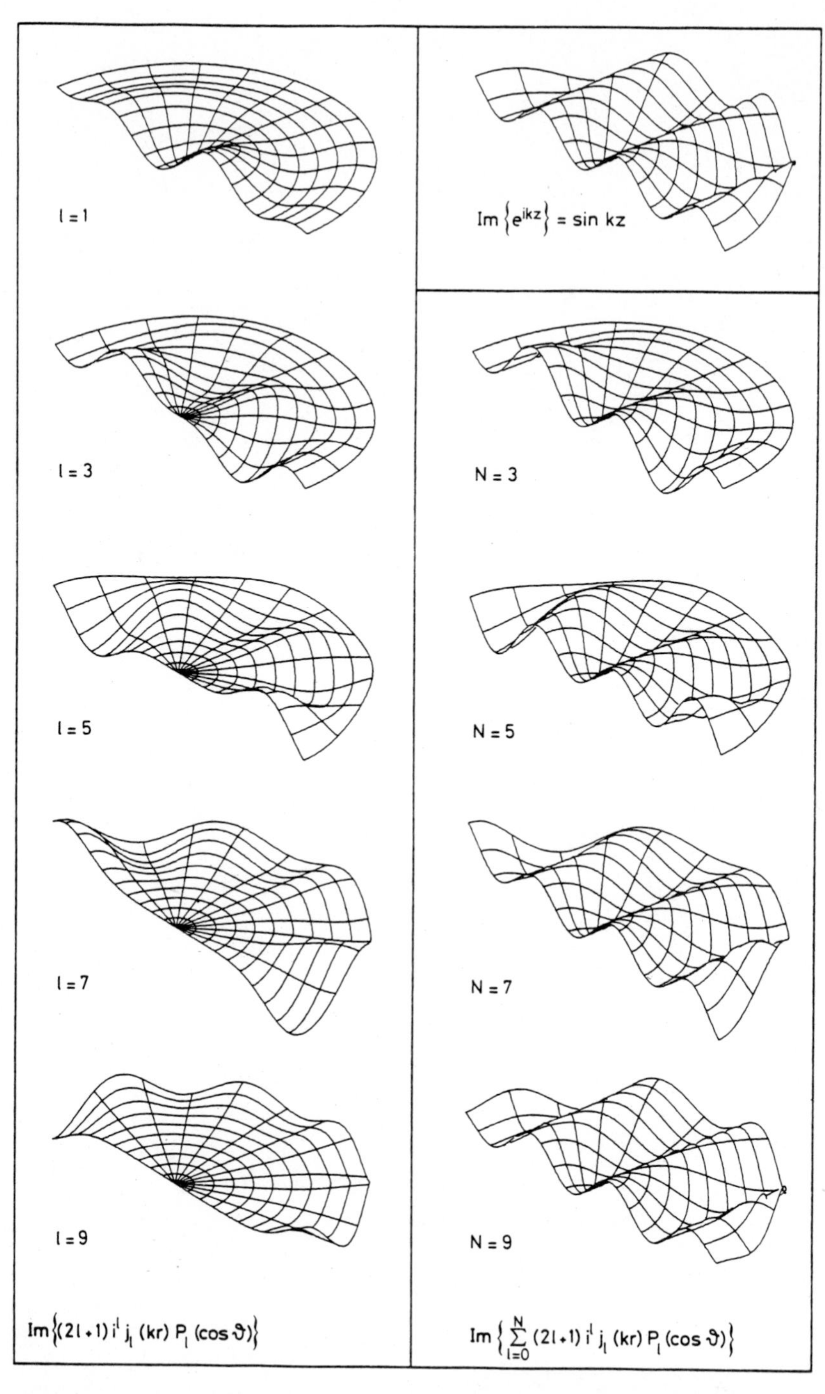

Figure 9.10 **Decomposition of a plane wave into spherical waves. The imaginary part Im$\{e^{ikz}\}$ = sin(kz) of a plane wave is shown in the top right corner. The left column contains the terms of the decomposition that are purely imaginary. The right column contains the sums of the first two terms ($N = 3$), three terms ($N = 5$), and so on, of the left column.**

which are the first few real terms in this decomposition. The right column shows the sums of the first two terms, three terms, and so on. In the neighborhood of the origin, the plane wave is described well by the first few terms of the sum. Farther away from the origin, more terms have to be added. Near the origin the first few terms are adequate because the functions $j_l(\rho)$ are suppressed there for increasing l (see Figure 9.6). A similar illustration for the imaginary part of the plane wave is given in Figure 9.10.

9.7 Free Wave Packet and Angular Momentum of the Classical Particle

In Section 9.1 we discussed a three-dimensional unaccelerated wave packet moving with group velocity $\mathbf{v}_0 = \mathbf{p}_0/M$. The wave packet was represented as a superposition of plane waves that are eigenfunctions for the momentum operator. The details of the superposition were determined by the spectral function $f(\mathbf{p})$ which specifies the contribution of the plane wave with wave vector $\mathbf{k} = \mathbf{p}/\hbar$. Analogously, the same wave packet can be understood as a superposition of the eigenfunctions $Y_{lm}(\vartheta, \phi)$ for angular momentum multiplied by appropriately chosen weight functions $a_{lm}(r, 0)$ for the radius variable r. In this kind of representation, the weight function regulates the relative weight contributed by the various angular momenta.

Representation of the wave packet at initial time has the form

$$\psi(\mathbf{r}, 0) = \sum_{l=0}^{\infty} \sum_{m=-l}^{l} a_{lm}(r, 0) Y_{lm}(\vartheta, \phi)$$

In an additional step we may decompose the radial functions $a_{lm}(r, 0)$ into purely wave number, that is, energy-dependent, coefficients,

$$b_{lm}(k) = \int_0^{\infty} j_l(kr) a_{lm}(r, 0) r^2 \, dr$$

$$a_{lm}(r, 0) = \frac{2}{\pi} \int_0^{\infty} b_{lm}(k) j_l(kr) k^2 \, dk,$$

so that the free wave packet at $t = 0$ is now

$$\psi(\mathbf{r}, 0) = \frac{2}{\pi} \sum_{l=0}^{\infty} \sum_{m=-l}^{l} \int b_{lm}(k) j_l(kr) Y_{lm}(\vartheta, \phi) k^2 \, dk$$

In this decomposition of the free wave packet in terms of the eigenfunctions of the free Schrödinger equation for the eigenvalues E, l, and m, the functions $b_{lm}(k)$ play the role of spectral coefficients for angular momentum and spectral functions for energy $E = \hbar^2 k^2/2M$. In Section 9.1 the spectral function $f(\mathbf{p})$ played a similar role in decomposition of the wave packet in terms of eigenfunctions of the three momentum components.

The moving wave packet is described by the time-dependent wave function $\psi(\mathbf{r}, t)$ which is obtained from the initial wave function by taking into account the time-dependent phase factor $\exp(-iEt/\hbar)$, that is

$$\psi(\mathbf{r}, t) = \frac{2}{\pi} \sum_{l=0}^{\infty} \sum_{m=-l}^{l} \int b_{lm}(k) \times \exp\left[-\frac{i}{\hbar} Et\right] j_l(kr) Y_{lm}(\vartheta, \phi) k^2 \, dk$$

The angular momentum content of the free wave packet is given by the spectral coefficients $b_{lm}(k)$. They are time-independent because angular momentum is conserved.

If we ask for the contribution having angular momentum quantum number l and magnetic quantum number m irrespective of wave number k, we have to integrate the probabilities $b_{lm}^*(k) b_{lm}(k) k^2 \, dk$ over all wave numbers:

$$W_{lm} = \frac{2}{\pi} \int b_{lm}^*(k) b_{lm}(k) k^2 \, dk$$

The probabilities W_{lm} fulfill the normalization condition:

$$\sum_{l=0}^{\infty} \sum_{m=-l}^{l} W_{lm} = 1$$

As an example, we consider the wave packet shown in Figure 9.11a. Its center moves with constant velocity in the negative x-direction keeping a constant distance b from the x-axis. That is, it behaves like a classical particle with time-dependent position vector

$$\mathbf{r}(t) = [x(t), b, 0]$$

and constant momentum vector

$$\mathbf{p}(t) = (-p, 0, 0)$$

The angular momentum vector of the classical particle,

$$\mathbf{L} = \mathbf{r} \times \mathbf{p} = (0, 0, bp)$$

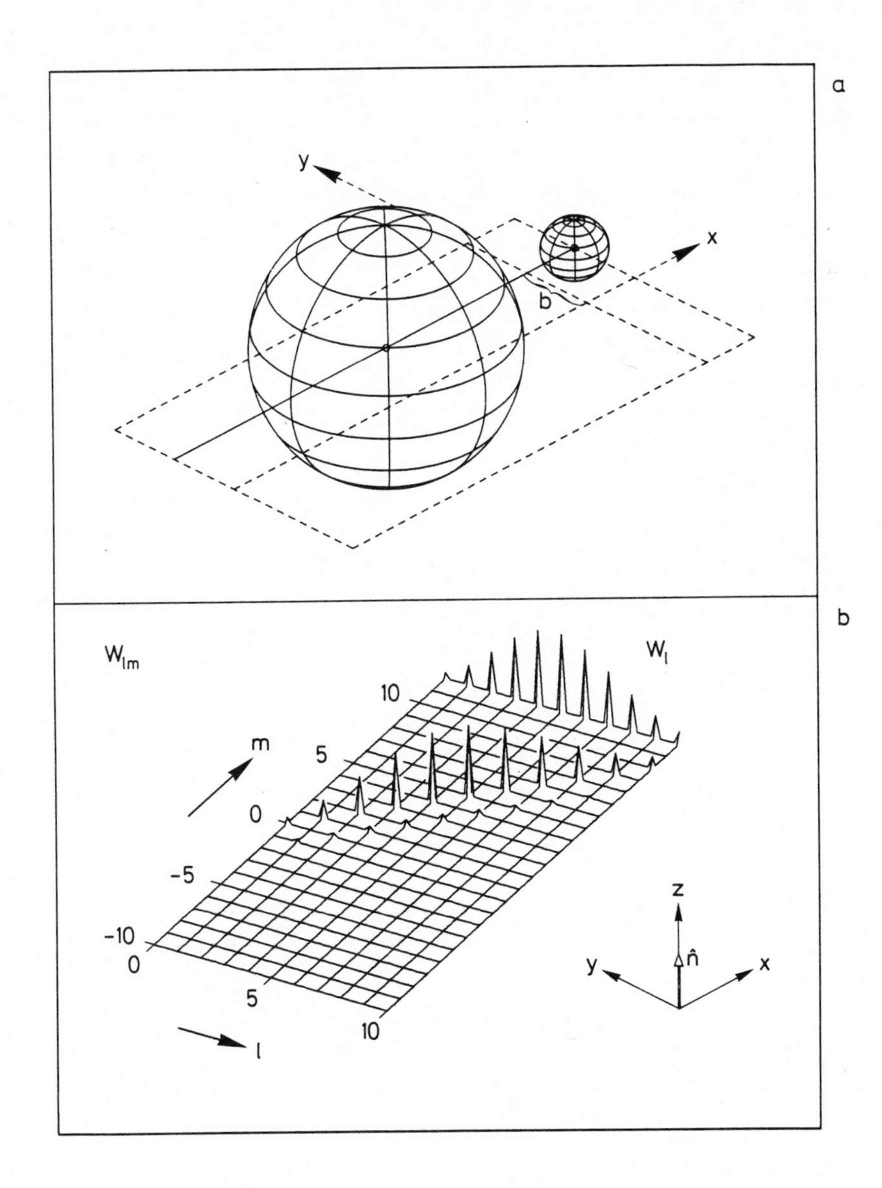

Figure 9.11 **(a) The probability ellipsoid, here a sphere, of a free wave packet moving in the x, y-plane antiparallel to the x-axis, shown at two moments of time. The dispersion of the wave packet is apparent through the growth of the sphere with time. (b) Decomposition of the wave packet shown in part a into angular momentum states. The height of the spike drawn at point l, m is proportional to the probability W_{lm} that the particle which is described by the wave packet has angular momentum quantum number l and quantum number m for the component of angular momentum along the quantization axis $\hat{\mathbf{n}}$. In this figure $\hat{\mathbf{n}}$ was chosen to be the z-axis. Also shown, on the upper margin, are the probabilities W_l that the particle possesses quantum number l irrespective of the value of m.**

is independent of time and is oriented along the z-direction. The absolute value of the angular momentum is

$$L = |\mathbf{L}| = bp$$

We now consider a particle of constant momentum $\mathbf{p}$ which travels along an arbitrary straight line. The shortest distance of this line from the origin is called the *impact parameter* b. Obviously, for reasons of symmetry, the absolute value of angular momentum for this particle is again $L = bp$.

Let us now study the probabilities W_{lm} for the wave packet of Figure 9.11a. We quantize the angular momentum in z-direction, that is, we use the eigenfunctions of $\hat{L}^2$ and $\hat{L}_z$ in decomposing the wave packet. In Figure 9.11b the probabilities W_{lm} are plotted for various values of l and m. In the graphical representation each probability is proportional to the height of the pyramid sitting on top of point l, m in the coordinate grid. Obviously, the probabilities can be different from zero only in points lying within a sector between two straight lines, for which $m = l$ and $m = -l$. We note that, in contrast to the classical point particle, various angular momenta contribute to the wave packet. In fact, for the quantization axis chosen, the probabilities at points $l = m$ are by far the largest for every l. This is not surprising since the angular momentum of the corresponding particle has only a z-component. Nevertheless, values $m < l$ also contribute. The contributions W_{lm} for $m = l - 1, l - 3, \ldots$ vanish. Because of the mirror symmetry of the wave packet with respect to the x, y-plane, functions $Y_{lm}(\vartheta, \phi)$ with $m = l - 1, l - 3, \ldots$ do not contribute. They are antisymmetric in ϑ with respect to point $\vartheta = \pi/2$.

The probabilities that a certain quantum number l will contribute irrespective of m are

$$W_l = \sum_{m=-l}^{l} W_{lm}$$

They are plotted on the upper margin of Figure 9.11b. As a function of l, the probabilities W_l have a bell-shaped envelope reminiscent of a Gaussian. The maximum of the marginal distribution corresponds to the angular momentum of the classical particle.

We study now the dependence of the W_{lm}-distribution on the quantization axis. Instead of the z-axis, we first choose an axis $\hat{\mathbf{n}}$ that forms an angle of $\pi/4$ with the y-axis in the z, y-plane. Figure 9.12a shows that many more m-values now participate in the superposition of the wave packet. The marginal distribution, however, remains unchanged. There are changes in the m-distribution because the new quantization axis does not point in the direction of the classical angular momentum vector. The distribution W_l of the modulus of angular momentum is independent of the quantization axis.

Finally, Figure 9.12b shows the probabilities W_{lm} for the y-axis as the quantization direction of angular momentum, and Figure 9.12c shows them for the x-axis as the quantization direction. Since in both figures the quantization direction is perpendicular to the classical angular momentum vector, we foresee that the expectation value of m will vanish. Indeed, the two distributions are symmetric around $m = 0$.

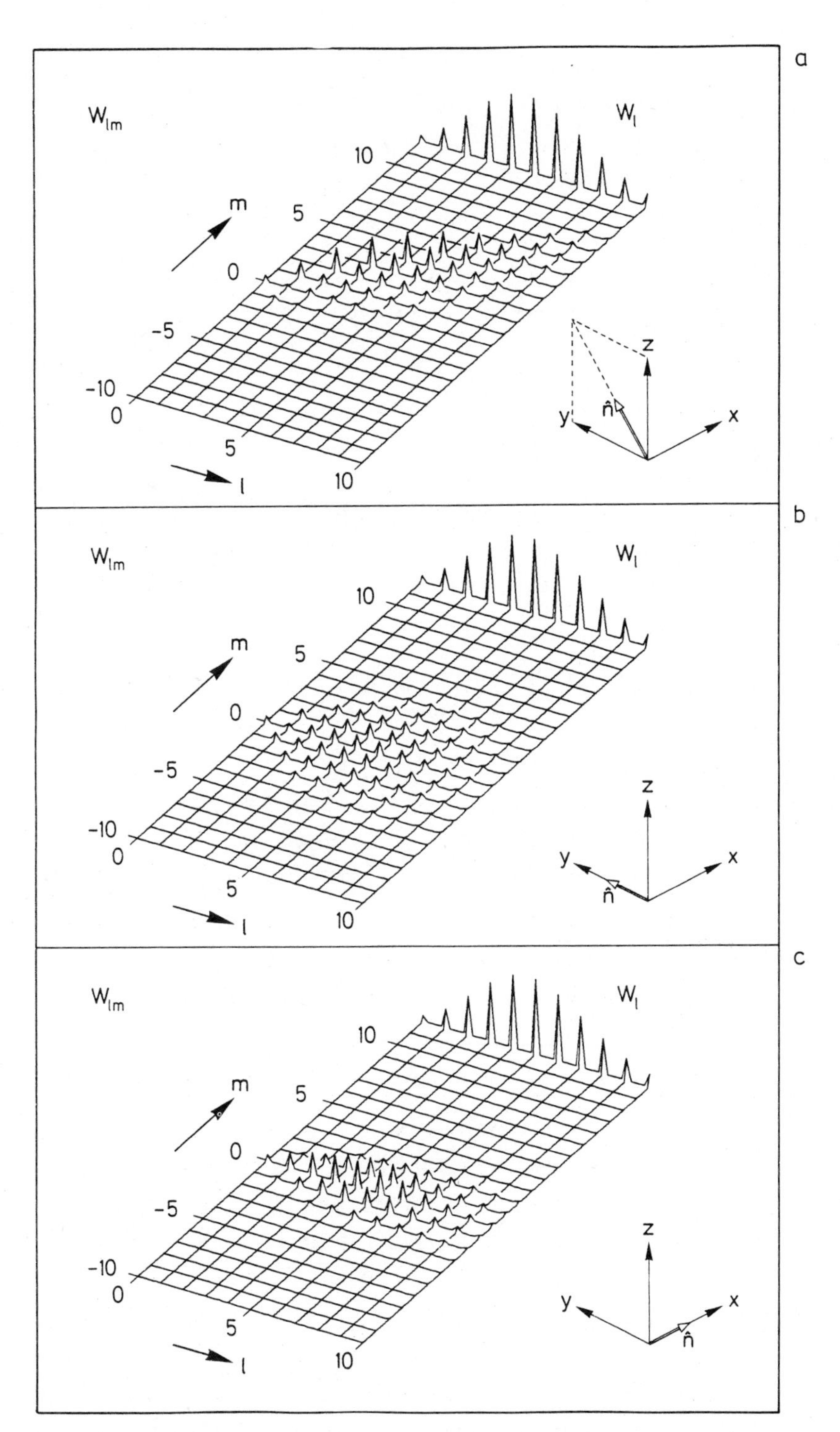

Figure 9.12 **All three figures apply to the situation of Figure 9.11a. Like Figure 9.11b, they show the decomposition of the wave packet into angular momentum states. The quantization axes are different, however.**

Problems

9.1 Assuming that the components of the momentum operator in the three spatial dimensions are given by

$$\hat{p}_i = \frac{\hbar}{i}\frac{\partial}{\partial x_i}, \qquad \hat{\mathbf{p}} = (\hat{p}_1, \hat{p}_2, \hat{p}_3)$$

show that the simultaneous stationary eigenfunction for the three momentum operators $\hat{p}_i$ is the product of three one-dimensional momentum eigenfunctions.

9.2 Calculate the probability density $\rho(\mathbf{r}, t) = \psi^*(\mathbf{r}, t)\psi(\mathbf{r}, t)$ of the three-dimensional Gaussian wave packet of Section 9.1, using the explicit form of $M(x, t)$ as given in Section 3.2. In which direction does the wave packet move? What is the square of its velocity? What determines in which direction the wave packet disperses fastest?

9.3 Verify the commutation relations of the components of angular momentum as given at the beginning of Section 9.2.

9.4 The spatial reflection is the transformation $\mathbf{r} \to -\mathbf{r}$. How is this transformation expressed in spherical coordinates? How do the spherical harmonics $Y_{lm}(\vartheta, \phi)$ behave under reflections?

9.5 Calculate the commutators of the angular momentum component operators $\hat{L}_x$, $\hat{L}_y$, and $\hat{L}_z$ with the coordinate operators x, y, and z and with the momentum component operators $\hat{p}_x$, $\hat{p}_y$, and $\hat{p}_z$.

9.6 Calculate the commutators of L_x, L_y, L_z, and $\hat{\mathbf{L}}^2$ with $r = \sqrt{x^2 + y^2 + z^2}$ and with $\hat{\mathbf{p}}^2$. Use the results to compute the commutators of angular momentum with a Hamiltonian for a spherically symmetric potential,

$$H = \frac{\hat{\mathbf{p}}^2}{2m} + V(r)$$

9.7 Show that the three-dimensional free wave packet, as given by its spectral representation in Section 9.1,

$$\psi(\mathbf{r}, t) = \int f(\mathbf{p})\psi_{\mathbf{p}}(\mathbf{r} - \mathbf{r}_0, t)\, d^3\mathbf{p}$$

is a solution of the Schrödinger equation for three-dimensional free motion.

9.8 What is the difference between the classical and the quantum-mechanical centrifugal term $\mathbf{L}^2/2Mr^2$ in the Hamiltonian for a given angular momentum?

9.9 Verify that the explicit expressions of the spherical Bessel functions $j_0(\rho)$ and $j_1(\rho)$ and the spherical Neumann functions $n_0(\rho)$ and $n_1(\rho)$ satisfy the free radial Schrödinger equation given at the beginning of Section 9.5.

9.10 The explicit form of the spherical Hankel functions $h_l^{(\pm)}(\rho)$ is given at the beginning of Section 9.5. Show that the asymptotic forms of the spherical Bessel and Neumann functions for $\rho \to 0$ and $\rho \to \infty$ are as given in this section.

9.11 Calculate the expression $\hat{\mathbf{L}}\varphi_{\mathbf{p}}(\mathbf{r})$. Explain why the result does not imply that $\varphi_{\mathbf{p}}(\mathbf{r})$ is a simultaneous eigenfunction of the angular momentum operators $\hat{L}_x, \hat{L}_y, \hat{L}_z$.

9.12 Calculate the expressions $\hat{\mathbf{L}}\{j_l(kr)Y_{lm}(\vartheta, \phi)\}$ for $l = 0, 1$. What distinguishes the two cases $l = 0$ and $l = 1$?

9.13 What is the expectation value of angular momentum of a Gaussian wave packet as given in Section 9.1? Explain why the result is time-independent.

9.14 Why is the m-distribution in Figure 9.12b wider than that in Figure 9.12c? To find the answer, consider the y- and z-components of angular momentum for the classical assembly of particles imitating the wave packet.

10. Solution of the Schrödinger Equation in Three Dimensions

In Section 9.3 the time-dependent Schrödinger equation for three-dimensional motion under the influence of a potential was separated with respect to time and space coordinates with the help of the *ansatz*

$$\psi(\mathbf{r}, t) = \exp\left[-\frac{i}{\hbar}Et\right]\varphi_E(\mathbf{r})$$

The three-dimensional stationary Schrödinger equation for the function $\varphi_E(\mathbf{r})$ obtained at the end of that section is

$$\left[-\frac{\hbar^2}{2M}\nabla^2 + V(\mathbf{r})\right]\varphi_E(\mathbf{r}) = E\varphi_E(\mathbf{r})$$

We now restrict ourselves to spherically symmetric systems, those in which the potential $V(\mathbf{r})$ depends only on the radial coordinate r. Following the same line of thought used in Section 9.4, we separate radial and angular coordinates,

$$\varphi_{Elm}(\mathbf{r}) = R(r)Y_{lm}(\vartheta, \phi)$$

and arrive at the radial Schrödinger equation for the radial wave functions $R_l(k, r)$:

$$-\frac{\hbar^2}{2M}\left[\frac{1}{r}\frac{d^2}{dr^2}r - \frac{l(l+1)}{r^2} - \frac{2M}{\hbar^2}V(r)\right]R_l(k, r) = ER_l(k, r)$$

Because the potential has spherical symmetry, this equation does not depend on quantum number m of the z-component of angular momentum. Therefore the $R_l(k, r)$ do not depend on m. Besides the kinetic and potential energies, the terms on the left-hand side of this equation represent the *centrifugal potential*

$$\frac{\hbar^2}{2M}\frac{l(l+1)}{r^2}$$

which is attributable to the angular momentum. This and the potential term $V(r)$ are often combined to give the *effective potential* for a given angular momentum l,

$$V^{\text{eff}}(r) = \frac{\hbar^2}{2M}\frac{l(l+1)}{r^2} + V(r)$$

The radial Schrödinger equation then reads

$$\left[-\frac{\hbar^2}{2M}\frac{1}{r}\frac{d^2}{dr^2}r + V^{\text{eff}}(r)\right]R_l(k,r) = ER_l(k,r)$$

This equation is a differential equation with one variable. Its solution for potential functions that are simple in structure proceeds along the same lines used to solve the one-dimensional Schrödinger equation in Chapter 4. Since the radius variable r assumes positive values only, here we are looking for solutions $R_l(k,r)$ only on the positive half-axis. At the origin the solution $R_l(k,r)$ must be finite. Again, we have to distinguish the two types of solutions, those for scattering processes and those for bound states.

In contrast to the three-dimensional Schrödinger equation, which does not refer to a particular angular momentum, the radial Schrödinger equation describes a particle of a given angular momentum quantum number l. The centrifugal potential acts as a repulsive potential, also called a *centrifugal barrier*, and keeps the particle of momentum p sufficiently distant from the origin of the polar coordinate system. This way the impact parameter b—see Figure 9.11—remains sufficiently large to guarantee that angular momentum $L = bp$ is conserved.

10.1 Stationary Scattering Solutions

As in Section 4.2, we have to formulate the boundary conditions for the solutions that describe elastic scattering of a particle on a potential. In Sections 9.4 and 9.5 we have seen that the solutions of the radial Schrödinger equation of free motion are the spherical Bessel and Neumann functions $j_l(kr)$

and $n_l(kr)$. From Section 4.2 we learned that for forces of finite range the particles move force-free at distances far from the range of the force. For elastic scattering on a potential of finite range d, the radial wave functions $R_l(k,r)$ must therefore approach a linear combination of spherical Bessel and Neumann functions for values of r large compared to range d:

$$R_l(k,r) \to A_l j_l(kr) + B_l n_l(kr), \qquad r \gg d$$

For some potentials the solution of the free radial Schrödinger equation can be given explicitly. As a particularly instructive example, we consider a square-well potential:

$$V(r) = \begin{cases} V_{\mathrm{I}}, & 0 \le r \le d_1, \quad \text{region I} \\ V_{\mathrm{II}}, & d_1 \le r \le d_2, \quad \text{region II} \\ V_{\mathrm{III}} = 0, & d_2 \le r < \infty, \quad \text{region III} \end{cases}$$

Since the potential vanishes in region III, we say that it has the finite range $d = d_2$.

Scattering solutions of the radial Schrödinger equation have energy $E > 0$. The solution in inner region I consists of $j_l(k_{\mathrm{I}} r)$ only, since $n_l(k_{\mathrm{I}} r)$ is singular for $r = 0$. In regions II and III the solution can be written as a superposition of j_l and n_l:

$$R_l(k,r) = \begin{cases} R_{l\mathrm{I}}(k,r) = A_{l\mathrm{I}} j_l(k_{\mathrm{I}} r) \\ R_{l\mathrm{II}}(k,r) = A_{l\mathrm{II}} j_l(k_{\mathrm{II}} r) + B_{l\mathrm{II}} n_l(k_{\mathrm{II}} r) \\ R_{l\mathrm{III}}(k,r) = A_{l\mathrm{III}} j_l(kr) + B_{l\mathrm{III}} n_l(kr) \end{cases}$$

Here the wave numbers k_i in regions $i =$ I, II are

$$k_i = \frac{1}{\hbar}\sqrt{2M(E - V_i)}$$

In region III

$$k = \frac{1}{\hbar}\sqrt{2ME}$$

is the wave number of the incident particles.

For every value of l, four of the coefficients A_{lN} and B_{lN} are determined in terms of the fifth by the continuity conditions for the wave function and its derivative at $r = d_1$ and $r = d_2$:

$$R_{l\mathrm{I}}(k,d_1) = R_{l\mathrm{II}}(k,d_1), \qquad \frac{dR_{l\mathrm{I}}}{dx}(k,d_1) = \frac{dR_{l\mathrm{II}}}{dx}(k,d_1)$$

and

$$R_{l\mathrm{II}}(k,d_2) = R_{l\mathrm{III}}(k,d_2), \qquad \frac{dR_{l\mathrm{II}}}{dx}(k,d_2) = \frac{dR_{l\mathrm{III}}}{dx}(k,d_2)$$

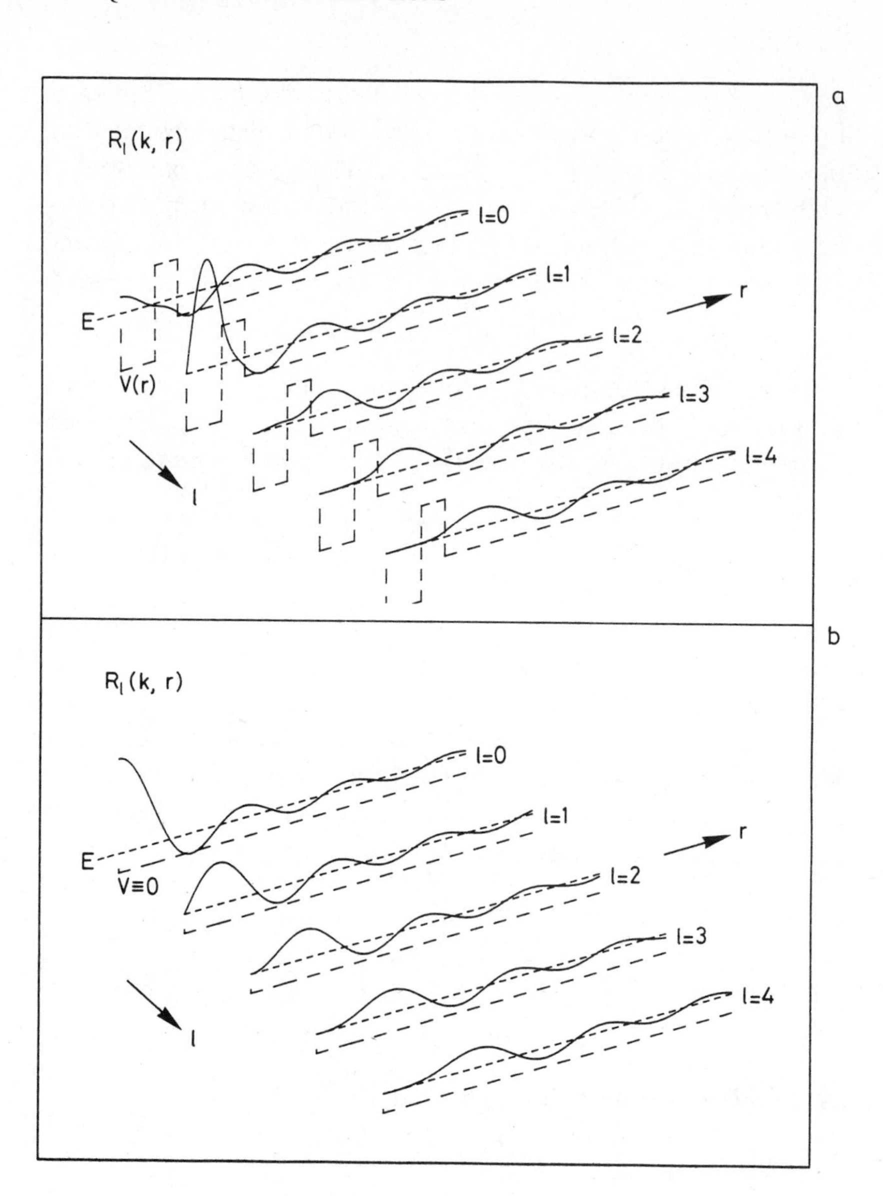

Figure 10.1 **(a) Solutions $R_l(k, r)$ of the radial Schrödinger equation for a potential that is negative in region I, $V_{\mathrm{I}} < 0$; is positive and larger than the particle energy in region II, $V_{\mathrm{II}} > E$; and vanishes in region III. The shape of the potential $V(r)$ is indicated by the long-dash line, the particle energy E by the short-dash line. The short-dash lines also serve as zero lines for the functions $R_l(k, r)$. The energy is kept constant. The various curves correspond to different angular momentum quantum numbers l. (b) The situation is the same as that in part a except that the potential is zero everywhere, $V(r) \equiv 0$. Here the solutions $R_l(k, r)$ are identical to the spherical Bessel functions $j_l(kr)$, $k = \sqrt{2ME}\,/\hbar$.**

The coefficient $A_{l\mathrm{III}}$ can be chosen equal to unity, thus fixing a normalization for the incident wave. For this choice the four coefficients $A_{l\mathrm{I}}$, $A_{l\mathrm{II}}$, $B_{l\mathrm{II}}$, and $B_{l\mathrm{III}}$ calculated from the continuity conditions as functions of the incident wave number k are real coefficients. Therefore the radial wave function $R_l(k, r)$ is real. Figure 10.1a presents the solutions $R_l(k, r)$ as functions of r for a fixed-energy value, that is, a fixed value of E_0, and for a number of angular momentum values l.

Figure 10.1b shows for comparison the functions $j_l(kr)$, which are in fact the functions $R_l(k, r)$ for a vanishing potential, that is, for the undisturbed plane waves. Because $j_l(kr) \sim (kr)^l$ is strongly suppressed near the origin for high l, the term in $R_l(k, r)$ proportional to $j_l(kr)$ has a numerically small coefficient. Therefore the cases with and without a potential do not differ substantially for high enough l. We obtain a rough idea of the size of the value of l above which the radial function R_l is only slightly changed by the potential.

The argument rests on the discussion in Section 9.7. In Figure 9.11b we showed the distribution of the angular momentum components of a Gaussian wave packet representing a classical particle moving with momentum $p = \hbar k$ and impact parameter b. The classical angular momentum has the value $L = pb$. We found that the spectral distribution of the angular momenta of the wave packet peaks at this classical value L. If the impact parameter b is larger than the range d of the potential, $b > d$, that is, if the classical angular momentum L is high enough,

$$L > L_0, \qquad L_0 = \hbar kd$$

the trajectory of the classical particle will not be changed by the potential. By implication, the radial wave functions $R_l(k, r)$ with angular momenta $\hbar l > L_0$, that is, $l > kd$, are essentially unaffected by the potential. Comparing the wave functions of Figures 10.1a and 10.1b shows that they are very similar for high values of l.

10.2 Stationary Bound States

The bound-state solutions occur for discrete values of negative energies E. Let us study the "spherical square well" potential, the simplest situation:

$$V(r) = \begin{cases} V_{\mathrm{I}} < 0, & 0 \le r \le d \quad \text{region I} \\ V_{\mathrm{II}} = 0, & d \le r < \infty, \quad \text{region II} \end{cases}$$

The wave number

$$k_i = \sqrt{2M(E - V_i)}\Big/\hbar$$

is real in region I for $E > V_{\mathrm{I}}$ and imaginary in region II for $E < 0$:

$$k_{\mathrm{II}} = i\varkappa_{\mathrm{II}}, \qquad \varkappa_{\mathrm{II}} = \sqrt{-2ME}\Big/\hbar$$

The wave function has to be proportional to $j_l(k_{\mathrm{I}}r)$ in region I, again because $n_l(k_{\mathrm{I}}r)$ is singular at $r = 0$. In region

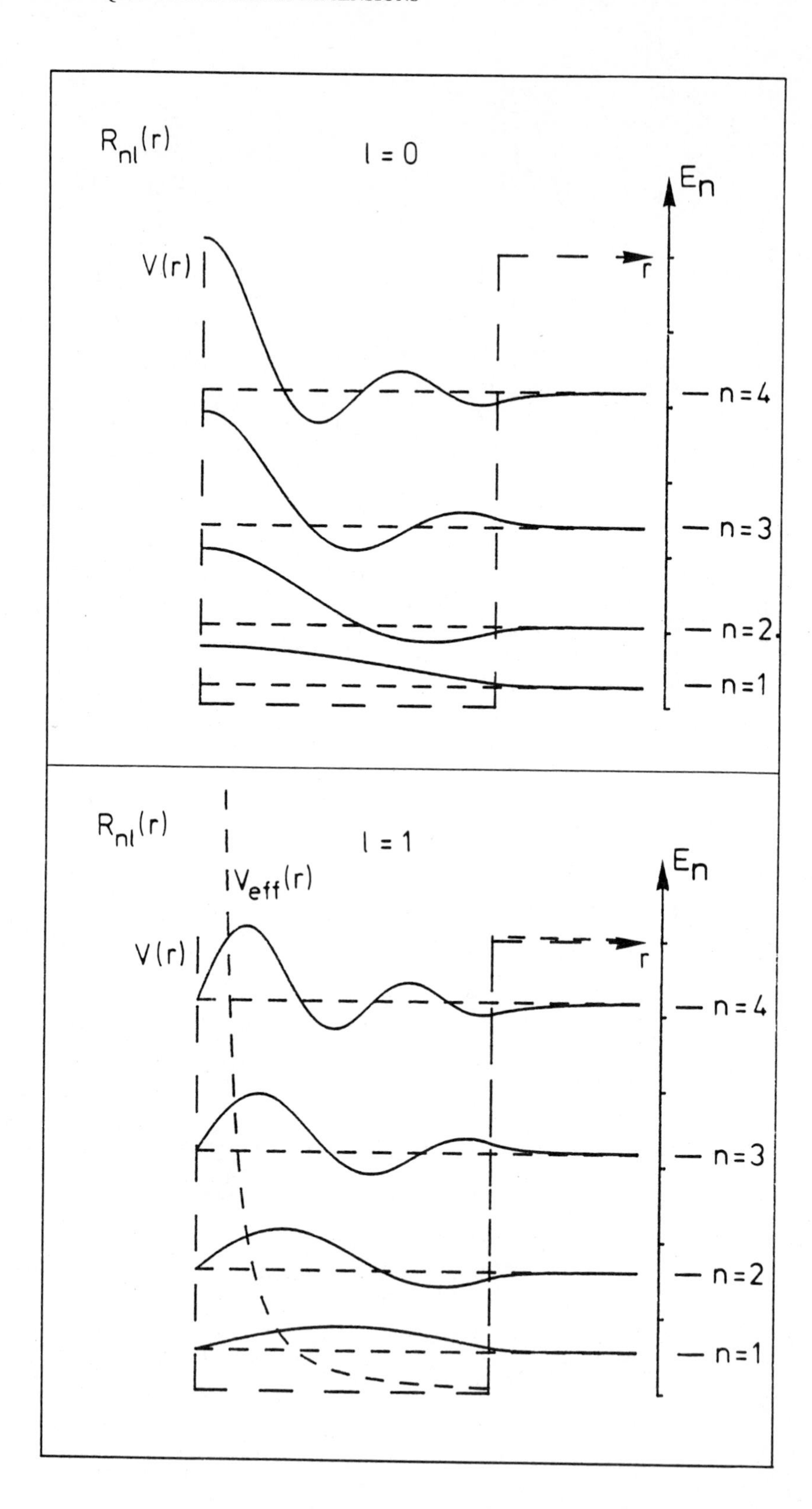

Figure 10.2 **Bound-state solutions $R_{nl}(r)$ of the radial Schrödinger equation for a square-well potential for two angular momentum values, $l = 0$, $l = 1$. The form of the potential $V(r)$ is indicated by the long-dash line. On the right side an energy scale is drawn, and to the right of it the energies E_n of the bound states are indicated by horizontal lines. These lines are repeated as short-dash lines on the left. They serve as zero lines for the solutions $R_{nl}(r)$. For $l \neq 0$ the radial dependence of the "effective potential" $V_{\mathrm{eff}}(r)$ indicates the influence of angular momentum (see Section 12.1).**

II the solution has to be proportional to $h_l^{(+)}(i\varkappa_{\rm II} r)$, for only this function converges toward zero for large distances r:

$$R_l(k,r) = \begin{cases} R_{l\rm I}(k,r) = A_{l\rm I} j_l(k_{\rm I} r) \\ R_{l\rm II}(k,r) = A_{l\rm II} h_l^{(+)}(i\varkappa_{\rm II} r) \end{cases}$$

The coefficient $A_{l\rm II}$ is determined in terms of $A_{l\rm I}$ as a function of energy by the two continuity conditions for the wave function and its derivative at $r = d$. The continuity can be achieved only for certain discrete bound-state energies. The constant $A_{l\rm I}$ is fixed by the normalization of the wave function,

$$\int_0^\infty |R_l(k,r)|^2 r^2\,dr = 1$$

Because the wave function falls off exponentially in region II, the particle is essentially confined to region I, the region of the potential. This confinement is the typical signature of a bound state. Figure 10.2 shows for low angular momenta the wave functions for the bound states in the square-well potential described.

Problems

10.1 Explain by a wave-mechanical argument resting on the potential of the centrifugal barrier why the radial wave functions $R_l(k,r)$ for higher values of l do not penetrate into the potential region in Figure 10.1

10.2 Show by direct calculation that the spherical wave $\varphi(r) = [\sin(kr)]/r$ is a solution of the three-dimensional Schrödinger equation

$$-\frac{\hbar^2}{2m}\nabla^2\varphi(r) = E\varphi(r), \qquad E = \frac{\hbar^2 k^2}{2m}$$

10.3 A bound-state solution of vanishing angular momentum in a square-well potential of finite depth V_0 is given by

$$\varphi(\mathbf{r}) = \frac{A}{2i}\left(\frac{e^{ikr}}{r} - \frac{e^{-ikr}}{r}\right) = A\frac{\sin(kr)}{r}$$

$$k = \frac{1}{\hbar}\sqrt{2m(V_0 - E)}$$

Outside the well the r dependence of the wave function is given by $\exp(-\varkappa r)/r$, $\varkappa = \sqrt{2ME}$. Therefore the function $\sin(kr)$ must have negative or zero slope at the edge $r = d$ of the well. Use this information about the slope to find a minimum value for the potential V_0 within the well so that there is at least one bound state. Explain why there is always at least one bound state in a one-dimensional square well.

11. Three-Dimensional Quantum Mechanics: Scattering by a Potential

11.1 Diffraction of a Harmonic Plane Wave, Partial Waves

In Section 10.1 we found the solutions $R_l(k, r)$ of the radial stationary Schrödinger equation for spherical square-well potentials. Since the radial Schrödinger equation is linear, its solutions are determined up to an arbitrary complex normalization constant, which has to be inferred from the boundary conditions of the three-dimensional problem we want to solve. As we have found in Section 5.4, a harmonic plane wave is an appropriately chosen idealization of an incoming wave packet representing a particle with sharp momentum. We want to apply this finding to the three-dimensional case, that is, the scattering or *diffraction* of a three-dimensional harmonic plane wave which represents a particle of sharp momentum. Then the normalization of the radial wave function has to be chosen in such a way that, for great distances from the region of the potential, the three-dimensional wave function consists of an incoming plane wave $\exp(i\mathbf{k}\cdot\mathbf{r})$ and an outgoing wave.

In Section 10.1 the solutions $R_l(k, r)$ of the radial Schrödinger equation for spherical well potentials were chosen to be real so that in particular the coefficients $A_{l\mathrm{III}}$ and $B_{l\mathrm{III}}$ are real. To help us find the correct normalization, we turn to the physical interpretation of the solution $R_l(k, r)$ in region III,

$$R_{l\mathrm{III}}(k, r) = A_{l\mathrm{III}} j_l(kr) + B_{l\mathrm{III}} n_l(kr)$$

using the decomposition of the spherical Bessel functions j_l

and n_l into the spherical Hankel functions $h_l^{(\pm)}$ of Section 9.5:

$$j_l(kr) = \frac{1}{2i}\left[h_l^{(+)}(kr) - h_l^{(-)}(kr)\right]$$

$$n_l(kr) = \tfrac{1}{2}\left[h_l^{(+)}(kr) + h_l^{(-)}(kr)\right]$$

The spherical Hankel functions have the asymptotic behavior of complex spherical waves:

$$h_l^{(\pm)}(kr) \xrightarrow[kr\to\infty]{} \frac{1}{kr}\exp\left[\pm i(kr) - l\frac{\pi}{2}\right] = (-i)^l\frac{1}{kr}\exp(\pm ikr)$$

In Section 4.2 we learned that wave packets formed with a stationary wave $\exp(ikx)$ move in the direction of increasing x, whereas those with $\exp(-ikx)$ move in the direction of decreasing x values. For spherical waves this implies that a stationary wave $\exp(-ikr)$ describes a particle moving from large values of r toward the origin $r = 0$, that is, an incoming particle. By the same token $\exp(ikr)$ describes an outgoing particle. Thus, except for an r-independent factor, the decomposition of $R_{l\mathrm{III}}$ into spherical Hankel functions

$$R_{l\mathrm{III}} = \frac{i}{2}\left[(A_{l\mathrm{III}} - iB_{l\mathrm{III}})h_l^{(-)} - (A_{l\mathrm{III}} + iB_{l\mathrm{III}})h_l^{(+)}\right]$$

describes an incoming, $h_l^{(-)}$, and an outgoing, $h_l^{(+)}$, part of the wave function.

We now divide the radial functions R_l by $A_{l\mathrm{III}} - iB_{l\mathrm{III}}$ and obtain

$$R_l^{(+)}(k,r) = \frac{1}{A_{l\mathrm{III}} - iB_{l\mathrm{III}}}R_l(k,r)$$

which takes in region III the explicit form

$$R_{l\mathrm{III}}^{(+)}(k,r) = -\frac{1}{2i}h_l^{(-)}(kr) + \frac{1}{2i}S_l(k)h_l^{(+)}(kr)$$

Here $S_l(k)$ is the *scattering matrix element* of the lth partial wave:

$$S_l(k) = \frac{A_{l\mathrm{III}} + iB_{l\mathrm{III}}}{A_{l\mathrm{III}} - iB_{l\mathrm{III}}}$$

Now, finally, we have achieved the decomposition of $R_{l\mathrm{III}}^{(+)}$ into the lth component $j_l(kr)$ of the plane wave and the outgoing spherical wave $h_l^{(+)}(kr)$. This structure becomes obvious if we add to the first term and subtract from the second $(1/2i)h_l^{(+)}(kr)$:

$$R_{l\mathrm{III}}^{(+)} = \frac{1}{2i}\left[h_l^{(+)}(kr) - h_l^{(-)}(kr)\right] + \frac{1}{2i}(S_l - 1)h_l^{(+)}(kr)$$
$$= j_l(kr) + f_l(k)h_l^{(+)}(kr)$$

Here f_l is the *partial scattering amplitude*

$$f_l(k) = \frac{1}{2i}(S_l(k) - 1)$$

It determines the amplitude of the outgoing spherical wave in relation to the lth component $j_l(kr)$ of the incoming plane wave.

The recipe for constructing the three-dimensional stationary wave function is indicated by the formula for decomposing the plane wave $\exp(i\mathbf{k}\cdot\mathbf{r})$ into partial waves:

$$e^{i\mathbf{k}\cdot\mathbf{r}} = \sum_{l=0}^{\infty}(2l+1)i^l j_l(kr)P_l(\cos\vartheta), \qquad \cos\vartheta = \mathbf{k}\cdot\mathbf{r}/(kr)$$

Replacing the free radial wave function $j_l(kr)$ by the solution $R_l^{(+)}(k,r)$ of the radial Schrödinger equation for a potential $V(r)$, we obtain

$$\varphi_{\mathbf{k}}^{(+)}(\mathbf{r}) = \sum_{l=0}^{\infty}(2l+1)i^l R_l^{(+)}(k,r)P_l(\cos\vartheta)$$

Figure 11.1 gives the real and imaginary parts and the absolute square of $\varphi_{\mathbf{k}}^{(+)}$ for the scattering of a plane wave from a repulsive potential that is constant within a sphere around the origin:

$$V(r) = \begin{cases} V_0 > 0, & 0 \le r \le d \\ 0, & r > d \end{cases}$$

The energy E of the wave is two-thirds of the height of the potential, that is, $E = 2V_0/3$. Figures 11.1a and b for the real and the imaginary parts show that the plane wave coming in from the left is strongly suppressed within the sphere of the repulsive potential and that its pattern is modified, particularly in the forward direction, by interference with the outgoing scattered spherical wave. The patterns in these figures bear a certain resemblance to those of water waves diffracted on a cylindrical obstacle in a ripple tank. The real and imaginary parts of $\varphi_{\mathbf{k}}^{(+)}$ are dominated by the incident plane wave $\exp(i\mathbf{k}\cdot\mathbf{r})$. The pattern of the absolute square $|\varphi_{\mathbf{k}}^{(+)}|^2$, how-

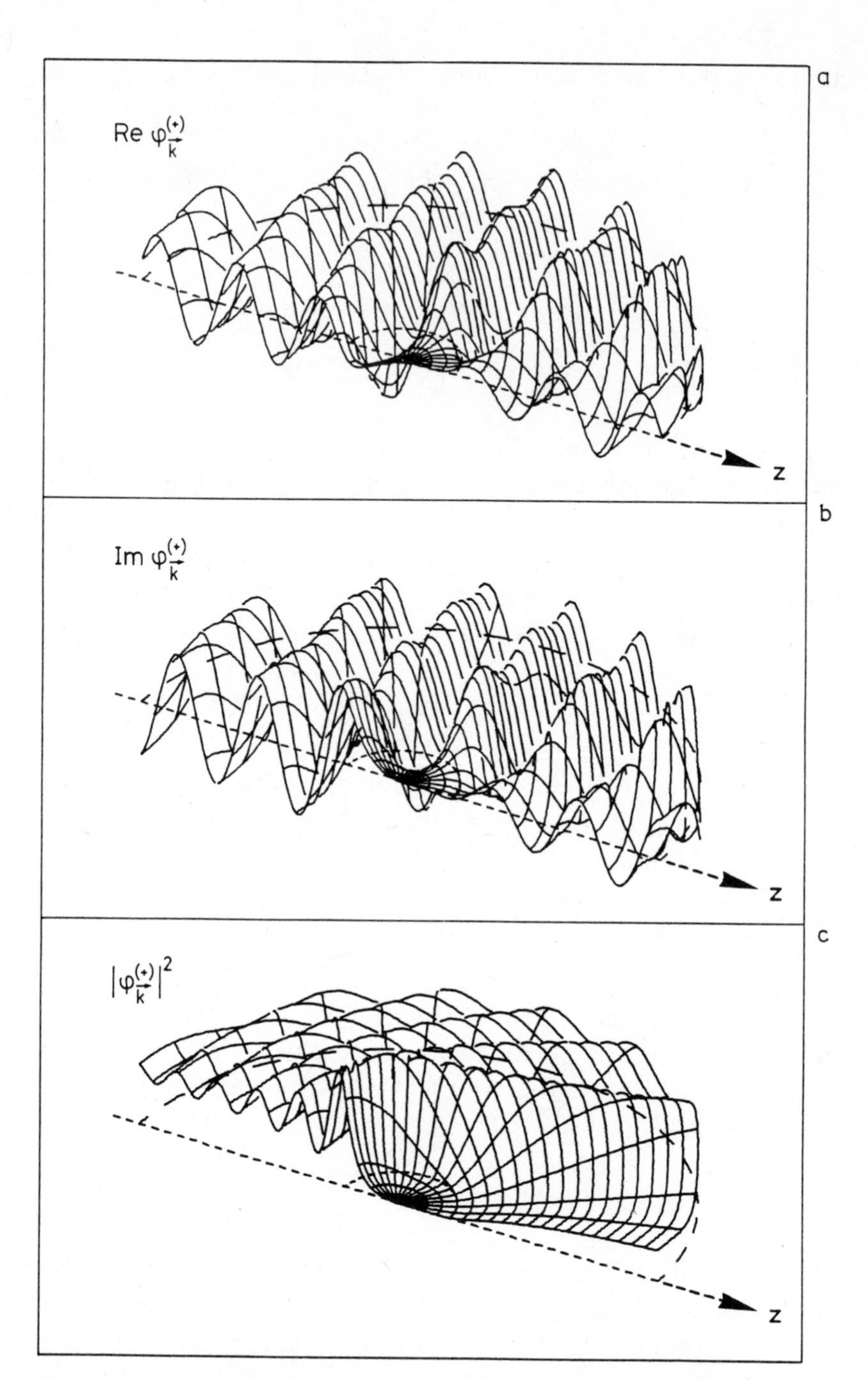

Figure 11.1 **Scattering of a plane wave incident from the left, that is, along the *z*-direction, by a repulsive potential. The potential is confined to region $r < d$, indicated by the small half circle marked off by a short-dash line. The energy E of the plane wave is two-thirds the height of the potential in this region. Shown are (a) the real part, (b) the imaginary part, and (c) the absolute square of the wave function $\varphi^{(+)}_{\mathbf{k}}$.**

ever, stems entirely from the superposition of the incident and scattered waves, since the absolute square of the unscattered incident wave $|\exp(i\mathbf{k}\cdot\mathbf{r})|^2 = 1$ would produce a flat sheet. In particular, the ripples to the left of center in Figure 11.1c are caused by the interference of the incident wave and the scattered wave in the backward direction. This interference pattern accordingly exhibits a wavelength half that of the incident wave. It tapers off with $1/r$ because the outgoing spherical wave itself falls off with $1/r$. There are no such ripples in the forward direction because the exponentials in the scattered spherical wave and the incident wave are identical.

11.2 Scattered Wave and Scattering Cross Section

If we insert into the right-hand side of the formula for $\varphi_{\mathbf{k}}^{(+)}(\mathbf{r})$ the function $R_{l\mathrm{III}}^{(+)}$ in terms of $j_l(kr)$ and the outgoing spherical wave $h_l^{(+)}$, we obtain the superposition

$$\varphi_{\mathbf{k}}^{(+)}(\mathbf{r}) = e^{i\mathbf{k}\cdot\mathbf{r}} + \eta_{\mathbf{k}}(\mathbf{r})$$

of the incoming plane wave and the scattered spherical wave

$$\eta_{\mathbf{k}}(\mathbf{r}) = \sum_{l=0}^{\infty} \eta_l$$

where η_l is the lth *scattered partial wave*:

$$\eta_l = (2l+1)i^l\left[R_l^{(+)}(kr) - j_l(kr)\right]P_l(\cos\vartheta)$$

In region III this scattered partial wave has the explicit form

$$\eta_l = (2l+1)i^l f_l(k)h_l^{(+)}(kr)P_l(\cos\vartheta)$$

which, for far-out distances, $kr \gg 1$, is dominated by the asymptotic term for $h_l^{+}(kr)$,

$$\eta_l = (2l+1)i^l f_l(k)\frac{e^{ikr}}{r}P_l(\cos\vartheta)$$

In external region III the scattered spherical wave has the explicit representation

$$\eta_{\mathbf{k}\mathrm{III}}(\mathbf{r}) = \sum_{l=0}^{\infty}(2l+1)i^l f_l(k)h_l^{(+)}(kr)P_l(\cos\vartheta)$$

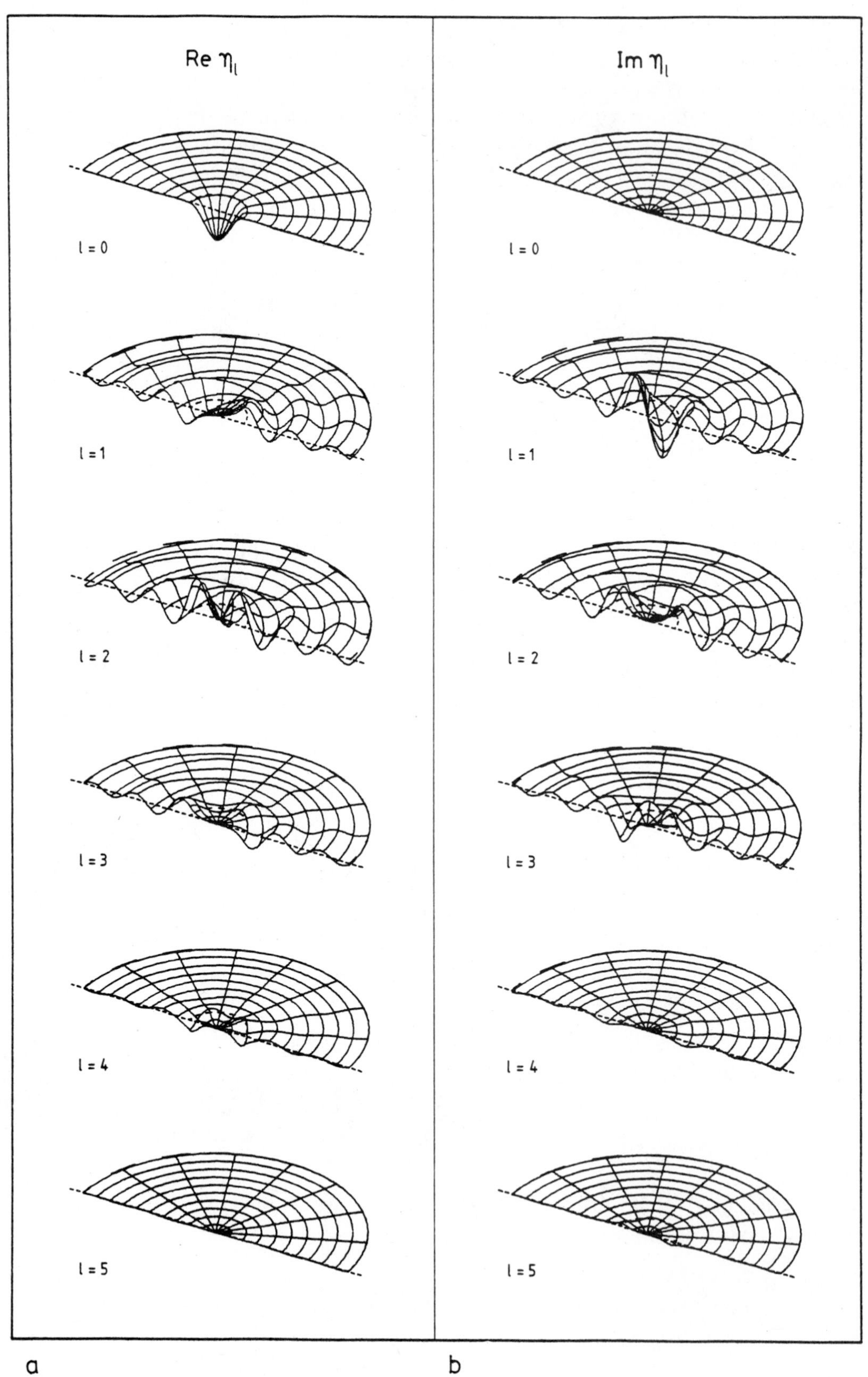

Re η_l
Im η_l
l = 0
l = 0
l = 1
l = 1
l = 2
l = 2
l = 3
l = 3
l = 4
l = 4
l = 5
l = 5
a
b

For far-out distances, $kr \gg 1$, this expression becomes

$$\eta_{\mathbf{k}\mathrm{III}}(\mathbf{r}) \xrightarrow[kr \gg 1]{} f(\vartheta) \frac{e^{ikr}}{r}$$

where the *scattering amplitude*

$$f(\vartheta) = \frac{1}{k} \sum_{l=0}^{\infty} (2l+1) f_l(k) P_l(\cos\vartheta)$$

modulates the amplitude of the scattered spherical wave for the various polar angles ϑ.

Figure 11.2 plots the real and imaginary parts of the lth scattered partial wave η_l for values $l = 0, 1, 2, 3, 4, 5$. This partial wave is the product of an r-dependent factor responsible for the variation along lines ϑ = const and a ϑ-dependent factor responsible for the variation along lines r = const. As expected from the Legendre polynomials $P_l(\cos\vartheta)$, there is no ϑ-variation for $l = 0$, whereas the increasing complexity of the higher P_l is signaled by their l nodes in ϑ. The pictures indicate a $1/r$ falloff for large values of r, as expected from the asymptotic form of the η_l. As already mentioned in Section 10.1, the deviations of the radial wave functions from the free radial wave function j_l are substantial only for $l \le kd$. Indeed, we observe that η_5 is essentially zero, as are η_6, η_7, and so on. In our example kd equals 4. We may wonder why the scattered partial waves η_l for low l have important contributions within the potential sphere. They are expected to contribute, however, for the superposition of the η_l has to compensate for the harmonic plane wave in this region since $\varphi_{\mathbf{k}}^{(+)}(\mathbf{r})$ is small in the sphere of a repulsive potential.

Figures 11.3a and b give the real and imaginary parts of the scattered spherical wave $\eta_{\mathbf{k}}(\mathbf{r})$ obtained by summing the scattered partial waves for $0 \le l \le 5$. The η_l essentially vanish for $l > 5$. Whereas the scattered partial waves η_l have the symmetry of the corresponding P_l, their superposition $\eta_{\mathbf{k}}(\mathbf{r})$ shows a definite forward structure, indicating that the scatter-

Figure 11.2 **Real and imaginary parts of the scattered partial waves η_l resulting from the scattering of a plane wave by a repulsive potential, as shown in Figure 11.1.**

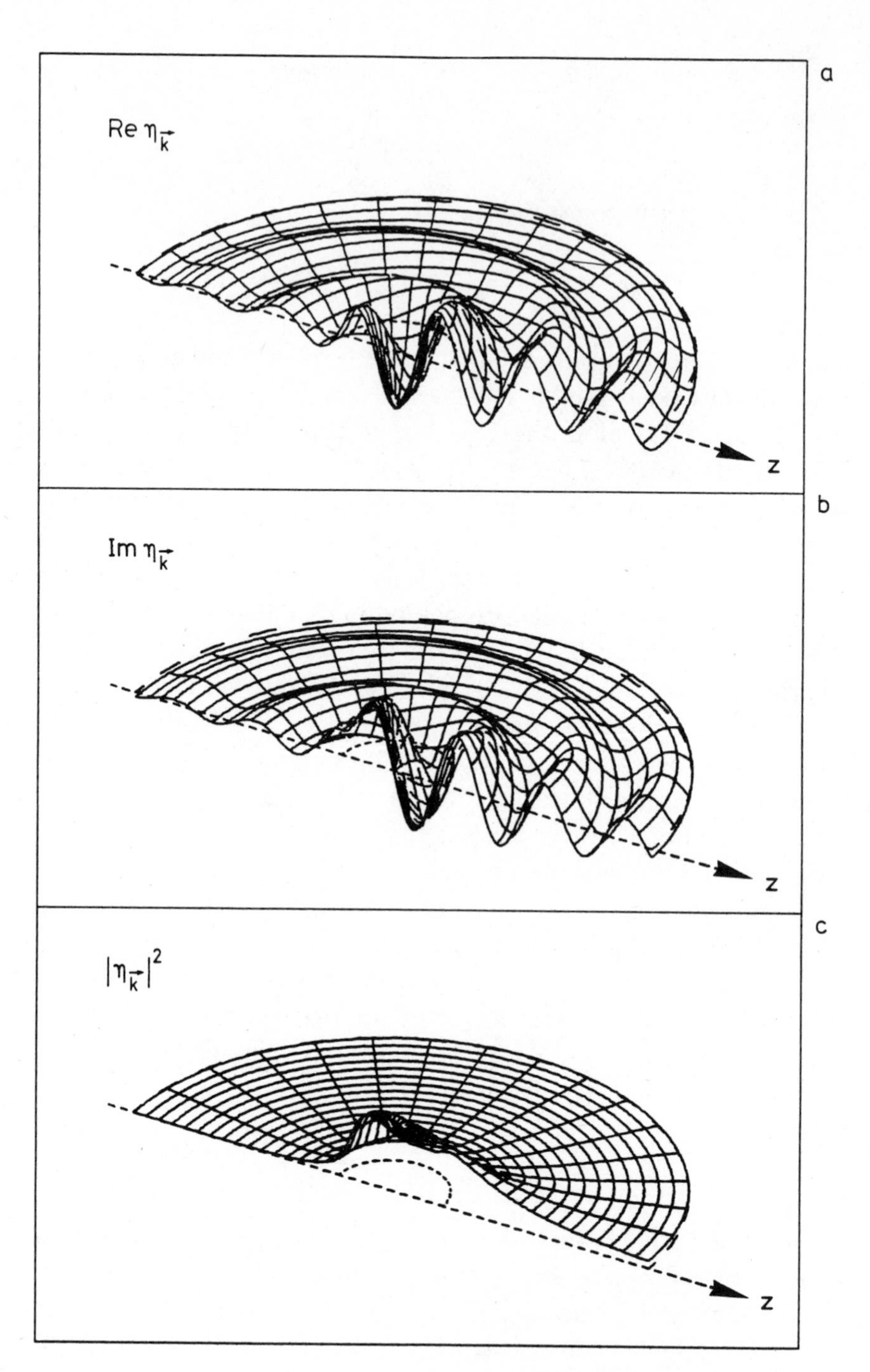

Figure 11.3 **(a) Real part, (b) imaginary part, and (c) absolute square of the scattered spherical wave η_k resulting from the scattering of a plane wave by a repulsive potential, as shown in Figure 11.1.**

ing occurs for the most part in the forward direction. Obviously, $\eta_{\mathbf{k}}(\mathbf{r})$ also falls off with $1/r$ for large r.

Figure 11.3b gives the absolute square $|\eta_{\mathbf{k}}(\mathbf{r})|^2$. This function falls off asymptotically with $1/r^2$. The physical significance of $|\eta_{\mathbf{k}}|^2$ is the average particle density for the scattered particles moving with velocity $v = \hbar k/M$ radially away from the center. In experiments the scattered particles can be detected only at distances that are large compared to the size of the scattering center. The average number Δn of scattered particles passing through the sensitive area Δa of the detector during the time interval Δt is the quantity usually measured. For a given sensitive area Δa, this number is the product of the current density $|\eta_{\mathbf{k}}(\mathbf{r})|^2 v$ of the particles and the area Δa times Δt:

$$\Delta n = v|\eta_{\mathbf{k}}(\mathbf{r})|^2 \Delta a \, \Delta t$$

The detector is located at $\mathbf{r}$.

For fixed experimental conditions Δa, Δt, and v, the quantity $|\eta_{\mathbf{k}}(\mathbf{r})|^2$ is directly proportional to the number of scattered particles observed. We assume that many detectors are distributed evenly along a half circle of radius r around the scattering center. The direction of the incident particles forms the diameter of the half circle. Then we have only to compute $|\eta_{\mathbf{k}}(\mathbf{r})|^2$ to predict the counting rates in all detectors. Figure 11.4a illustrates this situation. The function $|\eta_{\mathbf{k}}(\mathbf{r})|^2$ is plotted in a half-circle band in the region where the detectors could be placed. To overcome the $1/r^2$ suppression in $|\eta_{\mathbf{k}}|^2$, the values of this function have been blown up by a scaling factor. It becomes obvious from this figure that $|\eta_{\mathbf{k}}|^2$ depends to a considerable extent on the scattering angle ϑ, that is, the angle between the incident and the scattered particle.

Actually, the asymptotic form of $\eta_{\mathbf{k}}$, that is, of $\eta_{\mathbf{k}\mathrm{III}}$, which we found to be

$$\eta_{\mathbf{k}\mathrm{III}} \xrightarrow[kr \gg 1]{} f(\vartheta)\frac{e^{ikr}}{r}$$

shows that the quantity

$$|r\eta_{\mathbf{k}\mathrm{III}}|^2 \xrightarrow[kr \gg 1]{} |f(\vartheta)|^2$$

is dependent only on the scattering angle ϑ. Its physical interpretation in terms of the counting rate Δn becomes clear if we observe that $\Delta a/r^2 = \Delta\Omega$ is the sensitive solid angle of the detector. Furthermore, the incident current density is equal to the incident average particle density times the average velocity:

$$j = |e^{i\mathbf{k}\cdot\mathbf{r}}|^2 v = v$$

Thus the number of scattered particles Δn can be reexpressed in the form

$$\Delta n = j \frac{\Delta a}{r^2} |r\eta_{\mathbf{k}\mathrm{III}}|^2 \Delta t$$
$$= j|f(\vartheta)|^2 \Delta\Omega\, \Delta t$$

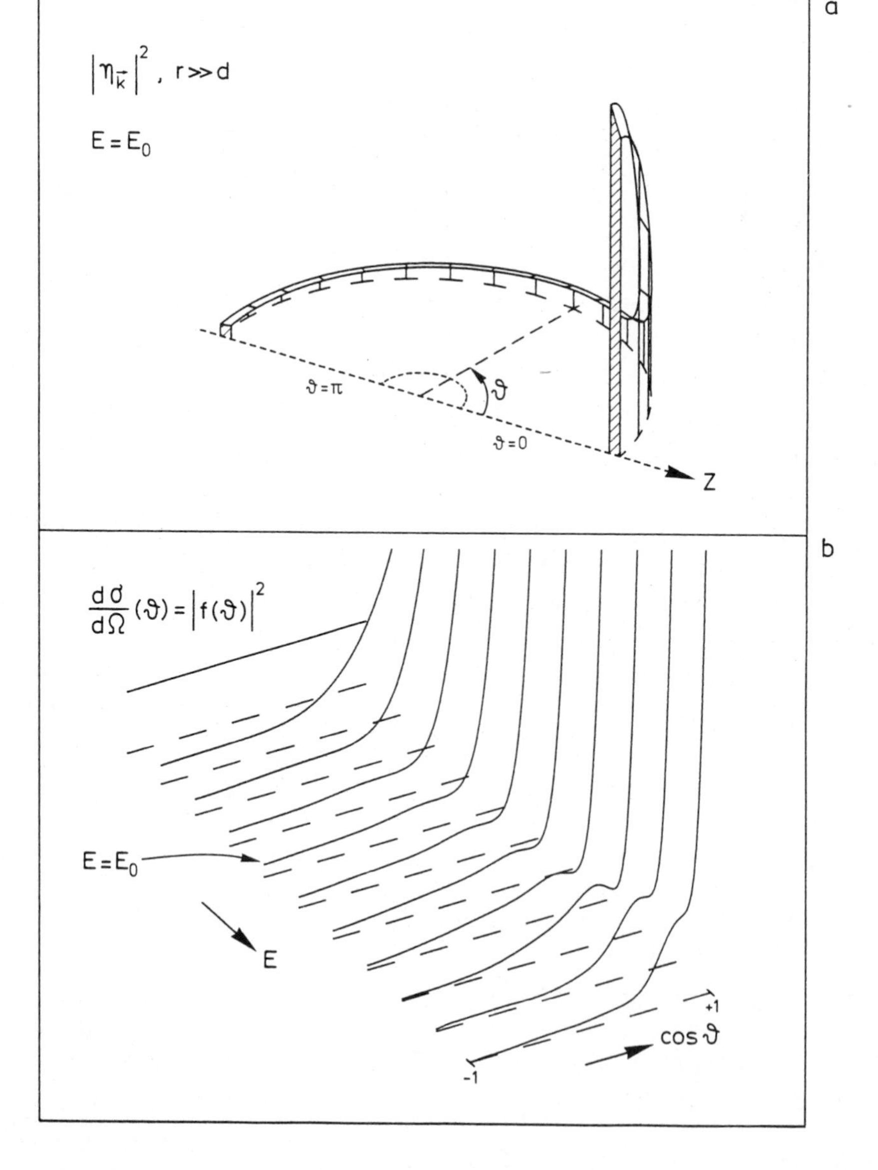

Figure 11.4 **(a) Intensity of the scattered spherical wave resulting from the scattering of a plane wave by a repulsive potential, as shown in Figure 11.1. The intensity at a fixed radius far outside the scattering region and for a given scattering angle ϑ is indicated by the height of the band. The band corresponds to the outer rim of Figure 11.3c, blown up by a scale factor.**

(b) Energy dependence of the differential scattering cross section $d\sigma(\vartheta)/d\Omega$ for the scattering of a plane wave by a repulsive potential. The differential cross section is proportional to the intensity of the scattered wave, as we can see by comparing the curve marked $E = E_0$ with the band in part a. Both correspond to the same energy.

which shows that $|f(\vartheta)|^2$ has the following physical meaning. It is the average number of particles from an incident particle current of density 1 scattered per second at angle ϑ into the solid angle $\Delta\Omega$:

$$|f(\vartheta)|^2 = \frac{1}{j}\frac{\Delta n}{\Delta\Omega\,\Delta t}$$

In a classical experiment in which a particle beam is incident on a hard sphere, the quantity on the right-hand side is the *differential scattering cross section*. This notion is derived from the elastic scattering of a beam of point particles with current density j incident on a rigid sphere of radius d. As Figure 11.5 indicates, the impact parameter b is related to the scattering angle ϑ by

$$b = d\cos\frac{\vartheta}{2}$$

The number Δn of particles incident during Δt in sector $\Delta\phi$ with an impact parameter between b and $b + \Delta b$ is

$$\Delta n = j\,\Delta t\, b\,\Delta b\,\Delta\phi$$

This number of particles is scattered into the solid angle $\Delta\Omega = \Delta\cos\vartheta\,\Delta\phi$ where $\Delta\cos\vartheta$ corresponds to Δb through the relation

$$\frac{db}{d\cos\vartheta} = \frac{d\vartheta}{d\cos\vartheta}\,\frac{db}{d\vartheta} = \frac{d}{4}\,\frac{1}{\cos\vartheta/2}$$

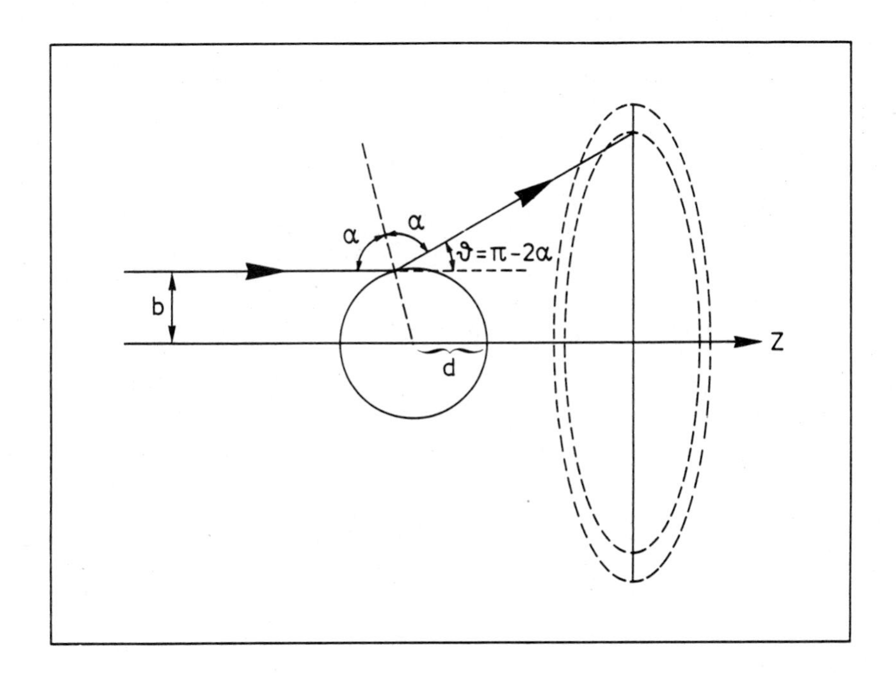

Figure 11.5 The classical elastic scattering of point particle by a rigid sphere.

The number of particles scattered at angle ϑ per unit solid angle and unit time is then

$$\frac{\Delta n}{\Delta\Omega\,\Delta t} = \tfrac{1}{4}d^2 j$$

This rate of particles for a current density one is completely determined by the properties of the scattering center, here a rigid sphere. It is the differential scattering cross section $d\sigma/d\Omega$. For a rigid sphere it is

$$\frac{d\sigma}{d\Omega} = \tfrac{1}{4}d^2$$

In general, the differential scattering cross section is not constant but depends on the direction of the scattered particle. When the scattering center is spherically symmetric, the differential scattering cross section is a function only of the scattering angle ϑ. Integration over the full solid angle 4π yields the *total scattering cross section*. When classical particles are scattered off a hard sphere, it is obtained by multiplying $\frac{1}{4}d^2$ by the full solid angle 4π:

$$\sigma_{\text{tot}} = \pi d^2$$

As expected, it is the geometrical cross section of the rigid sphere.

Coming back to our quantum-mechanical discussion, we can identify the differential scattering cross section as

$$\frac{d\sigma}{d\Omega} = |f(\vartheta)|^2$$

The function $f(\vartheta)$ as calculated earlier has the form

$$f(\vartheta) = \frac{1}{k}\sum_{l=0}^{\infty}(2l+1)f_l(k)P_l(\cos\vartheta)$$

which shows that it depends not only on the scattering angle but also on the energy $E = (\hbar k)^2/2M$ of the incident particles. It is customary to plot $d\sigma/d\Omega = |f(\vartheta)|^2$ as a function of $\cos\vartheta$ rather than ϑ. In Figure 11.4b this is done for a range of energies and for the potential used in the earlier figures of this chapter. For very low energy the differential cross section is constant in $\cos\vartheta$. With increasing energy it acquires a more complicated angular dependence. This dependence is easily explained by observing that for very low energy only the lowest partial wave, $l = 0$, contributes to the scattering amplitude $f(\vartheta)$ through the Legendre polynomial $P_0(\cos\vartheta)$, which is a constant. With increasing energy more and more partial waves contribute, allowing a richer structure in $\cos\vartheta$.

The total cross section is obtained by an integration over the full solid angle

$$\sigma_{\text{tot}} = \int \frac{d\sigma}{d\Omega}\,d\Omega = 2\pi\int_{-1}^{+1}|f(\vartheta)|^2\,d\cos\vartheta$$

For the following we need the orthogonality of different Legendre polynomials,

$$\int P_l(\cos\vartheta)P_{l'}(\cos\vartheta)\,d\cos\vartheta = \frac{2}{2l+1}\delta_{ll'}$$

which can be inferred from the orthonormality of the spherical harmonics Y_{l0} and their relation to the P_l, as discussed in Section 9.2. If we insert the series for $f(\vartheta)$ into the integral for σ_{tot}, we obtain

$$\sigma_{\text{tot}} = \frac{4\pi}{k^2}\sum_{l=0}^{\infty}(2l+1)|f_l(k)|^2 = \sum_{l=0}^{\infty}\sigma_l$$

The terms in this sum are called *partial cross sections*,

$$\sigma_l = \frac{4\pi}{k^2}(2l+1)|f_l(k)|^2$$

Figure 11.6 shows the various partial cross sections as functions of energy. We notice that the partial cross section for $l > 0$ starts at zero for $k = 0$. Furthermore, the contribution

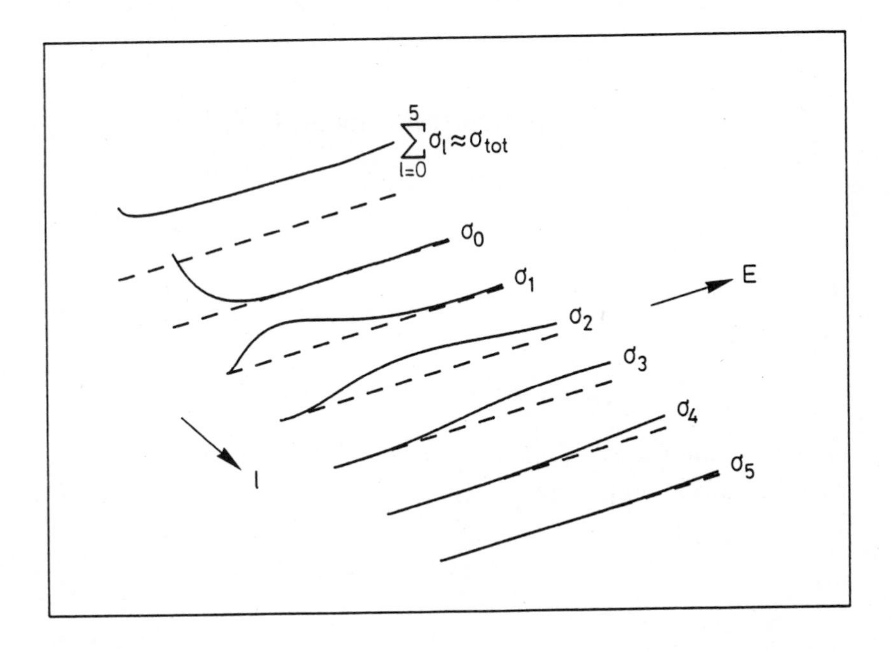

Figure 11.6 **The partial cross sections $\sigma_l(E)$ for $l = 0, 1, \ldots, 5$ and the total cross section $\sigma_{\text{tot}}(E)$, which is approximated by the sum over the first five partial cross sections for the scattering of a plane wave by a repulsive potential.**

of the cross sections for increasing l sets in with increasing energy so that for a given energy the sum over the partial cross sections can be truncated at $l_{max} \gtrsim kd$, the maximum value of the classical angular momentum at which scattering takes place. The total cross section obtained by the summation is plotted in the background of Figure 11.6.

11.3 Scattering Phase and Amplitude, Unitarity, and Argand Diagrams

In Section 11.1 we obtained as a representation for the radial wave function $R_{l\mathrm{III}}^{(+)}(k,r)$ the form

$$R_{l\mathrm{III}}^{(+)}(k,r) = \frac{i}{2}\left[h_l^{(-)}(kr) - S_l(k)h_l^{(+)}(kr)\right]$$

We interpreted this solution as the superposition of the incoming spherical wave $h_l^{(-)}$ and the outgoing spherical wave $h_l^{(+)}$, which is multiplied by the S-matrix element S_l. Potential scattering conserves particle number, angular momentum, and energy $E = (\hbar k)^2/2M$ so that the magnitude of velocity $\hbar k/M$ remains unaltered. Therefore the current density of the incoming spherical wave has the same size as the current density of the outgoing spherical wave. As a consequence, the particle densities in the incoming and outgoing spherical waves in region III have to be the same. Since particle densities are determined by the absolute squares of amplitudes, the scattering matrix element S_l representing the relative factor between the incoming and outgoing spherical waves must have absolute value one.

In fact, the representation for S_l found in Section 11.1,

$$S_l = \frac{A_{l\mathrm{III}} + iB_{l\mathrm{III}}}{A_{l\mathrm{III}} - iB_{l\mathrm{III}}}$$

satisfies this requirement,

$$S_l^* S_l = \frac{A_{l\mathrm{III}} - iB_{l\mathrm{III}}}{A_{l\mathrm{III}} + iB_{l\mathrm{III}}} \frac{A_{l\mathrm{III}} + iB_{l\mathrm{III}}}{A_{l\mathrm{III}} - iB_{l\mathrm{III}}} = 1$$

which is called the *unitarity relation for the S-matrix elements*. Thus S_l can be represented as a complex phase factor

$$S_l(k) = \frac{A_{l\mathrm{III}} + iB_{l\mathrm{III}}}{A_{l\mathrm{III}} - iB_{l\mathrm{III}}} = e^{2i\delta_l(k)}$$

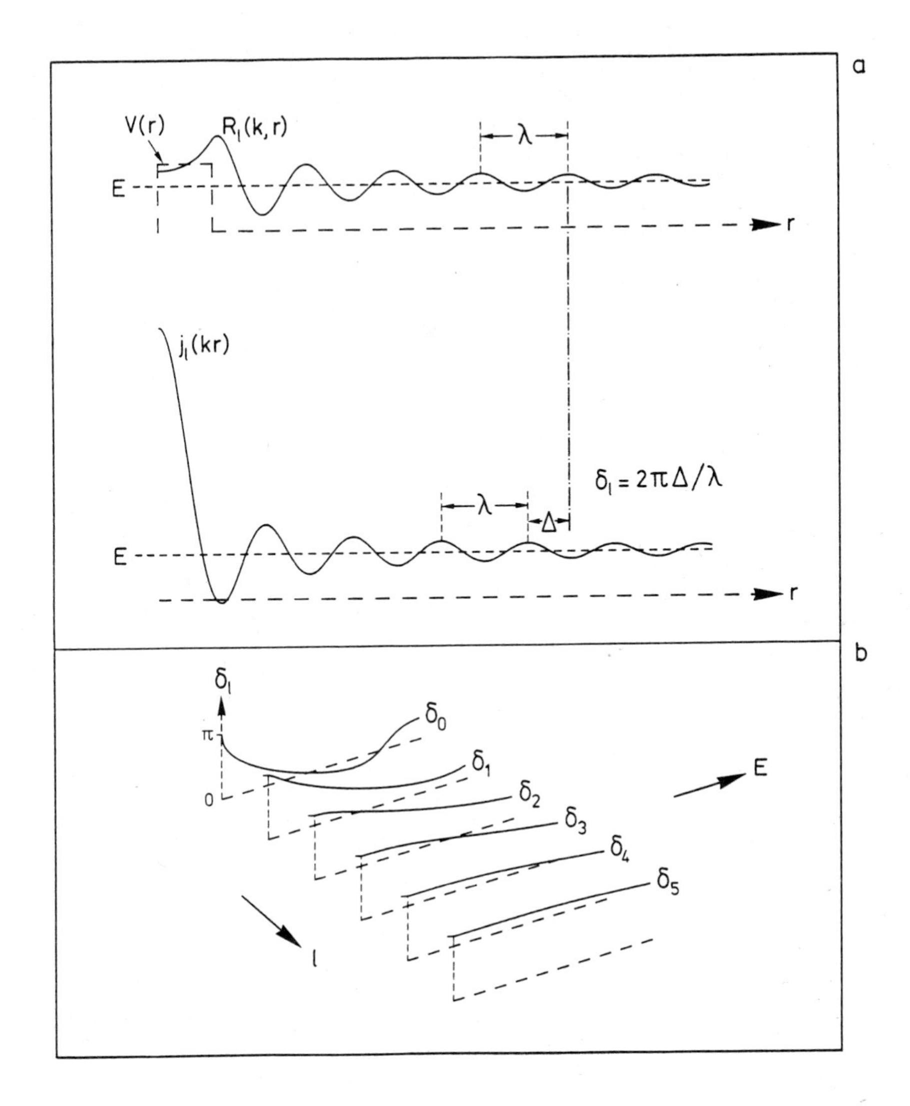

Figure 11.7 **(a) Definition of the scattering phase shift δ_l. The solution R_l of the radial Schrödinger equation for a given l, here $l = 0$, is shown for the scattering of a wave of energy E by a repulsive potential (top) and for vanishing potential (bottom). Asymptotically, that is, far outside the potential region, both solutions differ only by a phase shift δ_l.**

(b) Energy dependence of the phase shifts $\delta_0(E), \delta_1(E), \ldots, \delta_5(E)$ for scattering by a repulsive potential. There is an ambiguity in the definition of δ_l, which is resolved by choosing $\delta_l(0) = \pi$. All phase shifts vary slowly with energy for scattering by a repulsive potential.

The *scattering phase* δ_l determining S_l can be calculated directly from $A_{l\text{III}}$ and $B_{l\text{III}}$ if we observe that

$$e^{\pm i\delta_l} = \frac{A_{l\text{III}} \pm iB_{l\text{III}}}{\sqrt{A_{l\text{III}}^2 + B_{l\text{III}}^2}}$$

allows the identification

$$\cos\delta_l = \frac{A_{l\text{III}}}{\sqrt{A_{l\text{III}}^2 + B_{l\text{III}}^2}}, \qquad \sin\delta_l = \frac{B_{l\text{III}}}{\sqrt{A_{l\text{III}}^2 + B_{l\text{III}}^2}}$$

These relations can be used to show that δ_l is a phase shift produced by the potential. To this end we use the asymptotic representations, $kr \gg 1$, for $j_l(kr)$ and $n_l(kr)$, as given in Section 9.5. In fact, the solution $R_{l\text{III}}$ of the stationary Schrödinger equation presented at the beginning of section

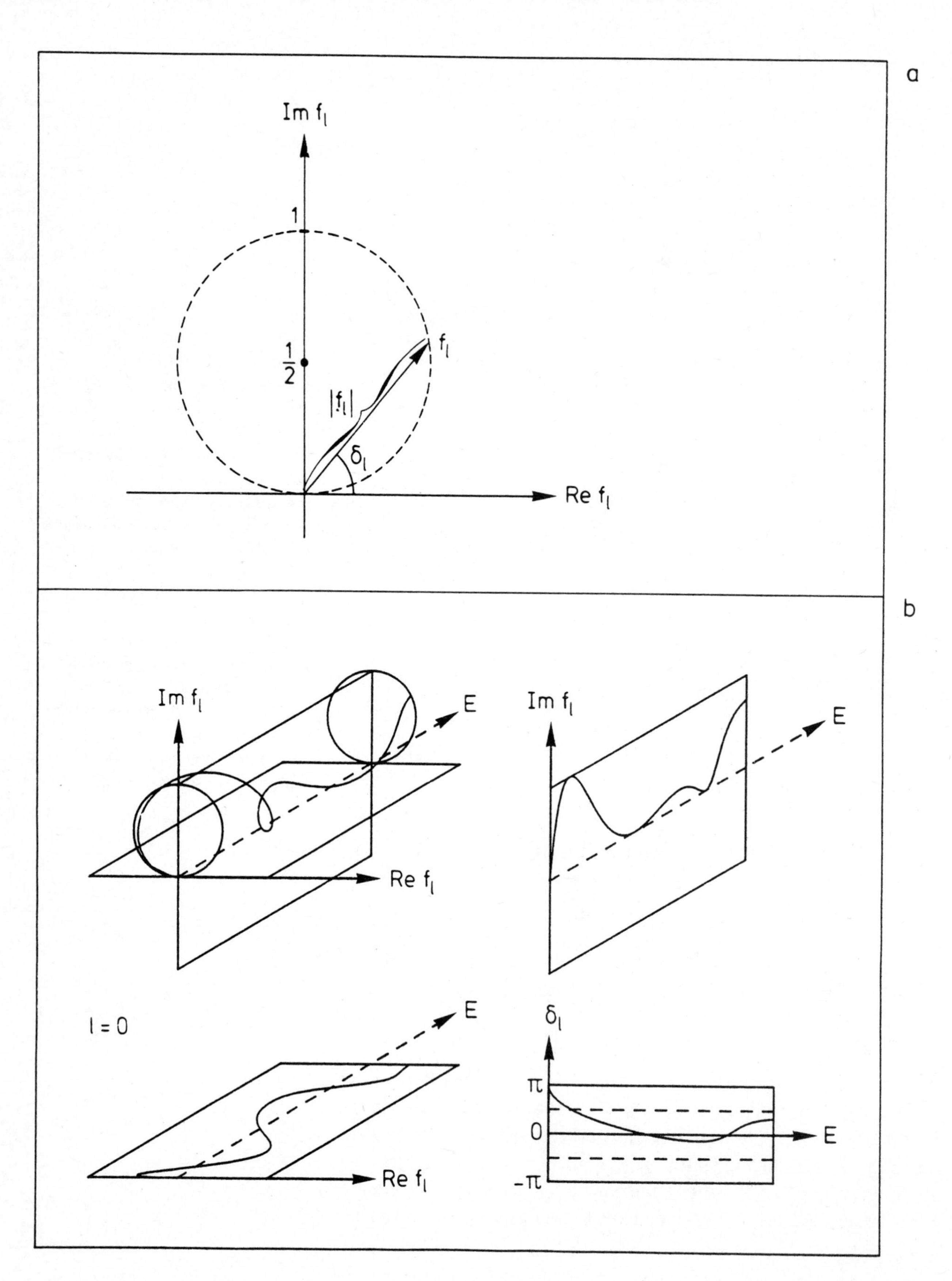

a
Im f_l
1
1/2
f_l
|f_l|
δ_l
Re f_l
b
Im f_l
E
Re f_l
Im f_l
E
l = 0
E
Re f_l
δ_l
π
0
E
-π

11.1 has the form

$$R_{l\mathrm{III}} = \sqrt{A_{l\mathrm{III}}^2 + B_{l\mathrm{III}}^2}\,[\cos\delta_l\, j_l(kr) + \sin\delta_l\, n_l(kr)]$$

which asymptotically becomes

$$R_{l\mathrm{III}} \xrightarrow[kr \gg 1]{} \sqrt{A_{l\mathrm{III}}^2 + B_{l\mathrm{III}}^2}\,\frac{1}{kr}\left[\cos\delta_l \sin\left(kr - l\frac{\pi}{2}\right) + \sin\delta_l \cos\left(kr - l\frac{\pi}{2}\right)\right]$$

$$\longrightarrow \sqrt{A_{l\mathrm{III}}^2 + B_{l\mathrm{III}}^2}\,\frac{1}{kr}\sin\left(kr - l\frac{\pi}{2} + \delta_l\right)$$

Figure 11.7a plots R_l together with j_l, the lth partial wave of the harmonic plane wave. The scattering phase shift δ_l is easily recognized as the phase difference between the two in the asymptotic region. In Figure 11.7b the energy dependence of the various phase shifts δ_l is shown for the repulsive square-well potential used as our example. We have chosen the phases δ_l so that they are equal to π for $E = 0$. For the potential of our example, they fall off smoothly with increasing energy.

In terms of the scattering phase $\delta_l(k)$ the partial scattering amplitude can be represented as

$$f_l(k) = \frac{1}{2i}[S_l(k) - 1] = \frac{1}{2i}(e^{2i\delta_l} - 1)$$

$$= e^{i\delta_l}\left[\frac{1}{2i}(e^{i\delta_l} - e^{-i\delta_l})\right]$$

$$= e^{i\delta_l}\sin\delta_l$$

The relation expressing that S_l has absolute value one, reflects itself in the equivalent *unitarity relation for the partial scattering amplitude* f_l:

$$\mathrm{Im}\, f_l(k) = |f_l(k)|^2$$

Figure 11.8 (a) Through the unitarity relation $\mathrm{Im}\, f_l = |f_l|^2$ the elastic partial-wave amplitude is confined to a circle in an Argand diagram. The angle between the vector f_l in the complex plane and the real axis is the phase shift δ_l. (b) Argand diagram for the elastic partial-wave amplitude f_l describing scattering by a repulsive potential. An energy axis is erected perpendicular to the complex f_l-plane of part a. The circle of part a turns into a cylinder. The point $f_l(E)$ describes a trajectory on this cylinder as the energy E changes (top left). Projections onto a vertical and horizontal plane yield graphs of $\mathrm{Im}\, f_l(E)$ (top right) and $\mathrm{Re}\, f_l(E)$ (bottom left), respectively. The function $\delta_l(E)$ is also shown (bottom right).

As a complex number, the partial scattering amplitude can be plotted in an Argand diagram similar to the one in Section 5.4. Here, however, f_l stays on the circumference of the circle with radius $\frac{1}{2}$ centered at $i/2$ in the complex plane because the unitarity relation can be written as

$$(\mathrm{Re}\, f_l)^2 + (\mathrm{Im}\, f_l - \tfrac{1}{2})^2 = \tfrac{1}{4}$$

This relation is the equation for a circle of radius $\frac{1}{2}$ centered at $i/2$ in the complex plane. It is shown in Figure 11.8a.

As the wave number $k = (1/\hbar)\sqrt{2ME}$ of the incident wave changes, f_l moves on the circle. The scattering phase δ_l is the angle between the arrow representing the complex number f_l and the real axis. The energy dependence of the complex scattering amplitude f_l is shown in detail in Figure 11.8b. We construct a three-dimensional diagram by erecting an energy axis perpendicular to the complex plane. The complex number $f_l(k)$ is then given by a point in this three-dimensional diagram. Since f_l is bound to lie on a circle in the complex plane, it is now confined to the *hull* of the corresponding cylinder in the plane passing through the energy value given by k. The real and imaginary parts of f_l as a function of the energy are then obtained by projecting the three-dimensional diagram onto the corresponding planes spanned by the energy axis and the real and imaginary axes. Finally, for completeness, we also show the energy dependence of the scattering phase shift δ_l.

There is an interesting relationship between the function $f(\vartheta)$ in the forward direction and the total cross section. We have

$$\sigma_{\mathrm{tot}} = \frac{4\pi}{k^2} \sum_{l=0}^{\infty} (2l+1) |f_l(k)|^2$$

Using the unitarity relation for the partial scattering amplitude

$$|f_l(k)|^2 = \mathrm{Im}\, f_l(k)$$

and the particular value of $P_l(\cos\vartheta)$ in the forward direction ($\vartheta = 0$),

$$P_l(1) = 1$$

we obtain

$$\sigma_{\mathrm{tot}} = \frac{4\pi}{k} \frac{1}{k} \sum_{l=0}^{\infty} (2l+1)\, \mathrm{Im}\, f_l(k) P_l(1)$$

$$\sigma_{\mathrm{tot}} = \frac{4\pi}{k}\, \mathrm{Im}\, f(0)$$

if we use the partial wave representation of $f(\vartheta)$ for $\vartheta = 0$.

This equation is called the *optical theorem*. It states that the total cross section is directly given by the imaginary part of the forward scattering amplitude. The optical theorem reflects the conservation of the particle current in the scattering process. In fact, the total current contained in the scattered wave has to supplied by the incident current. That is done through the interference between the incident and the scattered waves in the forward direction.

Problems

11.1 Why is the wave function $\varphi_{\mathbf{k}}^{(+)}(\mathbf{r})$ in Figure 11.1 suppressed beyond the potential region indicated by the dashed circle close to the center? Which effect makes it recover along the positive z-axis? Use Huygens's principle to draw an analogy to the scattering of light by a black disk.

11.2 Why must the scattered spherical wave $\eta_{\mathbf{k}}(\mathbf{r})$ as shown in Figure 11.3 be unequal to zero in the region of the repulsive potential and have a wave pattern there? What can be said about its wavelength within the potential region?

11.3 In Section 10.2 the classical elastic scattering of point particles by a rigid sphere of radius d was discussed. Replace the point particles by spheres of radius a. Show that the results for the differential and total cross sections stay valid if d is replaced by $d + a$.

11.4 Verify the unitarity relation for the partial scattering amplitude f_l,

$$\operatorname{Im} f_l = f_l f_l^*$$

using the unitarity relation for the scattering matrix element S_l,

$$S_l S_l^* = 1$$

as derived in Section 11.3. Put the unitarity relation for f_l into the form of an equation for the unitarity circle as given in Section 11.3.

12. Three-Dimensional Quantum Mechanics: Bound States

12.1 Bound States in a Spherical Square-Well Potential

Figure 10.2 has already shown the radial wave function of bound states in a three-dimensional square-well potential. Now in Figure 12.1 we plot the radial wave function R_{nl} together with its square R_{nl}^2 and the function $r^2R_{nl}^2$ for the low angular momentum quantum numbers $l = 0, 1, 2$. The reason for showing $r^2R_{nl}^2$ is that $r^2R_{nl}^2(r)\,dr$ represents the probability that a particle is within a spherical shell of radius r and thickness dr. Also shown in Figure 12.1 is the energy spectrum of the eigenvalues. We observe that the number of bound states is finite. The spacing between the different eigenvalues increases with increasing energy. For a given l-value the lowest-lying state has no node in r, the next one has one node, and so on. We can enumerate the eigenvalues E_{nl}, $n = 1, 2, \ldots,$ for a given l by the number $n - 1$ of nodes they possess. In Figure 12.1 the square-well potential $V(r)$ is drawn as a long-dash line, the effective potential as a short-dash line. The effective potential, as we have learned, is made up of the centrifugal potential and the square-well potential:

$$V_l^{\text{eff}}(r) = \frac{\hbar^2}{2M}\frac{l(l+1)}{r^2} + V(r)$$

The repulsive nature of the centrifugal potential suppresses the radial wave function,

$$R_{nl}(r) = A_{\text{I}}j_l(kr) \rightarrow \frac{A_{\text{I}}}{(2l+1)!!}(kr)^l, \qquad kr \ll 1$$

$$(2l+1)!! = 1\cdot 3\cdot 5\cdots(2l+1)$$

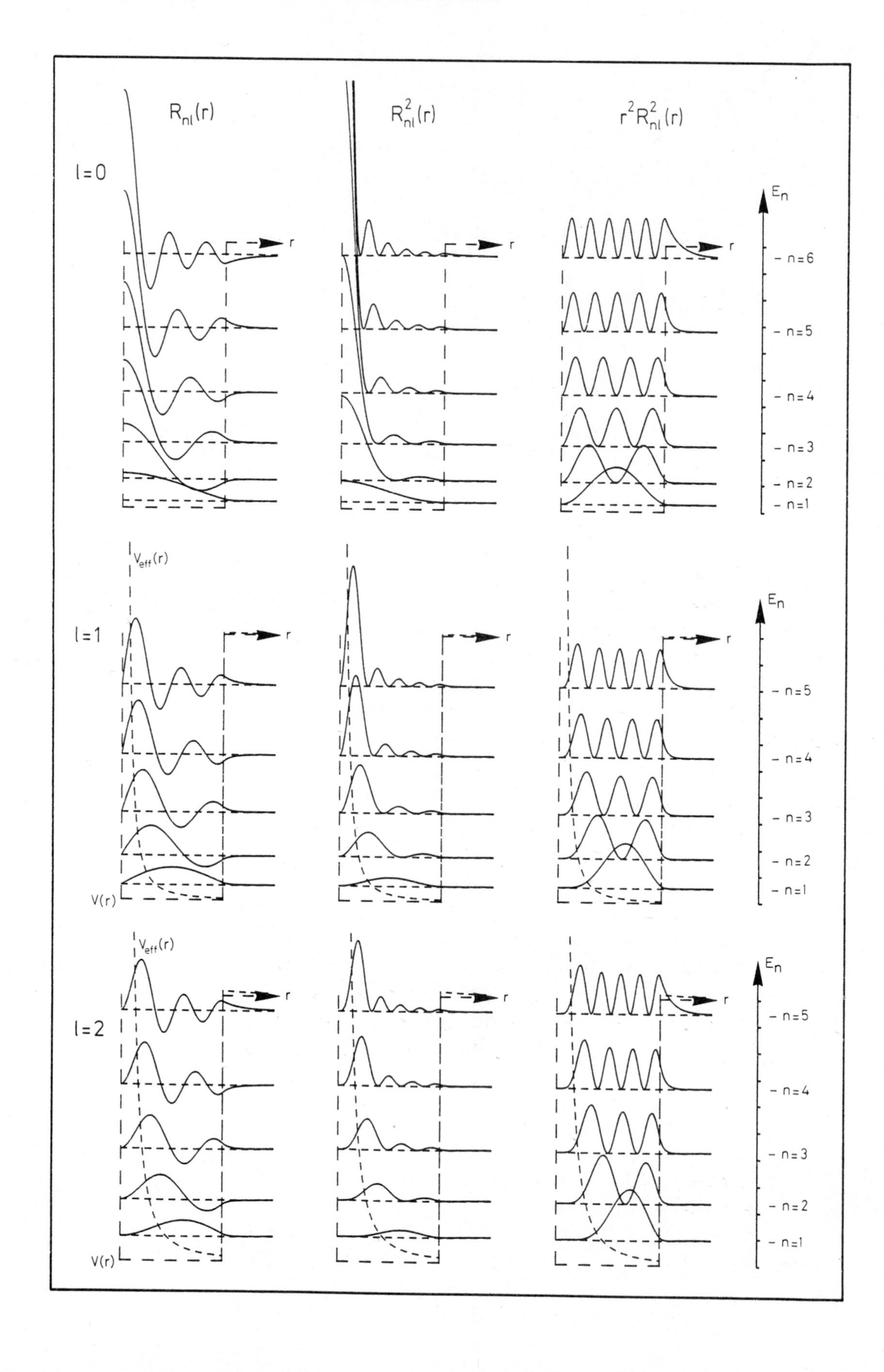

$R_{nl}(r)$
$R^2_{nl}(r)$
$r^2R^2_{nl}(r)$
l=0
l=1
l=2
E_n
r
n=6
n=5
n=4
n=3
n=2
n=1
$V_{eff}(r)$
V(r)

for small values of r and $l > 0$. It is also responsible for the increase in energy E_{nl} for given n and increasing l. This suppression of the wave function for $l \geq 1$ near the origin is easily verified in Figure 12.1. For $l = 0$ the wave functions start with values larger than zero at the origin. For $l = 1$ the wave function is zero at the origin; however, it increases linearily close to $r = 0$. For $l = 2$ the growth of the wave function from zero at the origin is only that of a parabola. The slopes close to the origin become steeper with higher quantum numbers at fixed l. Thus for fixed l the particle comes closer to the origin for higher values of n. Higher values of n correspond to higher values of the energy and of the momentum p. For a given angular momentum L the classical relation $L = bp$ shows that larger momenta p correspond to smaller impact parameters b. In this respect the quantum-mechanical behavior of the particle corresponds to classical mechanics.

The plot of $r^2R^2_{nl}(k, r)$ allows a particularly simple discussion. Let us start with the value $l = 0$. The radial wave functions within the potential region $r < d$ are,

$$R_{n0}(r) = A_{\mathrm{I}} j_0(k_{\mathrm{I}n} r) = A_{\mathrm{I}} \frac{\sin k_{\mathrm{I}n} r}{k_{\mathrm{I}n} r}$$

$$k_{\mathrm{I}n} = \sqrt{2m(E_{n0} - V_0)}$$

so that the function

$$r^2 R^2_{n0}(r) = \frac{A^2_{\mathrm{I}}}{k^2_{\mathrm{I}n}} \sin^2 k_{\mathrm{I}n} r$$

behaves in a simple sine-squared manner. For the higher

Figure 12.1 **The radial eigenfunctions $R_{nl}(r)$ of bound states in a square-well potential for three angular momentum values, $l = 0, 1, 2$, are shown as continuous lines in the left column. The form $V(r)$ of the potential is indicated by the long-dash line. Also shown for $l \neq 0$ is the effective potential $V_{\mathrm{eff}}(r)$ which contains the influence of angular momentum. On the far right is an energy scale and to the right of it the energy eigenvalues E_n are indicated by horizontal lines. These lines are repeated as short-dash lines on the left. They serve as zero lines for the plotted functions. In the middle column the squares $R^2_{nl}(r)$ of the radial eigenfunctions are shown. Along a fixed direction ϑ, ϕ away from the origin, this quantity is proportional to the probability that the particle will be observed within a unit volume element around point r, ϑ, ϕ. In the right column are the functions $r^2R^2_{nl}(r)$. Their values are a measure for the probability of observing the particle anywhere within a spherical shell of radius r and unit thickness.**

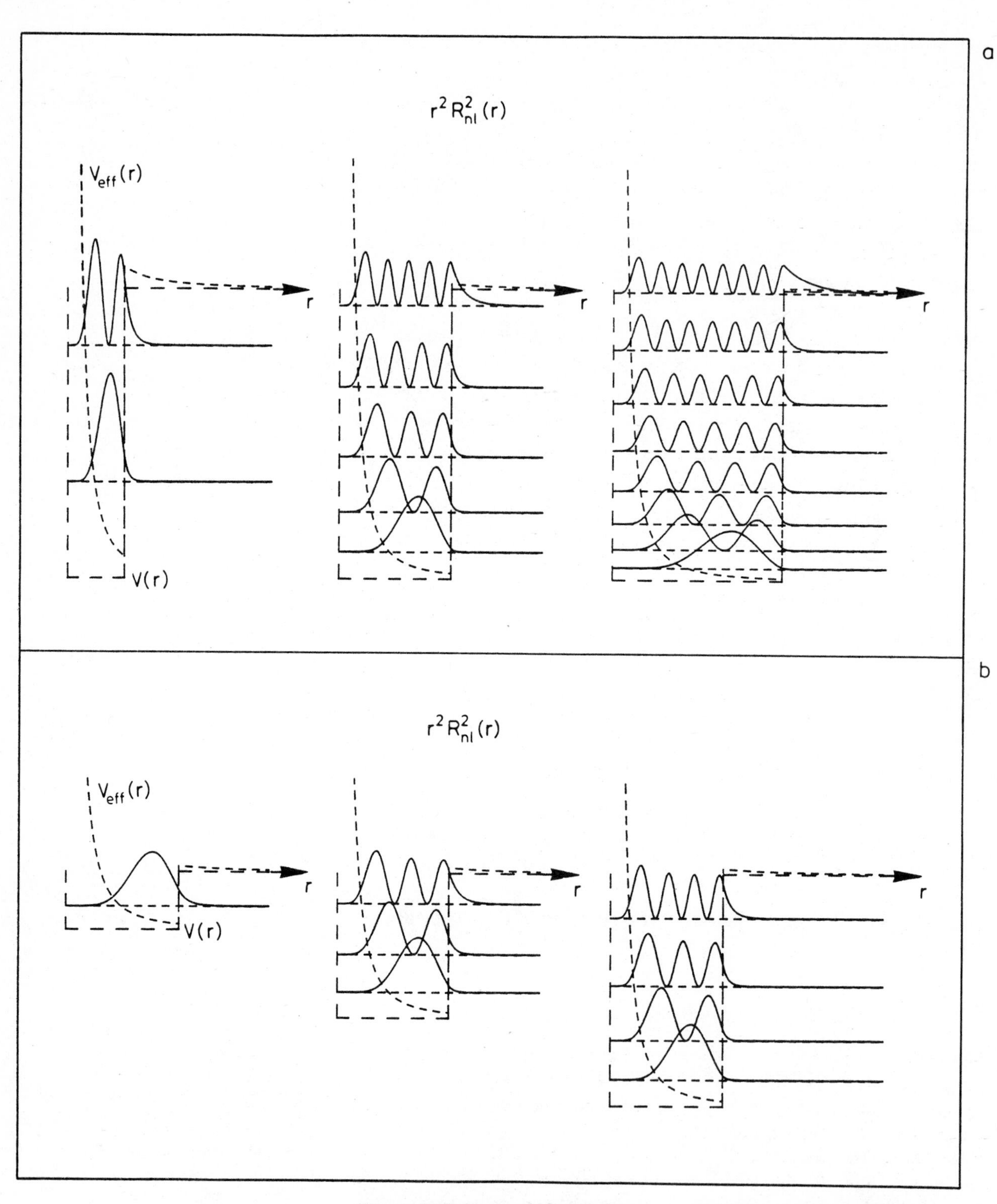

Figure 12.2 **Dependence of the eigenvalue spectrum of a square-well potential on (a) the width and (b) the depth of the well. The function shown is $r^2R_{nl}^2(r)$ for the fixed angular momentum quantum number $l = 2$.**

angular momenta we recall the asymptotic relation

$$j_l(kr) \to \frac{1}{kr}\sin\left(kr - l\frac{\pi}{2}\right), \qquad kr \gg 1$$

so that for $r \gg 1/k$ the behavior is again sine-squared:

$$r^2 R_{nl}^2 \to \frac{A_{\mathrm{I}}^2}{k^2}\sin^2\left(kr - l\frac{\pi}{2}\right), \qquad kr \gg 1$$

Again, looking at Figure 12.1, we recognize the approach of the quantity $r^2 R_{nl}^2$ toward this behavior. In region I close to the edge of the potential at $r = d$, the centrifugal barrier is low for low values of l; it can therefore be neglected in a coarse approximation. Thus, close to the outer rim of the potential, the wave functions for different l, but equal n should look almost alike and behave in a sine-squared manner. This is easily verified in Figure 12.1.

Figure 12.2 shows the dependence of the eigenvalue spectrum on the width and depth of the potential. The number of eigenvalues grows as the potential widens and deepens.

The full three-dimensional wave function is obtained by multiplying the radial wave function $R_{nl}(r)$ by the spherical harmonic $Y_{lm}(\vartheta, \phi)$:

$$\varphi_{nlm}(\mathbf{r}) = R_{nl}(r) Y_{lm}(\vartheta, \phi)$$

Since the absolute square $\rho_{nlm}(r, \vartheta)$ of this wave function is independent of ϕ,

$$\begin{aligned}\rho_{nlm}(r, \vartheta) &\equiv |\varphi_{nlm}(\mathbf{r})|^2 \\ &= R_{nl}^2(r)\frac{2l+1}{4\pi}\frac{(l-|m|)!}{(l+|m|)!}\left[P_l^{|m|}(\cos\vartheta)\right]^2\end{aligned}$$

it can be shown in an r, ϑ-plot. In Figures 12.3, 12.4, and 12.5 this function is plotted as a surface over a half circle ($0 \le r \le R$; $0 \le \vartheta \le \pi$) in the x, z-plane. It is the probability density for observing a particle at location r, ϑ, ϕ; that is to say

$$\begin{aligned}dw &= |\varphi_{nlm}(\mathbf{r})|^2\, dV \\ &= R_{nl}^2 \frac{2l+1}{4\pi}\frac{(l-|m|)!}{(l+|m|)!}\left[P_l^{|m|}(\cos\vartheta)\right]^2 r^2\, dr\, d\cos\theta\, d\phi\end{aligned}$$

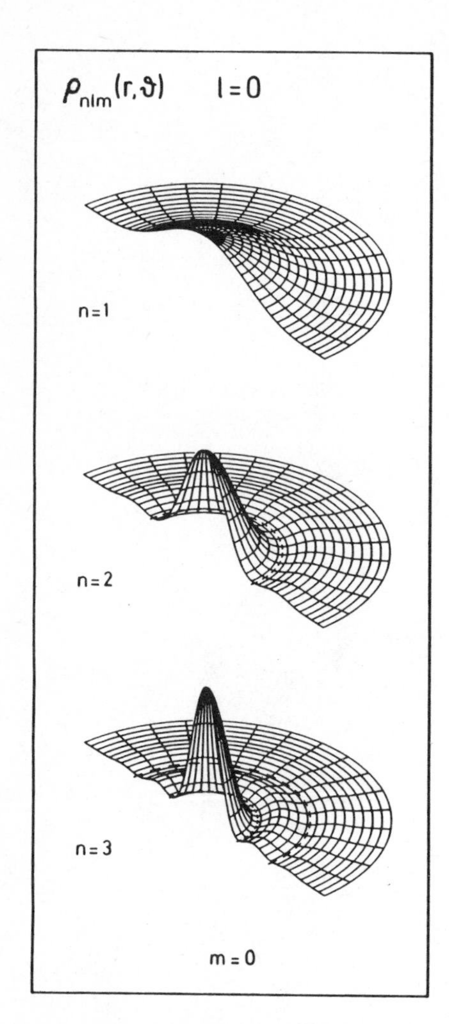

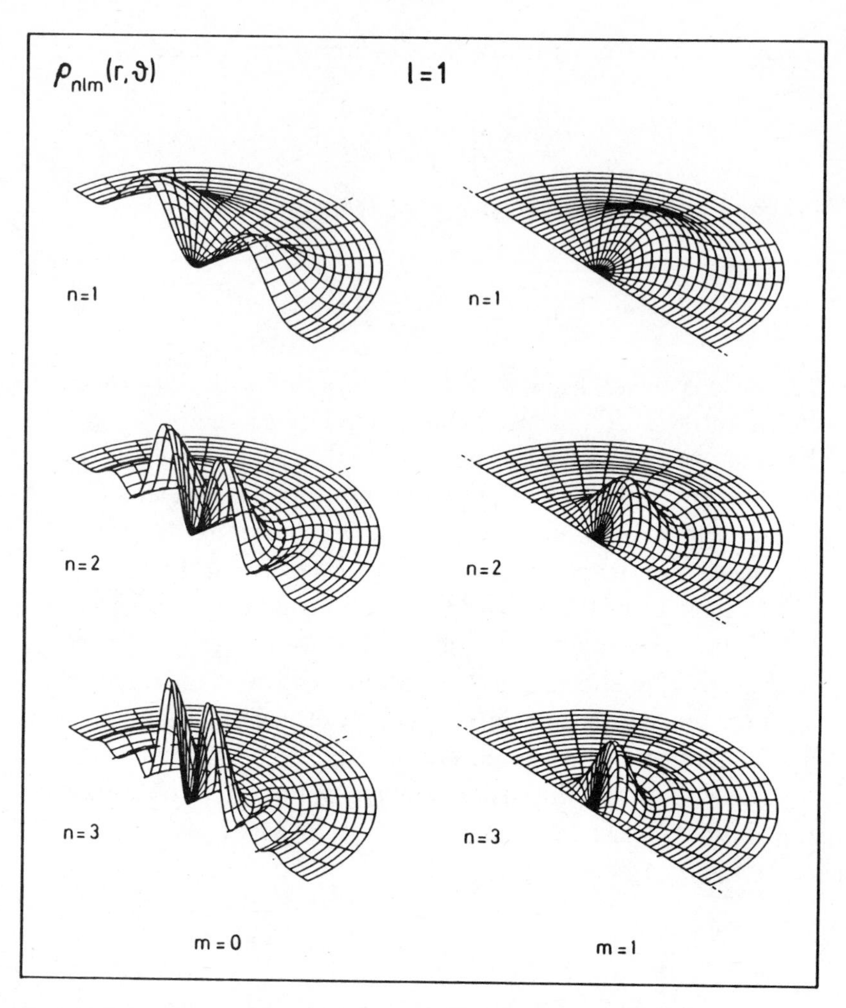

Figure 12.3 **Absolute squares $\rho_{nlm}(r, \vartheta) = |\varphi_{nlm}(r, \vartheta, \phi)|^2$ of the full three-dimensional eigenfunction of a square-well potential. Here $\rho_{nlm}(r, \vartheta)\,dV$ represents the probability of observing the particle in the volume element dV at location r, ϑ, ϕ. It is a function only of the distance r from the origin and of the polar angle ϑ. In this figure, which applies to zero angular momentum quantum number l, the function φ depends only on r. For values $n = 1, 2, 3$ of the principal quantum number it has $n - 1 = 0, 1, 2$ nodes in r indicated by the dashed half circles. Each figure gives the probability density for observing the particle at any point in a half-plane containing the z-axis. All pictures in Figures 12.3 through 12.5 have the same scale in r and ϑ. They do, however, have different scale factors in ρ.**

Figure 12.4 **The functions $\rho_{nlm}(r, \vartheta)$ as given in Figure 12.3 but for $l = 1$ and $m = 0, 1$. The ϑ-dependence is given by the associated Legendre functions $P_l^{|m|}(\cos\vartheta)$ which have $l - |m|$ nodes in ϑ, indicated by the dashed lines $\vartheta = \text{const}$.**

is the probability of finding the particle in the volume element $dV = r^2\,dr\,d\cos\vartheta\,d\phi$ at r, ϑ, ϕ. In Figures 12.3, 12.4, and 12.5 we recognize the nodes in r as half circles in the plane at which the probability density vanishes. They are attributable to the nodes in the radial wave function $R_{nl}(r)$. In addition, there are $l - |m|$ nodes in ϑ along rays $\vartheta = \text{const}$ in the plane originating from zeroes of $P_l^m(\cos\vartheta)$.

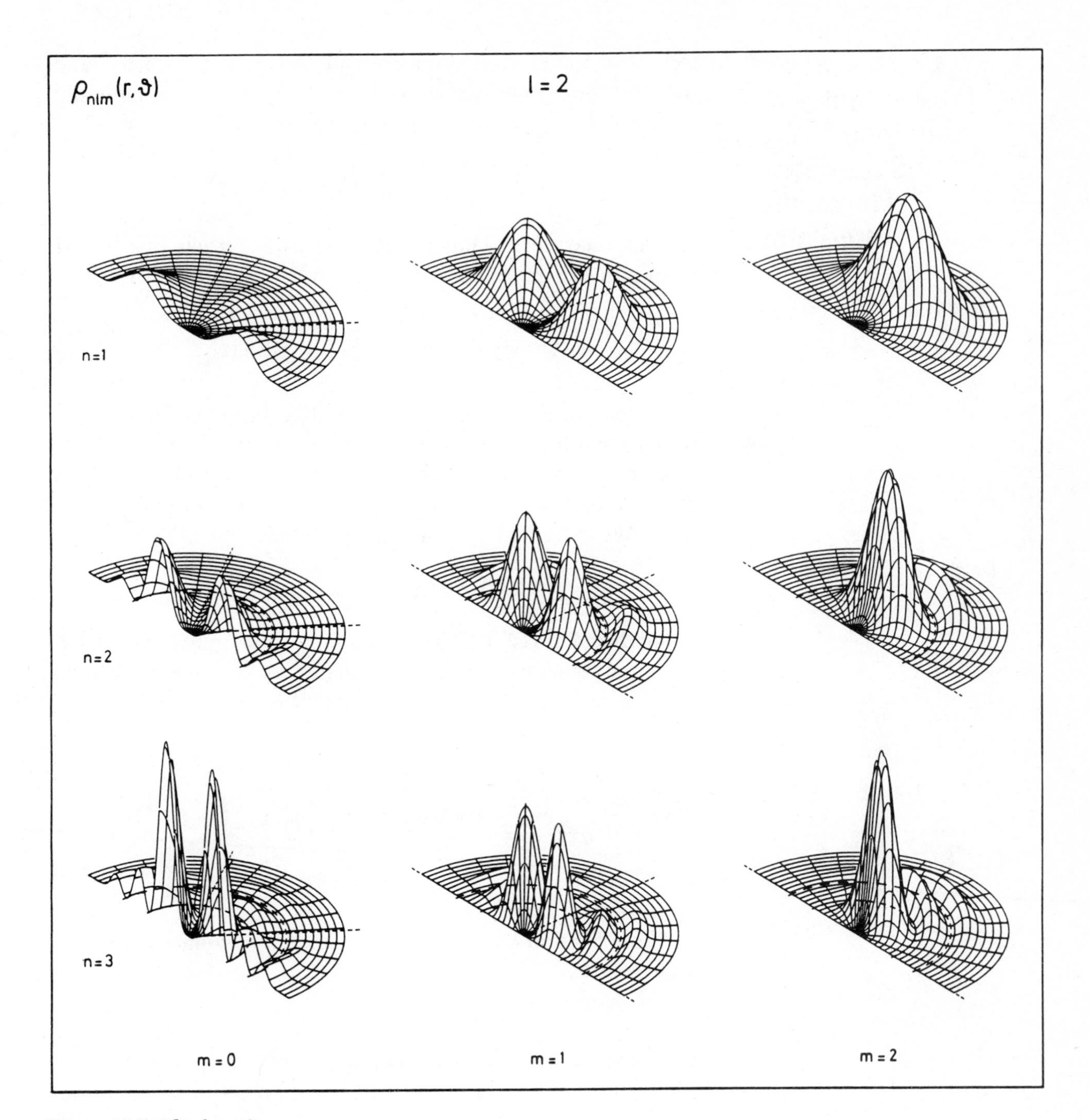

Figure 12.5 **The functions $\rho_{nlm}(r,\vartheta)$ as given in Figure 12.4 but for $l=2$ and $m=0,1,2$.**

12.2 Bound States of the Spherically Symmetric Harmonic Oscillator

For many model calculations in nuclear physics, a harmonic oscillator potential has proved to be useful. The potential energy of a spherically symmetric harmonic oscillator is

$$V(\mathbf{r}) = \frac{k}{2}r^2 = \frac{k}{2}(x_1^2 + x_2^2 + x_3^2)$$

The stationary Schrödinger equation for a particle of mass M moving in this potential has the form

$$\left(-\frac{\hbar^2}{2M}\nabla^2 + \frac{k}{2}r^2\right)\varphi(\mathbf{r}) = E\varphi(\mathbf{r})$$

Instead of the separation of variables in polar coordinates, as discussed in Section 9.4, we may just as well carry out the separation in Cartesian coordinates, for the potential is a sum of terms, each of which depends on only one of these coordinates. We start with the factorized *ansatz*

$$\varphi(\mathbf{r}) = \varphi_1(x_1)\varphi_2(x_2)\varphi_3(x_3)$$

and arrive at three Schrödinger equations for one-dimensional harmonic oscillators in the coordinates x_1, x_2, and x_3, which are identical to the equation discussed in Section 6.3 for the coordinate x,

$$\left(-\frac{\hbar^2}{2M}\frac{d^2}{dx_i^2} + \frac{M}{2}\omega^2 x_i^2\right)\varphi_i(x_i) = E_i\varphi_i(x), \qquad i = 1, 2, 3$$

$$\omega = \sqrt{k/M}$$

From Section 6.3 we know that the energy eigenvalues are

$$E_i = E(n_i) = (n_i + \tfrac{1}{2})\hbar\omega, \qquad n_i = 0, 1, 2, \ldots$$

with independent integer quantum numbers n_i for the three oscillators. The total energy E depends on the three quantum numbers n_1, n_2, and n_3:

$$E(n_1, n_2, n_3) = E(n_1) + E(n_2) + E(n_3) = (n_1 + n_2 + n_3 + \tfrac{3}{2})\hbar\omega$$

The eigenfunctions $\varphi_{n_i}(x_i)$ are normalized products of Hermite polynomials and Gaussians. They were shown in Figures 6.3 and 6.4.

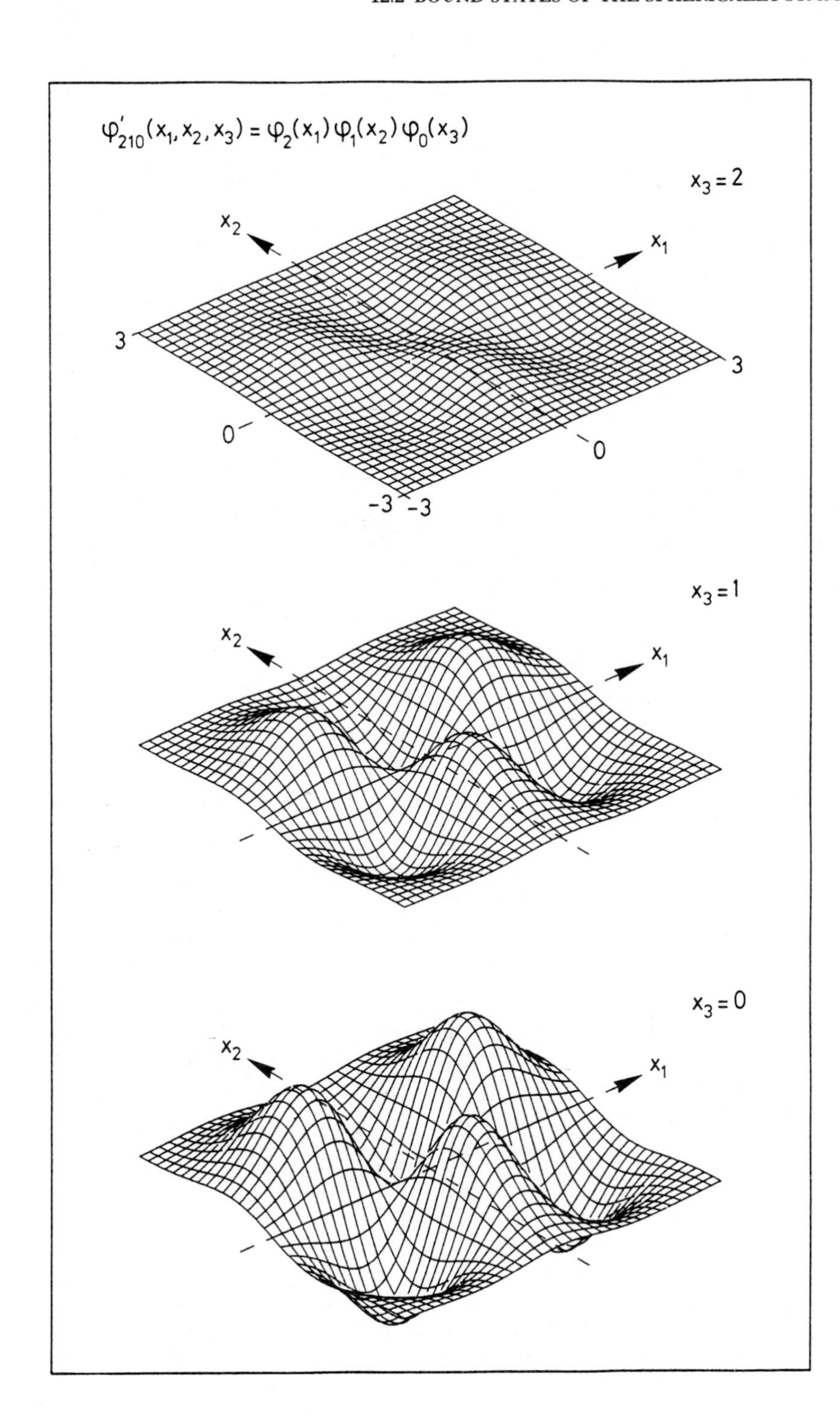

Figure 12.6 **Eigenfunction $\varphi'_{210}(x_1, x_2, x_3) = \varphi_2(x_1)\varphi_1(x_2)\varphi_0(x_3)$ of the three-dimensional harmonic oscillator expressed in Cartesian coordinates x_1, x_1, x_3 and written as a product of three one-dimensional harmonic oscillator eigenfunctions. For this figure the width parameter $\sigma_0 = 1$ was chosen. The function is plotted for three planes $x_3 = 0, 1, 2$. Because $\varphi_0(x_3)$ is symmetric, the plots remain unchanged if the substitution $x_3 \to -x_3$ is performed. This figure should be compared with Figure 6.3.**

The eigenfunctions of the three-dimensional harmonic oscillator are

$$\varphi'_{n_1 n_2 n_3}(x_1, x_2, x_3) = \varphi_{n_1}(x_1)\varphi_{n_2}(x_2)\varphi_{n_3}(x_3)$$

with the eigenvalue $E(n_1, n_2, n_3)$. Figure 12.6 shows, as an example, the eigenfunction

$$\varphi'_{210}(x_1, x_2, x_3) = \varphi_2(x_1)\varphi_1(x_2)\varphi_0(x_3)$$

Since it is a function of the independent coordinates x_1, x_2, and x_3 we represent it by plotting it for various planes $x_3 = \text{const}$ in x_1, x_2, x_3-space. Since the x_3-dependence is given by the simple Gaussian factor

$$\varphi_0(x_3) = \text{const} \cdot \exp\left(-\frac{x_3^2}{2\sigma_0^2}\right), \qquad \sigma_0^2 = \frac{\hbar}{M\omega}$$

the function is symmetric in x_3 and is damped away as x_3 increases in magnitude (see Section 6.3). It is also symmetric in x_1 and antisymmetric in x_2.

Obviously, all the different quantum triplets n_1, n_2, n_3 having the same sum correspond to different eigenfunctions $\varphi'_{n_1 n_2 n_3}$, that is, to different physical states of the system. All these physical states, however, have the same energy eigenvalue. They are therefore called *degenerate states*.

The usual separation of the three-dimensional Schrödinger equation in polar coordinates yields the radial Schrödinger equation

$$\left[-\frac{\hbar^2}{2M}\frac{1}{r}\frac{\partial^2}{\partial r^2}r + V_l^{\text{eff}}(r)\right]R_{nl}(r) = E_n R_{nl}(r)$$

with the effective potential

$$V_l^{\text{eff}}(r) = \frac{\hbar^2}{2M}\frac{l(l+1)}{r^2} + \frac{k}{2}r^2$$

The solutions of this equation are

$$R_{nl}(r) = N_{nl}\left(\frac{r^2}{\sigma_0^2}\right)^{l/2} \exp\left(-\frac{r^2}{2\sigma_0^2}\right) L_{n_r}^{l+1/2}\left(\frac{r^2}{2\sigma_0^2}\right)$$

where the functions $L_{n_r}^{l+1/2}$ are the *associated Laguerre polynomials*:

$$L_{n_r}^{l+1/2}(x) = \sum_{j=0}^{n_r} (-1)^j \binom{n_r + l + \frac{1}{2}}{n_r - j} \frac{x^j}{j!}, \qquad n_r = 0, 1, 2, \ldots$$

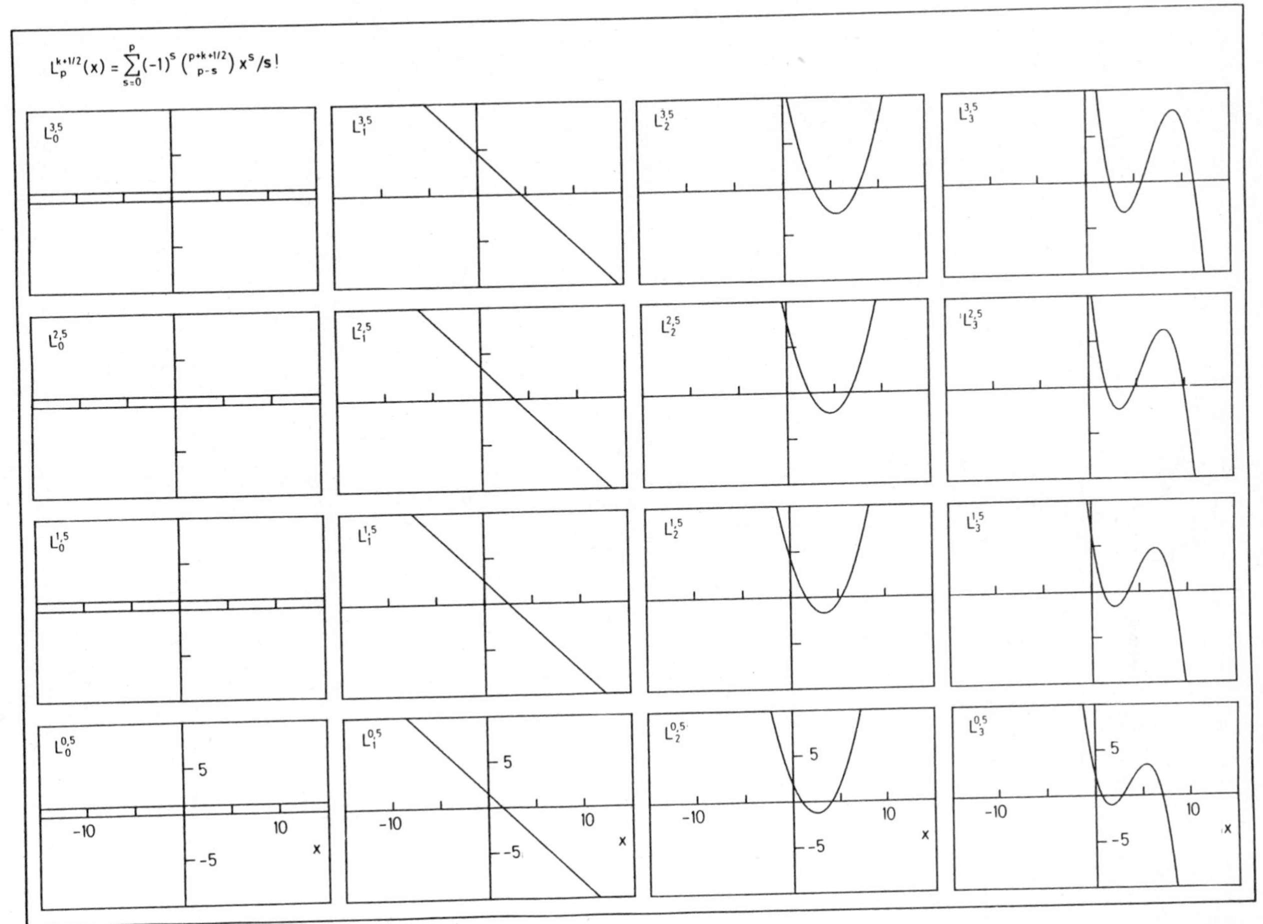

Figure 12.7 **Laguerre polynomials of half-integer upper index. The lower index is equal to the degree of the polynomial and to the number of its zeroes. All zeroes are at positive values of the argument x.**

The normalization constants are

$$N_{nl} = \sqrt{\frac{n_r!\,2^{n+2}}{[2(l+n_r)+1]!!\,\sqrt{\pi}\,\sigma_0^3}}$$

The energy eigenvalues are given by

$$E_n = \left(n + \tfrac{3}{2}\right)\hbar\omega$$

with

$$n = 2n_r + l$$

Before studying the radial solutions R_{nl}, we first present the associated Laguerre polynomials in Figure 12.7. The degree of the polynomial is equal to its lower index and to the number of zeroes on the positive x-axis.

The radial solutions R_{nl} are shown in Figure 12.8. Their zeroes are determined by the zeroes of the corresponding

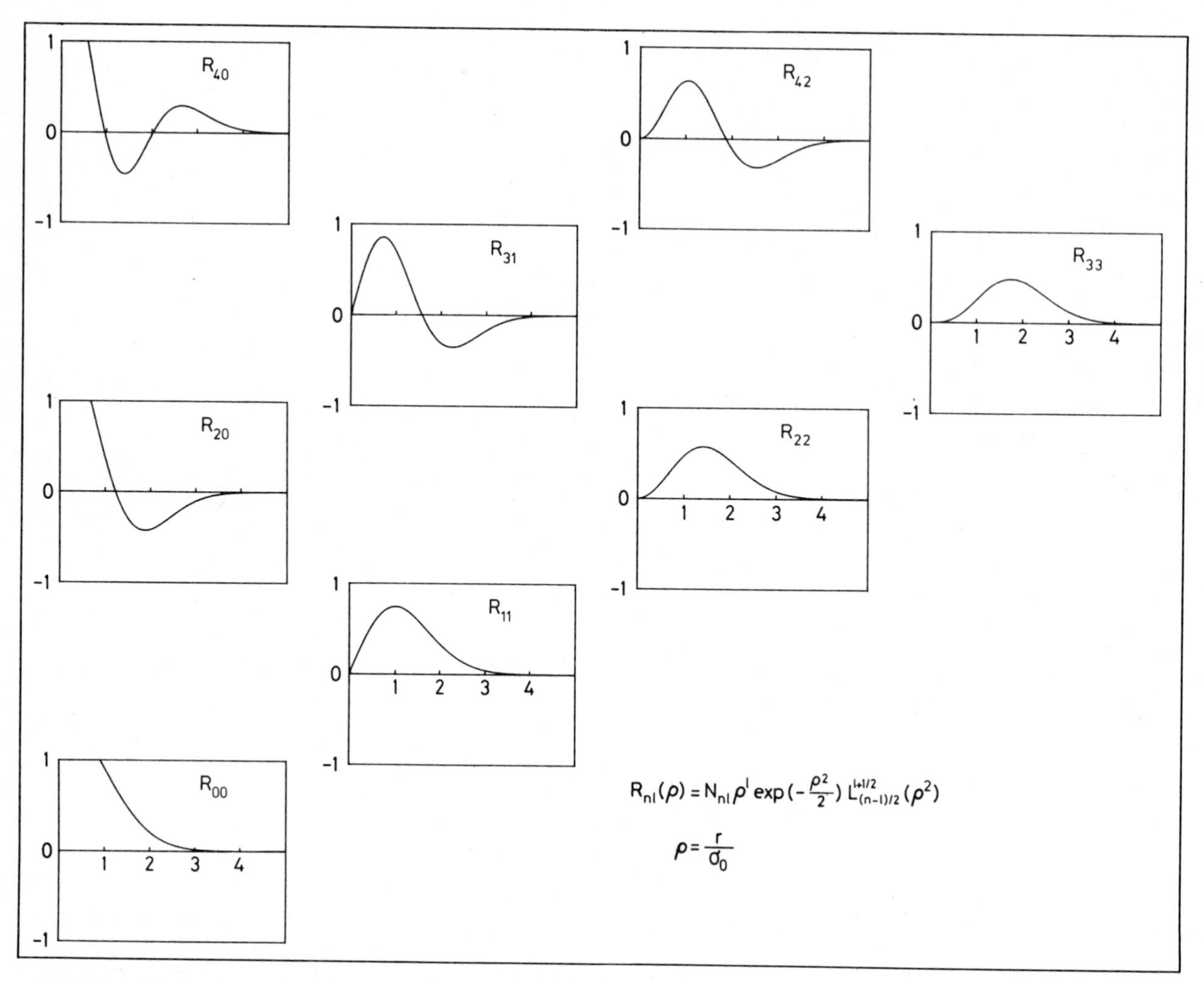

Figure 12.8 **Radial eigenfunctions $R_{nl}(\rho)$ for the three-dimensional harmonic oscillator. Their zeroes are the $(n - l)/2$ zeroes of the associated Laguerre polynomial $L^{l+1/2}_{(n-l)/2}(\rho^2)$. The argument ρ is the distance r from the origin divided by the width σ_0 of the ground state of the oscillator. Graphs in the same row belong to the same value of n. Graphs in the same column belong to the same value of l.**

Figure 12.9 **Radial eigenfunctions $R_{nl}(r)$, their squares $R^2_{nl}(r)$, and the functions $r^2R^2_{nl}(r)$ for the lowest eigenstates of the harmonic oscillator and the lowest angular momentum quantum numbers $l = 0, 1, 2$. On the far right side are the eigenvalue spectra. The form of the harmonic oscillator potential $V(r)$ is indicated by a long-dash line, and, for $l \neq 0$, that of the effective potential $V_{\text{eff}}(r)$ by a short-dash line. The eigenvalues have equidistant spacing. The eigenvalue spectra are degenerate for all even l values and all odd l values, except that the minimum value of the principal quantum number is $n = l$.**

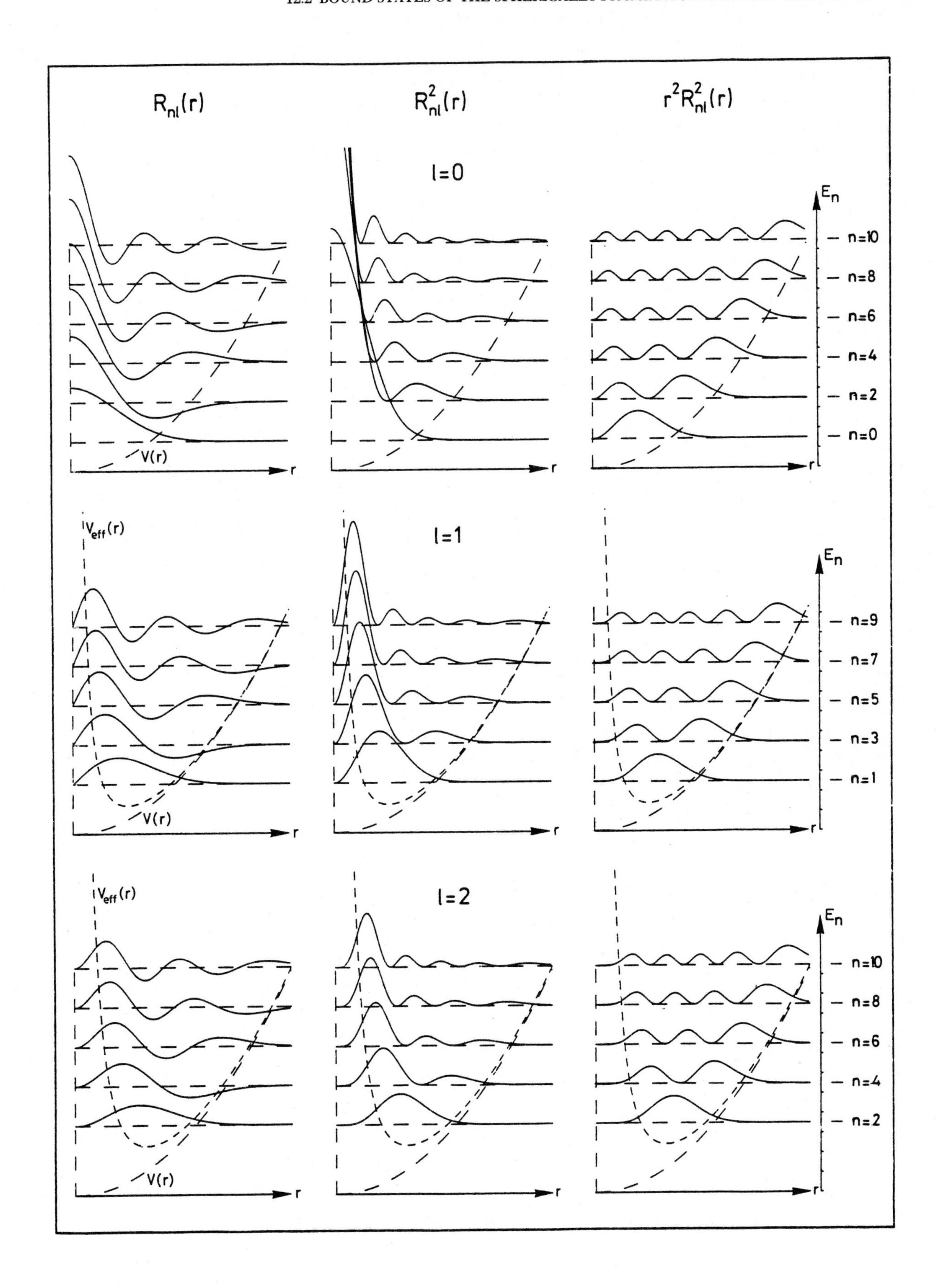

$R_{nl}(r)$
$R^2_{nl}(r)$
$r^2R^2_{nl}(r)$
l=0
E_n
n=10
n=8
n=6
n=4
n=2
n=0
V(r)
r
l=1
$V_{eff}(r)$
n=9
n=7
n=5
n=3
n=1
l=2

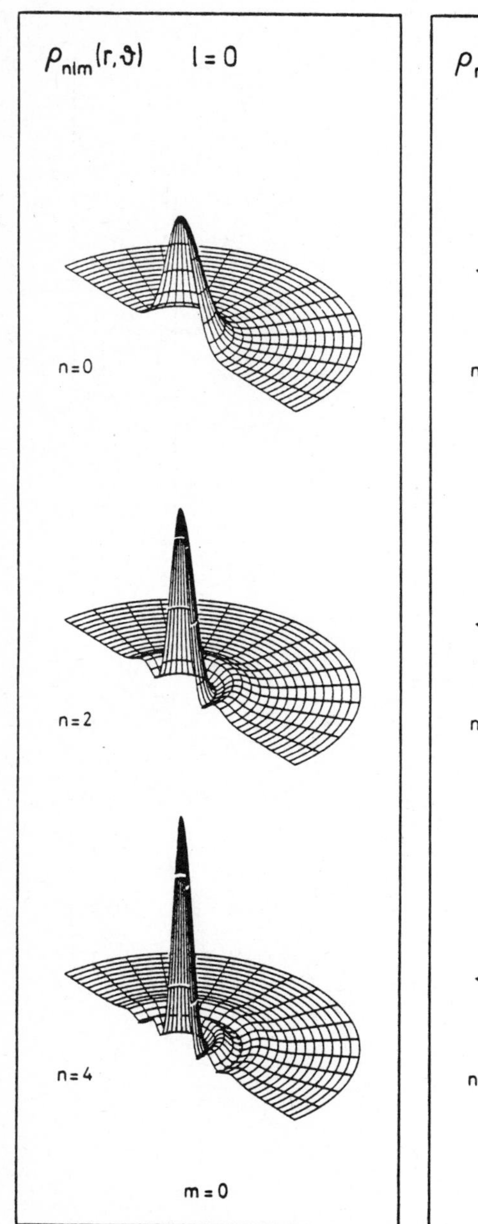

Figures 12.10–12.12 **The absolute squares $\rho_{nlm}(r, \vartheta) = |\varphi_{nlm}(r, \vartheta, \phi)|^2$ of the full three-dimensional eigenfunctions for the harmonic oscillator. The absolute squares are functions only of r and ϑ. There are $(n - l)/2$ radial nodes and $l - |m|$ polar nodes, indicated by dashed half circles and rays, respectively. Each figure gives the probability density for observing the particle at any point in a half-plane containing the z-axis. All pictures have the same scale in r and ϑ. They do, however, have different scale factors in ρ.**

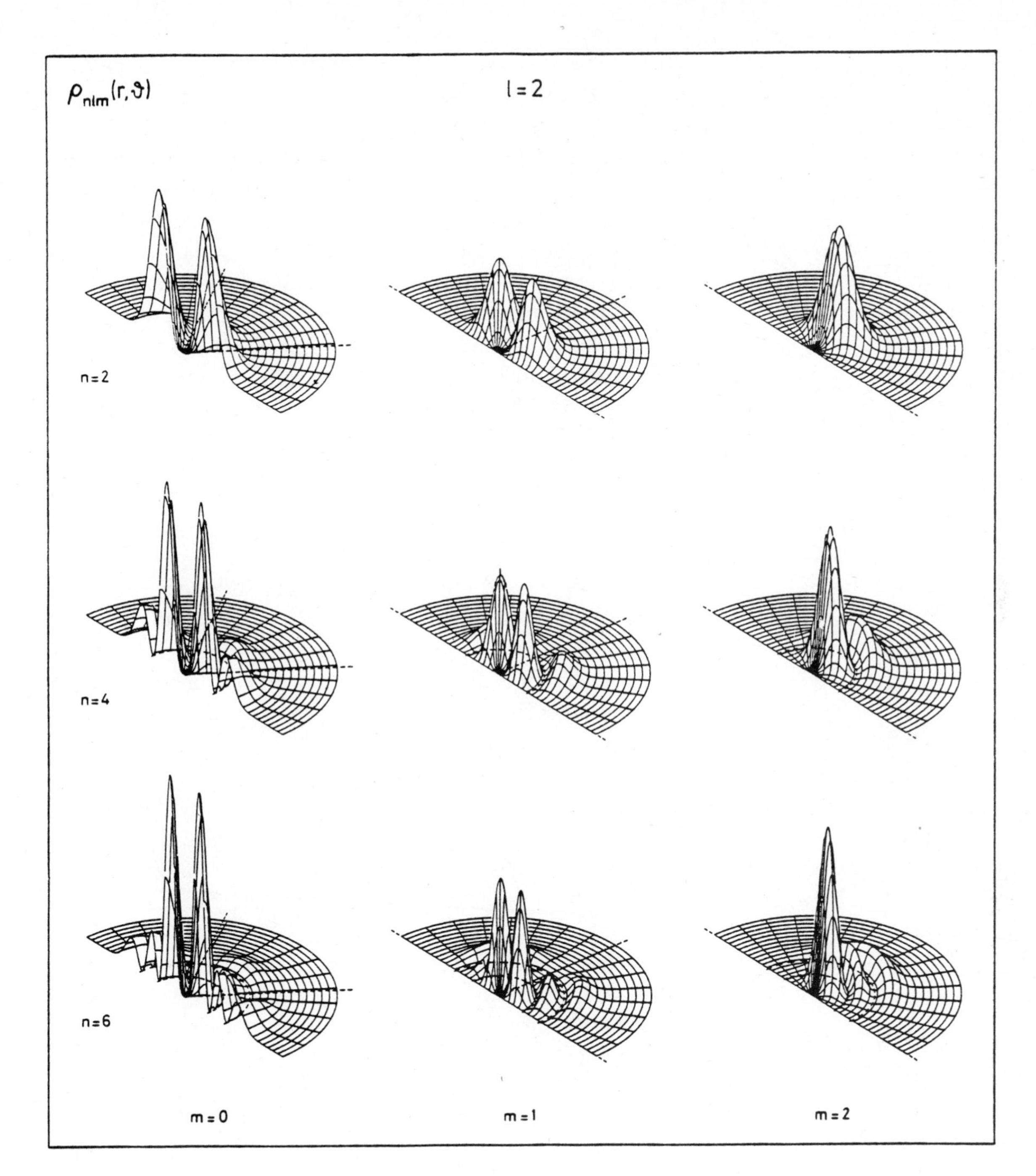

Figure 12.12

Laguerre polynomial. Because of the relation between the integer quantum numbers n, n_r, and l, quantum number n takes the values $l, l+2, l+4, \ldots$.

In Figure 12.9 the functions R_{nl}, R_{nl}^2, and $r^2 R_{nl}^2$ are shown together with the potential $V(r)$, the effective potential $V_{\text{eff}}(r)$, and the eigenvalue spectra for the lowest eigenstates of the harmonic oscillator and the lowest angular momentum quantum numbers $l = 0, 1, 2$. With increasing energy the functions reach out farther in r since the potential increases with r^2. The functions are again suppressed near $r = 0$ for $l \neq 0$ by the centrifugal barrier. The suppression is strongest for low energy E but high angular momentum quantum numbers l.

The three-dimensional stationary wave functions are

$$\varphi_{nlm}(\mathbf{r}) = R_{nl}(r) Y_{lm}(\vartheta, \phi)$$

Their absolute squares $|\varphi_{nlm}|^2$, which are independent of the azimuth ϕ, are plotted in Figure 12.10, 12.11, and 12.12 for low values of n and $l = 0, 1, 2$. Since the energy eigenvalues E_n depend on one quantum number only, there are again degenerate eigenfunctions. From the properties of the spherical harmonics, we know that for every l there are $2l+1$ states of different quantum number m. Moreover, for a given energy eigenvalue E_n there are eigenstates with different angular momenta l. Because of the relation $n = 2n_r + l$, the number n of quanta of energy $\hbar\omega$ above the energy $\frac{3}{2}\hbar\omega$ of the ground state is even or odd, depending on whether l is even or odd.

How are the two different sets of solutions $\varphi'_{n_1 n_2 n_3}$ and φ_{nlm} related? Obviously, we have to be able to describe the same physical states by either set. In fact, we are able to do so because most of the states are degenerate; that is, a large number of states have the same eigenvalue. Obviously too, a linear superposition of degenerate eigenstates is again an eigenstate of the same energy. Thus it is possible to express the eigenstates of a given energy in one set by a linear superposition of the eigenstates of the same energy in the other set. The only nondegenerate eigenstate is the ground-state

$$\varphi'_{000}(\mathbf{r}) = \pi^{-3/4} \sigma_0^{-3/2} \exp\left(- \frac{r^2}{2\sigma_0^2} \right) = \varphi_{000}(\mathbf{r})$$

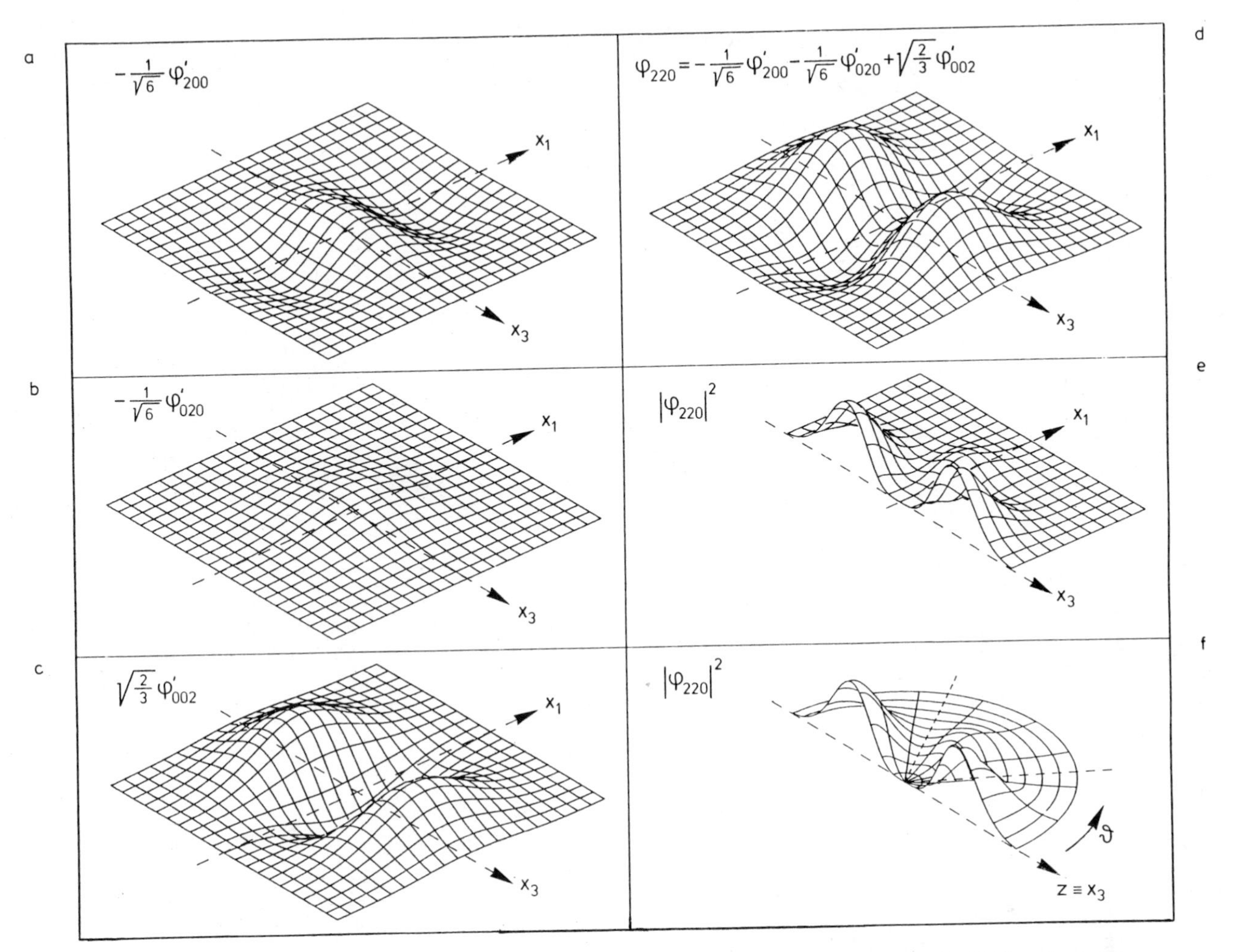

Figure 12.13 **An eigenstate $\varphi_{nlm}(r, \vartheta, \phi)$ of the harmonic oscillator can be written as a linear superposition of the degenerate eigenfunctions $\varphi'_{n_1n_2n_3}(x_1, x_2, x_3)$ having the same energy eigenvalue (a, b, c). The three eigenfunctions for $n = 2$ in the x_1, x_2-plane each multiplied by the appropriate factor; (d) the sum; (e) its square; (f) the function $|\varphi_{220}|^2$ in r, ϑ-representation as known from Figure 12.12. Parts e and f are identical except that part e has Cartesian coordinates, part f polar coordinates.**

which has the ground-state energy $E_0 = \frac{3}{2}\hbar\omega$. This eigenstate is the same in both sets. All other states are degenerate. As an example of a superposition of Cartesian eigenstates $\varphi'_{n_1n_2n_3}$ which forms an angular momentum eigenstate φ_{nlm}, we look at $n = 2$, $l = 2$, and $m = 0$. We have

$$\varphi_{220}(\mathbf{r}) = -\frac{1}{\sqrt{6}}\varphi'_{200}(\mathbf{r}) - \frac{1}{\sqrt{6}}\varphi'_{020}(\mathbf{r}) + \sqrt{\tfrac{2}{3}}\,\varphi'_{002}(\mathbf{r})$$

Figure 12.13 demonstrates this particular superposition. Figures 12.13a, b, and c give the three terms of this superposition in the x_1, x_3-plane. In Figure 12.13d the sum $\varphi_{220}(\mathbf{r})$ is shown. In Figure 12.13e its absolute square is plotted in the x_1, x_3-half-plane to facilitate comparision with the r, ϑ-plot of this same function $|\varphi_{220}(\mathbf{r})|^2$, which is given in Figure 12.13f.

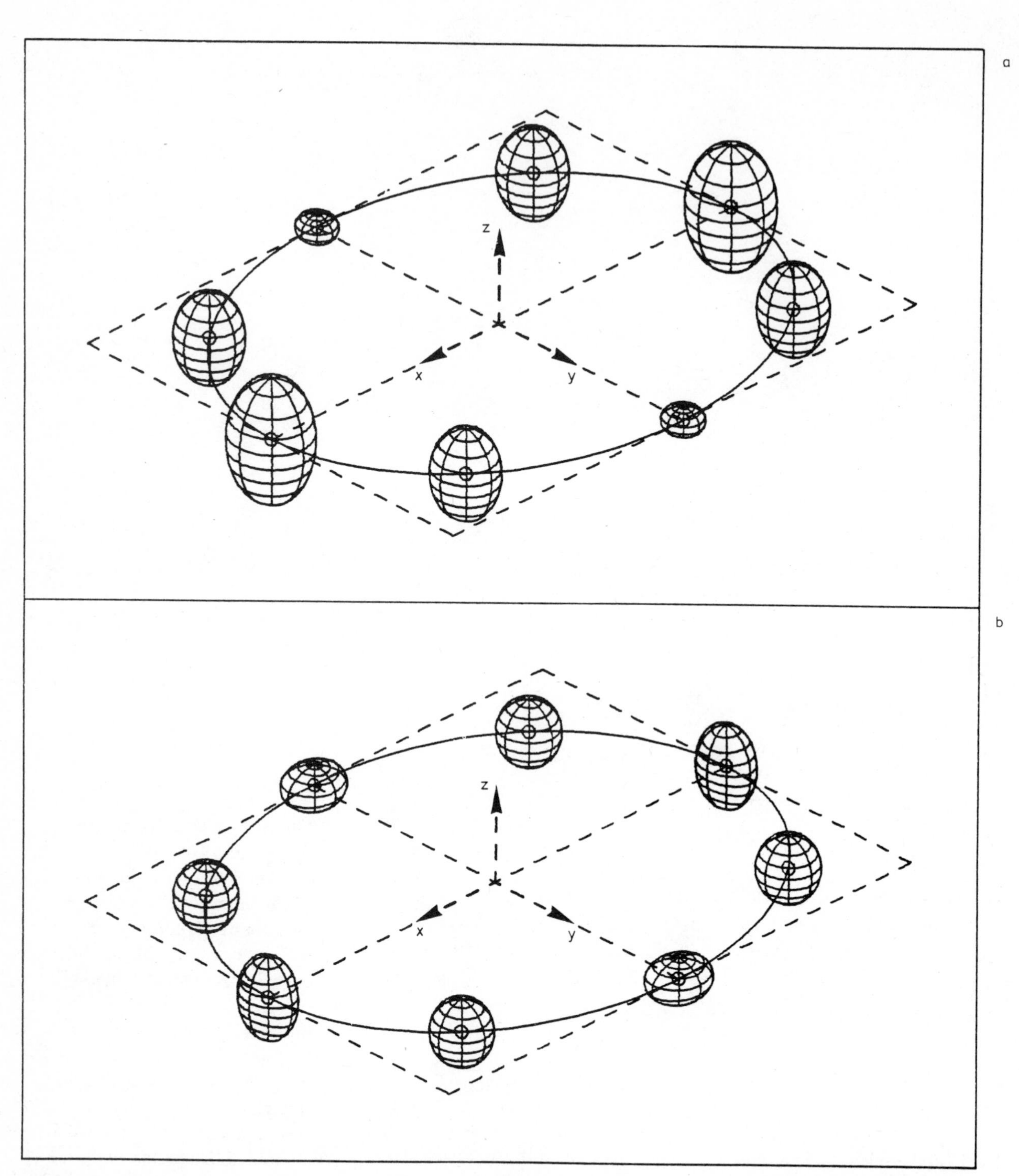

Figure 12.14 A three-dimensional Gaussian wave packet, represented by its probability ellipsoid, moves under the influence of an attractive force described by a harmonic oscillator potential. Its expectation value, that is, the center of the ellipsoid, describes an elliptical trajectory. The initial conditions were chosen so that the ellipsoid does not tumble, that is, its principal axes keep constant orientations. The magnitudes of the principal axes oscillate with twice the oscillator frequency. Two examples are shown. (a) The ellipsoid stays rotationally symmetric with respect to the z-axis. (b) All three principal axes of the ellipsoid are different.

12.3 Harmonic Particle Motion in Three Dimensions

In Section 6.4 we described the motion of a Gaussian wave packet in a one-dimensional harmonic oscillator potential. We obtained for the absolute square of the time-dependent wave functions a Gaussian distribution with an expectation value oscillating like the classical point particle. Its width oscillates with twice the oscillator frequency. We therefore anticipate that in the three-dimensional oscillator the expectation value of a three-dimensional Gaussian wave packet moves on an elliptical trajectory as a classical point particle does. The shape of the three-dimensional wave packet is completely described by its covariance ellipsoid, which we introduced in Section 9.1. The shape of the covariance ellipsoid itself oscillates, that is, it changes periodically with time, its frequency being twice the oscillator frequency.

Figure 12.14 shows two examples for such a motion. The classical trajectory is chosen to be identical for both. For simplicity the covariance ellipsoid has two of its principal axes in the plane of motion. Moreover, the initial conditions were chosen so that the directions of its principal axes do not change while the ellipsoid is moving. Because the harmonic oscillator is spherically symmetric, the oscillation in magnitude of all three principal axes has the same frequency but may have different phases. In Figure 12.14a the covariance ellipsoid stays rotationally symmetric with respect to the axis perpendicular to the plane of motion. The size of the ellipsoid changes dramatically with time. So does its shape: it oscillates between prolate and oblate. In Figure 12.14b all three principal axes of the ellipsoid are in general different: the ellipsoid does not have rotational symmetry.

12.4 The Hydrogen Atom

The most fundamental application of quantum mechanics is atomic physics. The simplest atom is that of *hydrogen*; it consists of a simple nucleus, the proton, and one electron bound by the electric force acting between them. Since the mass of the proton is nearly 2000 times that of the electron, M, the center of mass of the atom, for our purposes, coincides with the position of the proton. We choose it to be the origin of our polar coordinate system. The potential energy of the electron, which carries the charge $-e$ in the electric field of the proton of charge $+e$, is given by the *Coulomb potential* of the proton,

$$U(\mathbf{r}) = \frac{e}{4\pi\epsilon_0}\frac{1}{r}$$

multiplied by the charge of the electron:

$$E_{\text{pot}} = V(r) = -\frac{e^2}{4\pi\epsilon_0}\frac{1}{r}$$

The constant $e^2/(4\pi\epsilon_0)$ has the dimension of action times velocity. It can therefore be expressed by a multiple of two fundamental constants of nature, namely Planck's constant $\hbar$ and the speed of light c. Inserting numbers, we obtain

$$\frac{e^2}{4\pi\epsilon_0} = \alpha\hbar c, \qquad \alpha = \frac{1}{137}$$

The dimensionless proportionality constant α is called the *fine-structure constant*. It was introduced by Arnold Sommerfeld in 1916.

The stationary Schrödinger equation for the hydrogen atom then has the form

$$\left(-\frac{\hbar^2}{2M}\nabla^2 - \hbar c\frac{\alpha}{r}\right)\varphi(\mathbf{r}) = E\varphi(\mathbf{r})$$

with M the electron mass. We solve this equation with the separation *ansatz* in polar coordinates,

$$\varphi(\mathbf{r}) = R(r)Y_{lm}(\vartheta, \phi)$$

which yields the radial Schrödinger equation for the hydrogen atom,

$$\left[-\frac{\hbar^2}{2M}\frac{1}{r}\frac{\partial^2}{\partial r^2}r + V_l^{\text{eff}}(r)\right]R_{nl}(r) = E_n R_{nl}(r)$$

It is an eigenvalue equation for the radial eigenfunctions R_{nl} with the energy eigenvalues E_n. The effective potential is the sum of the centrifugal potential and the Coulomb potential:

$$V_l^{\text{eff}}(r) = \frac{\hbar^2}{2M}\frac{l(l+1)}{r^2} - \hbar c\frac{\alpha}{r}$$

The energy eigenvalues E_n depend on the *principal quantum number n* only. We have

$$E_n = -\tfrac{1}{2}Mc^2\frac{\alpha^2}{n^2}, \qquad n = 1, 2, \ldots$$

They form an infinite set of discrete energies. The coefficient in this equation has the value $Mc^2\alpha^2/2 = 13.61$ eV.

The normalized radial wave functions R_{nl} have the form

$$R_{nl} = N_{nl}\left(\frac{2r}{na}\right)^{l} e^{-r/na} L_{n-l-1}^{2l+1}\left(\frac{2r}{na}\right)$$

$$n = 1, 2, 3, \ldots, \qquad l = 0, 1, 2, \ldots, n-1$$

with the normalization factor

$$N_{nl} = \frac{1}{a^{3/2}} \frac{2}{n^2} \sqrt{\frac{(n-l-1)!}{[(n+l)!]^3}}$$

Here the parameter

$$a = \frac{\hbar}{\alpha M c} = 0.5292 \cdot 10^{-10} \text{ m}$$

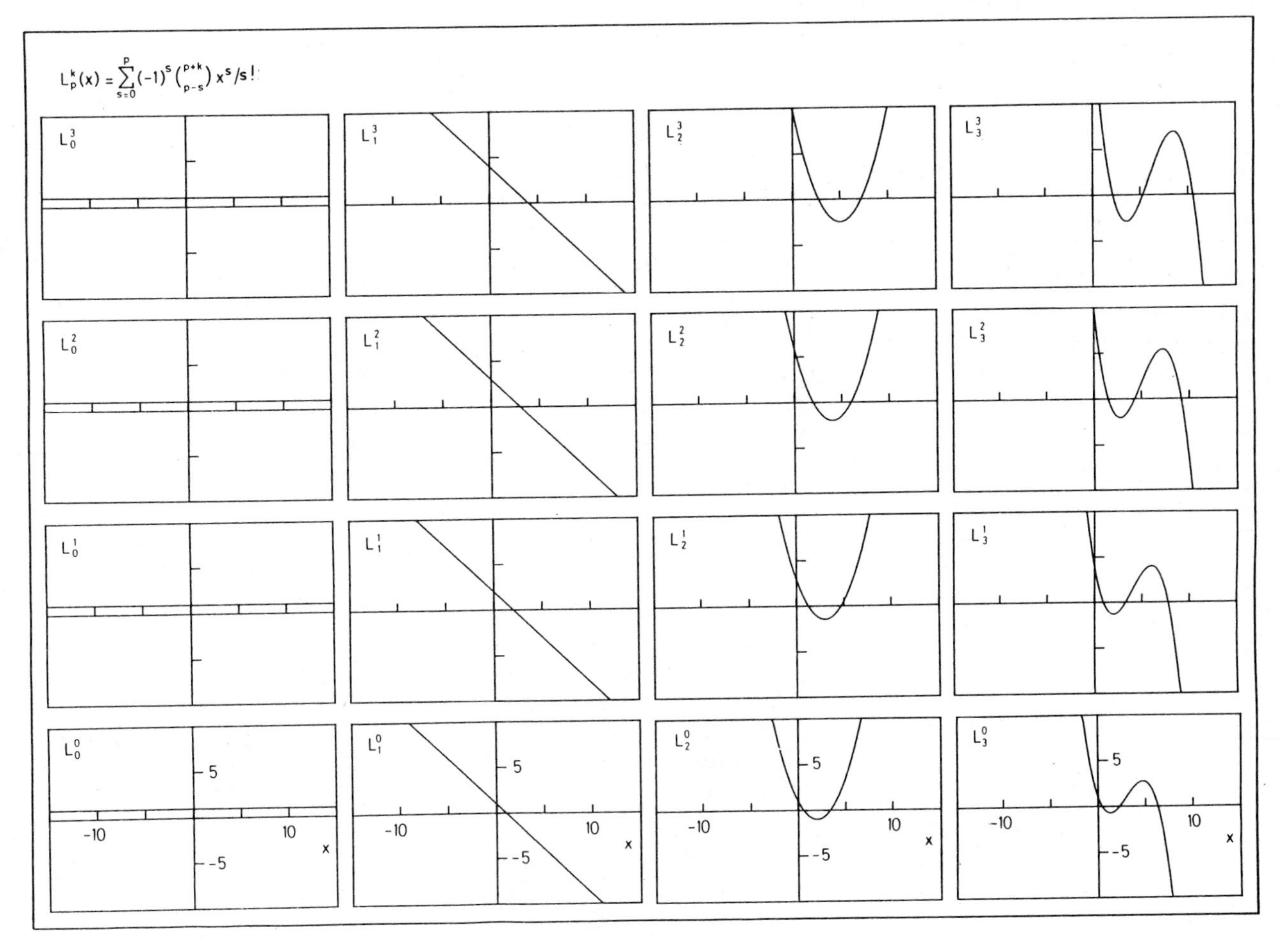

Figure 12.15 **Laguerre polynomials of integer upper index. The lower index is equal to the degree of the polynomial and to the number of its zeroes. All zeroes are at positive values of the argument *x*. The graphs look rather similar to those of Figure 12.7, which shows the Laguerre polynomials of half-integer upper index. That they are in fact different can be seen, for example, from the positions of the zeroes.**

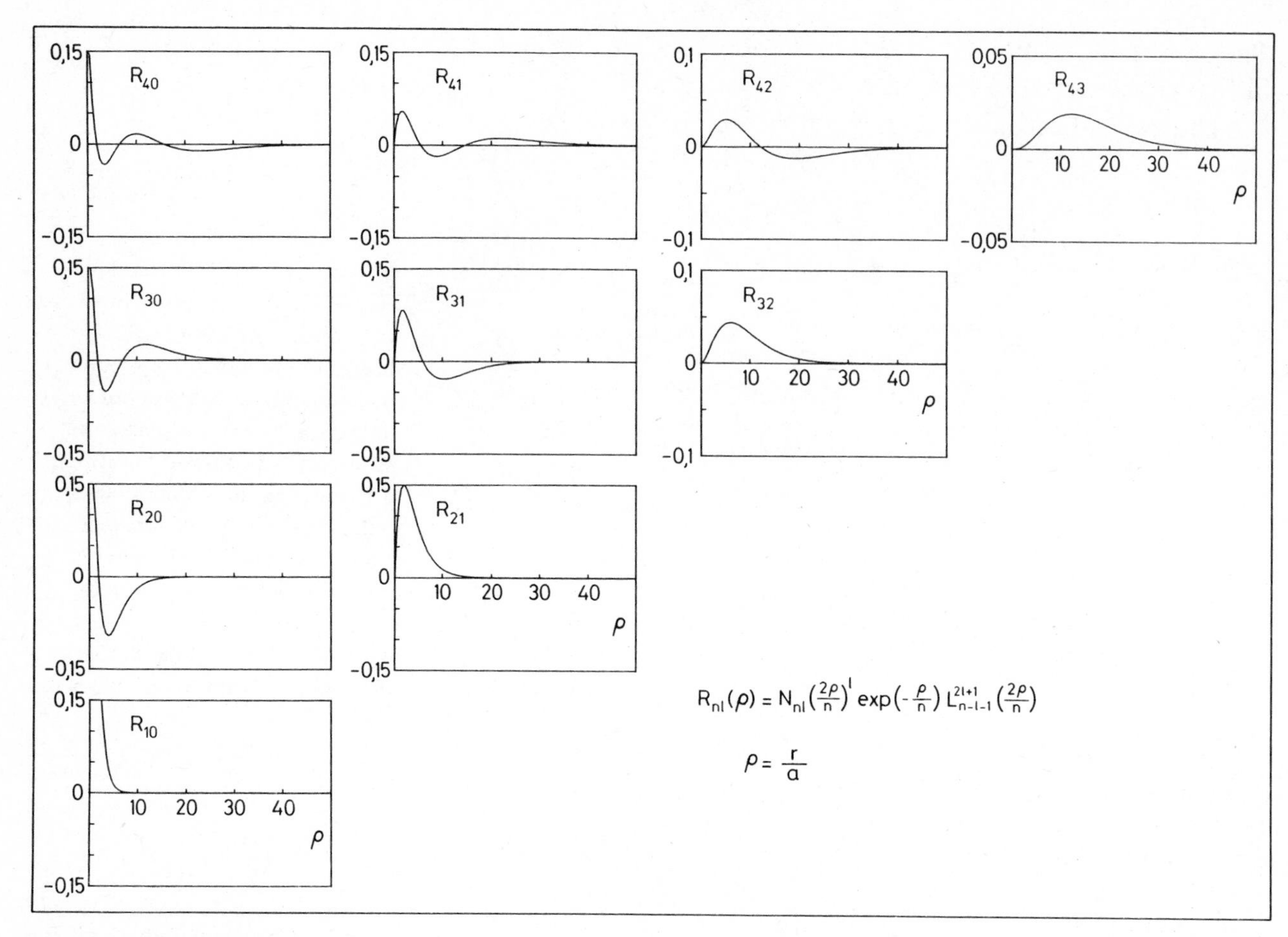

Figure 12.16 Radial eigenfunctions $R_{nl}(\rho)$ for the electron in the hydrogen atom. Their zeroes are in the $n - l - 1$ zeroes of the Laguerre polynomial $L^{2l+1}_{n-l-1}(2\rho / n)$. Here the argument of the Laguerre polynomial is $2\rho / n$ with n being the principal quantum number and $\rho = r / a$ the distance between electron and nucleus divided by the Bohr radius a.

is the Bohr radius of the innermost orbit. In the model of the hydrogen atom that was put forward by Niels Bohr in 1913, the electron can turn around the nucleus in circular orbits. These orbits can have only certain discrete radii $r_n = n^2\hbar/(\alpha Mc)$. The innermost orbit for $n = 1$ is $r_1 = a$.

The function $L^{2l+1}_{n-l-1}(x)$ is a particular *Laguerre polynomial*,

$$L^k_p(x) = \sum_{s=0}^{p} (-1)^s \binom{p+k}{p-s} \frac{x^s}{s!}$$

Figure 12.17 Radial eigenfunctions $R_{nl}(r)$, their squares $R^2_{nl}(r)$, and the functions $r^2R^2_{nl}(r)$ for the lowest eigenstates of the electron in the hydrogen atom and the lowest angular momentum quantum numbers $l = 0, 1, 2$. Also shown are the eigenvalue spectra, on the far right, the form of the Coulomb potential $V(r)$, and, for $l \neq 0$, the forms of the effective potential $V_{\text{eff}}(r)$. The eigenvalue spectra are degenerate for all l values, except that the minimum value of the principal quantum number is $n = l + 1$.

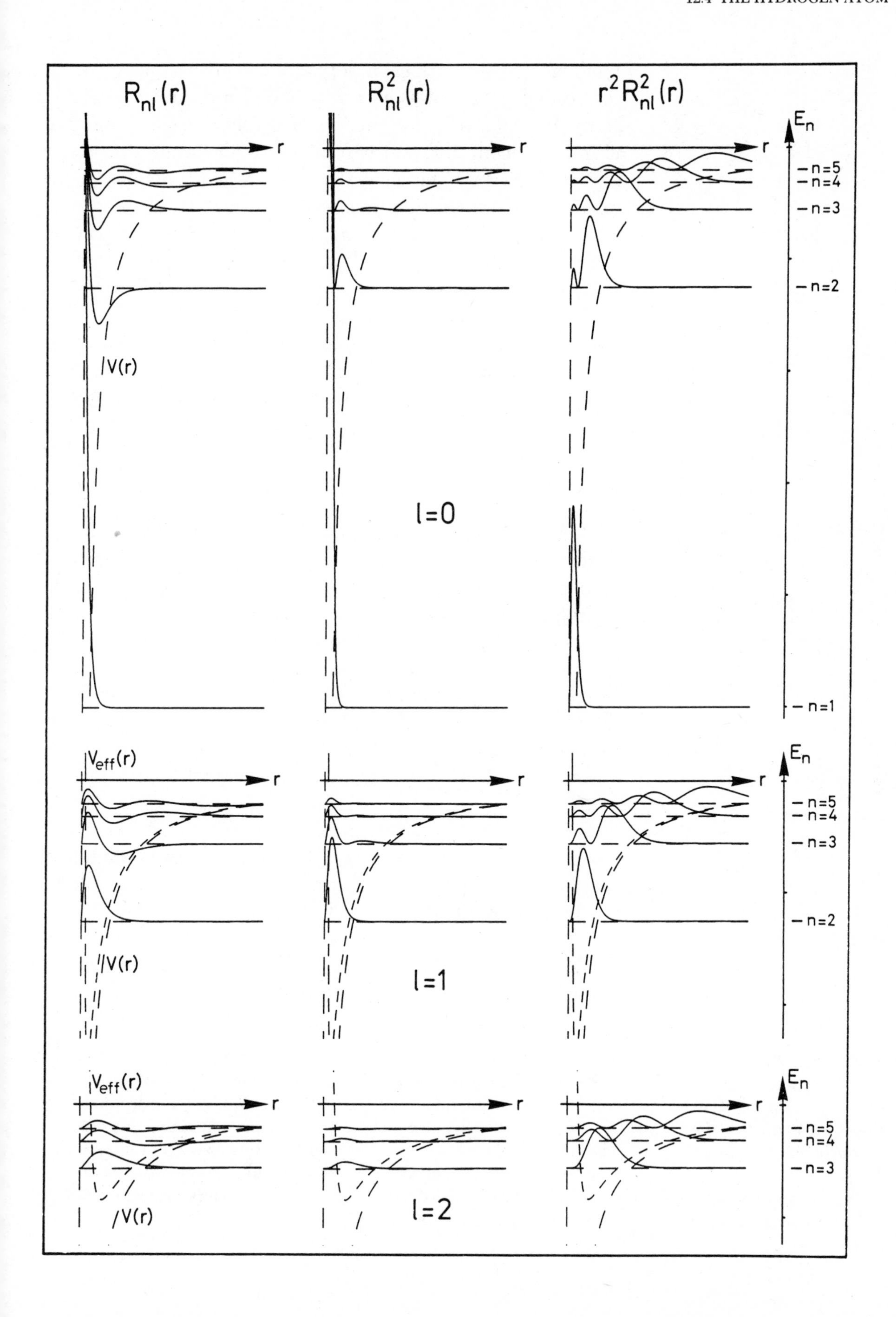
$R_{nl}(r)$
$R_{nl}^2(r)$
$r^2R_{nl}^2(r)$
r
E_n
n=5
n=4
n=3
n=2
n=1
V(r)
l=0
$V_{eff}(r)$
l=1
l=2

with the integer upper index $k = 2l + 1$. Some of these polynomials with low values of p and k are shown in Figure 12.15. We note that the number of zeroes equals p, and that all zeroes occur for positive values of the argument x. In Section 12.2 we found that the radial wave functions of the spherically symmetric harmonic oscillator contain the associated Laguerre polynomials $L_{n_r}^{l+1/2}(x)$ of half-integer upper index. They were plotted in Figure 12.7 for a few values of the indices. A comparison of Figures 12.15 and 12.7 reveals a strong similarity between the two sets of polynomials.

The radial wave functions R_{nl} for the electron in the hydrogen atom are shown in Figure 12.16. Their behavior for large values of r is dominated by the exponential $\exp[-r/(na)]$. Near $r = 0$ it is determined by the power $[2r/(na)]^l$. Their zeroes are those of the corresponding Laguerre polynomial, that is, the radial wave functions $R_{nl}(r)$ possess $n - l - 1$ zeroes.

Let us compare the radial wave functions of the hydrogen atom with those of the harmonic oscillator. We realize that with increasing energy eigenvalue the wave functions of the hydrogen atom spread much faster to larger radii than do those of the harmonic oscillator. Obviously, the reason is that the Coulomb potential becomes wide with total energy much more quickly than the harmonic oscillator potential does. This difference manifests itself in the analytic form of the wave functions of the two systems. The radial wave functions of the harmonic oscillator, as presented in Section 12.2, contain the factor $\exp[-r^2/(2\sigma_0^2)]$, which varies little for $r^2/(2\sigma_0^2) \ll 1$ and approaches zero very quickly for values $r^2/(2\sigma_0^2) > 1$. The radial wave functions of the hydrogen atom contain the factor $\exp[-r/(na)]$, which varies more strongly for $r/(na) \ll 1$ and falls off to zero much more slowly in region $r/(na) > 1$.

The spectrum of energy eigenvalues is highly degenerate because, for a given quantum number n, angular momentum

Figures 12.18 and 12.19 **The absolute squares $\rho_{nlm}(r,\vartheta) = |\varphi_{nlm}(r,\vartheta,\phi)|^2$ of the full three-dimensional wave functions for the electron in the hydrogen atom. They are functions only of r and ϑ. All eigenstates having the same principal quantum number have the same energy eigenvalue E_n. The possible angular momentum quantum numbers are $l = 0, 1, \ldots, n-1$. The wave functions have $n - l - 1$ nodes in r and $l - |m|$ nodes in ϑ, indicated by dashed half circles and rays, respectively.**

Each figure gives the probability density for observing the electron at any point in a half-plane containing the z-axis. All pictures have the same scale in r and ϑ. They do, however, have different scale factors in ρ.

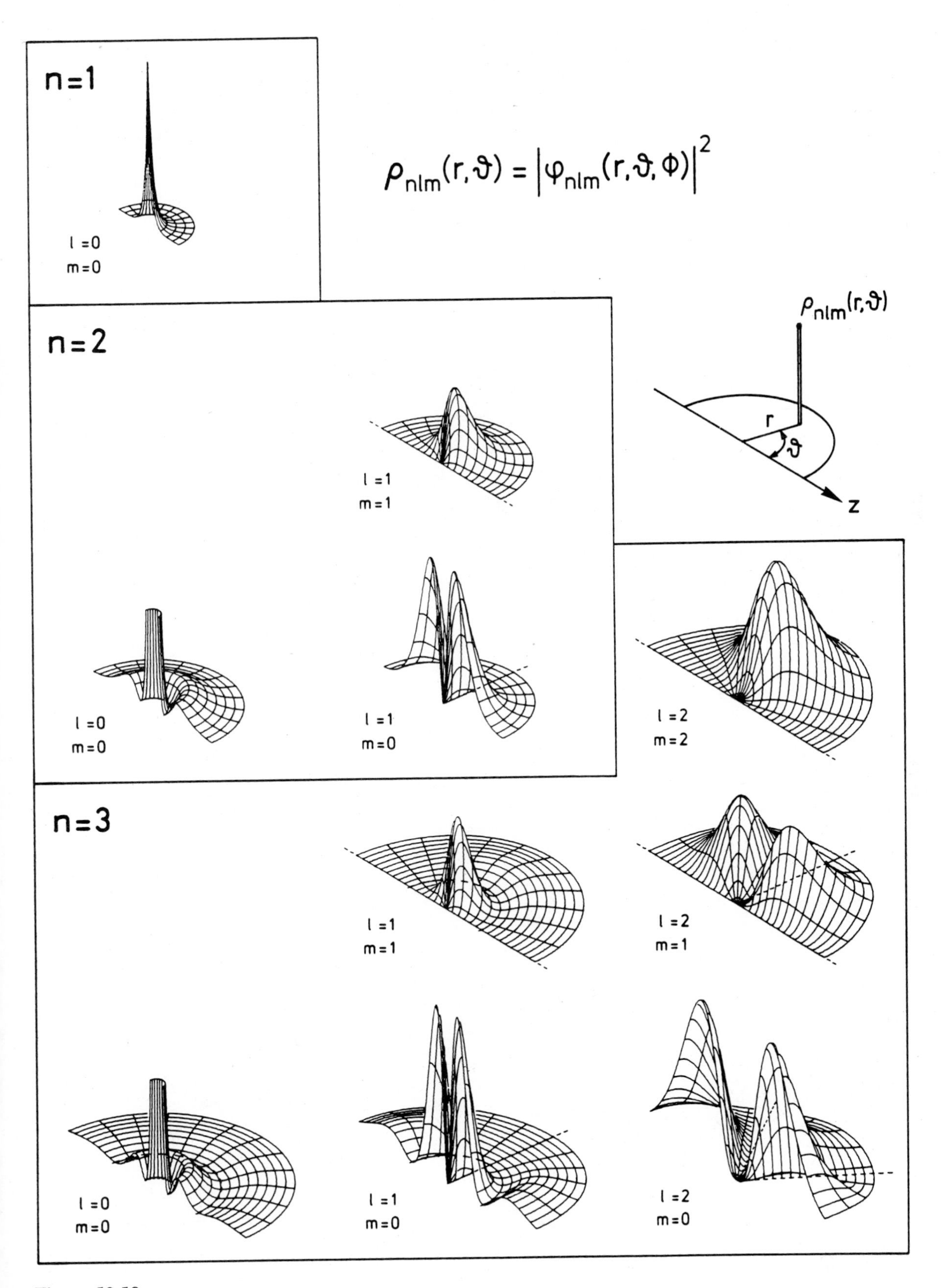
n = 1
l = 0
m = 0
$\rho_{nlm}(r,\vartheta) = \left|\varphi_{nlm}(r,\vartheta,\phi)\right|^2$
$\rho_{nlm}(r,\vartheta)$
r
ϑ
z
n = 2
l = 1
m = 1
l = 0
m = 0
l = 1
m = 0
l = 2
m = 2
n = 3
l = 1
m = 1
l = 2
m = 1
l = 0
m = 0
l = 1
m = 0
l = 2
m = 0

Figure 12.18

quantum number l can take any one of the values $0 \leq l \leq n-1$, and, for a given l, quantum number m of the z-component L_z of angular momentum runs between $-l \leq m \leq l$. Thus for a given n there are $\Sigma(2l+1) = n^2$ different states, all having the same eigenvalue E_n.

In Figure 12.17 the radial wave functions $R_{nl}(r)$ are shown together with the Coulomb potential $V(r)$ and the effective potential $V_{\text{eff}}(r)$ for the lowest values of the principal quantum number, $n = 1, \ldots, 5$, and for the lowest values of the angular momentum quantum number, $l = 0, 1, 2$. Figure 12.17 also contains the plots for $R_{nl}^2(r)$ and $r^2 R_{nl}^2(r)$.

It is interesting to compare the radial wave functions and the energy spectra for the three types of potentials discussed in this chapter, namely the square-well potential (Figure 12.1), the harmonic-oscillator potential (Figure 12.9), and the Coulomb potential (Figure 12.17). For the square-well potential, which vanishes in the external region, the wave functions fall off like an exponential function in this region. The Coulomb potential approaches zero with increasing r. The wave func-

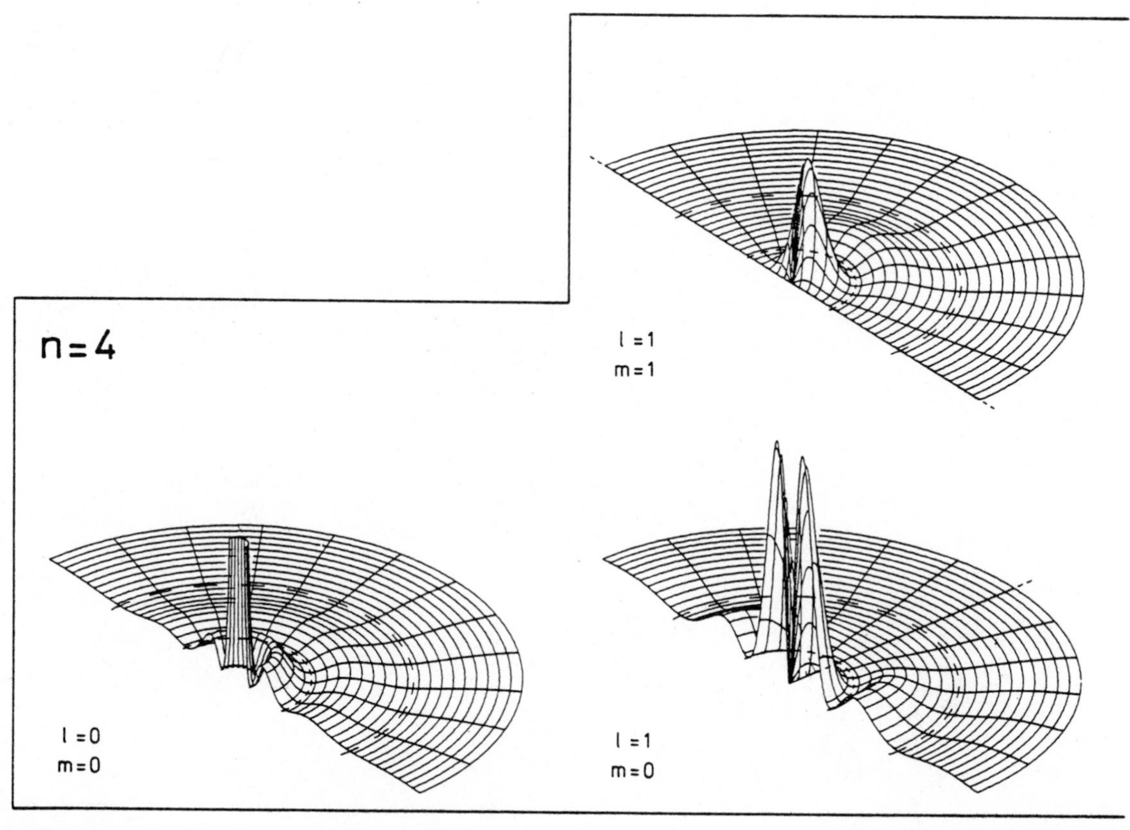

Figure 12.19

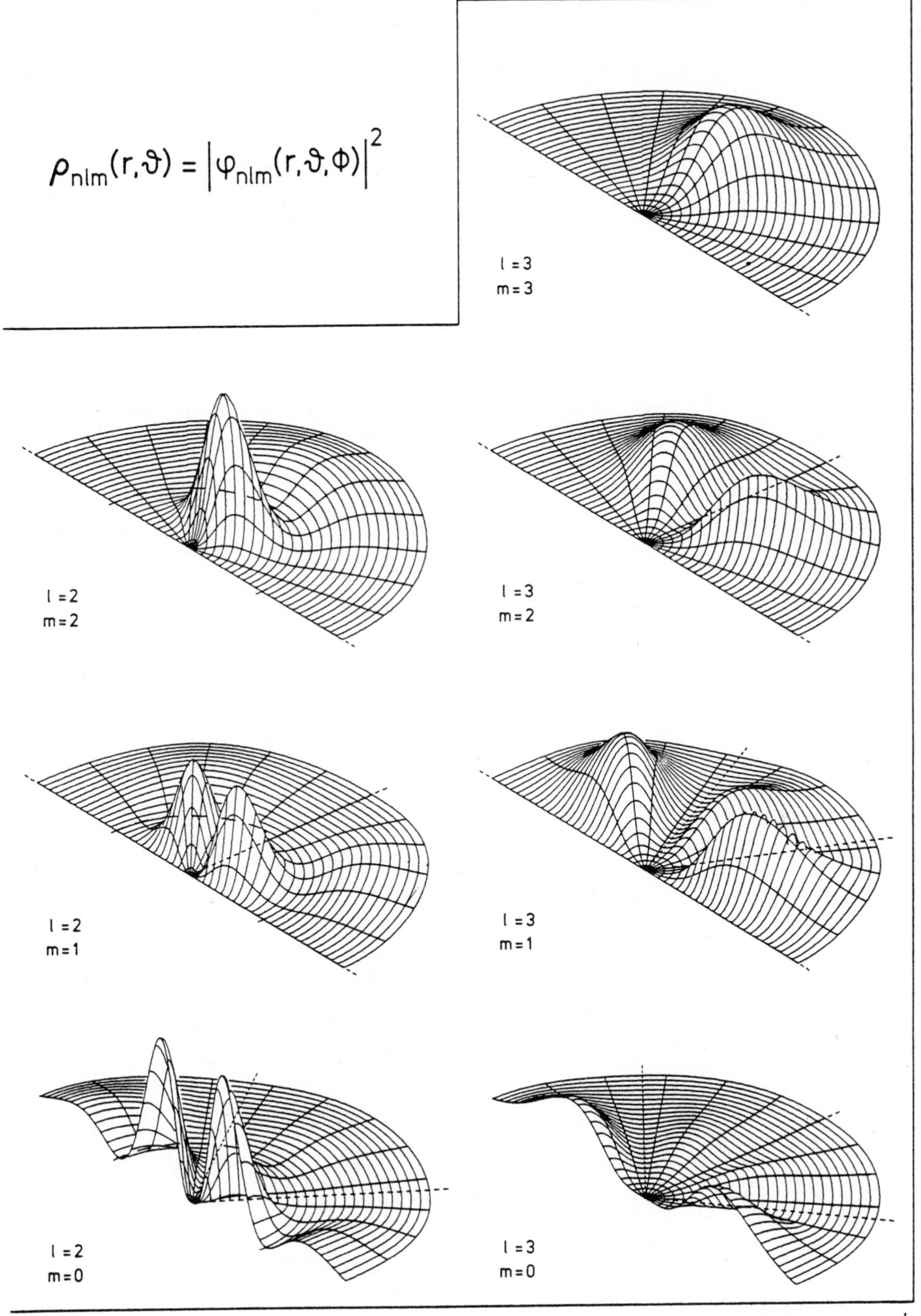
$\rho_{nlm}(r,\vartheta) = |\varphi_{nlm}(r,\vartheta,\Phi)|^2$
l = 3
m = 3
l = 2
m = 2
l = 3
m = 2
l = 2
m = 1
l = 3
m = 1
l = 2
m = 0
l = 3
m = 0
b

tions are products of polynomials and an exponential function, so that they taper off for large r like a power of r times the exponential function. Finally, for the quadratically increasing potential of the harmonic oscillator, the falloff of the wave functions is more pronounced. They behave like $r^n \exp(-r^2/2\sigma_0^2)$ for large r. This behavior reflects the intuitive expectation that an ever-increasing potential confines the particle best and that attractive potentials that approach zero better confine the particle the faster they reach zero.

Now, looking at the energy spectra, we observe that the spacing between levels increases with energy for the square-well potential, is equidistant for the harmonic oscillator, and decreases for the Coulomb potential.

Finally, we turn to the three-dimensional wave functions for the electron in the hydrogen atom,

$$\varphi_{nlm}(\mathbf{r}) = R_{nl}(r) Y_{lm}(\vartheta, \phi)$$

Figures 12.18 and 12.19 plot the absolute squares $|\varphi_{nlm}|^2$, which are independent of the azimuth. They show the probability densities for observing the electron as a function of r and the polar angle ϑ.

Problems

12.1 Calculate the energies E_n of the states of angular momentum zero for an infinitely deep potential well in three dimensions. Compare this spectrum with the one in Figure 12.1. Explain the deviations.

12.2 Why are the energies of the same quantum numbers n for $l = 1, 2$ in Figure 12.1 larger than those for $l = 0$?

12.3 Why does the energy of the lowest (in general, the nth) state decrease with increasing width of the spherical square-well potential of the same depth (Figure 12.2a)?

12.4 Why does the difference $E_{1l} - V_0$ of the state of lowest energy for a given angular momentum l increase as the potential well deepens?

12.5 Explain the structure of the product function $\varphi'_{210}(x_1, x_2, x_3)$, as plotted in Figure 12.6, in terms of the structures of the harmonic oscillator functions in one dimension, as plotted in Figure 6.3.

12.6 Why does the average probability density in a spherical shell of unit volume, given by $r^2 R_{nl}^2(r)$ as plotted in Figure 12.9, increase toward the harmonic oscillator wall?

12.7 What do you expect the harmonic oscillator probability densities for $n = 5$, $l = 0, 1, 2, 3$ to look like roughly? Describe their nodes in r and ϑ in analogy to those in Figures 12.10, 12.11, and 12.12.

12.8 Verify by explicit calculation that the angular momentum eigenstate $\varphi_{220}(\mathbf{r})$ of the harmonic oscillator can be decomposed as was done in Figure 12.13.

12.9 Let us consider time-dependent motion in a rotationally symmetric harmonic oscillator. The wave function of the initial state $\psi(\mathbf{r}, 0)$ at $t = 0$ is given explicitly in terms of a decomposition into eigenfunctions $\varphi'_{n_1 n_2 n_3}(x_1, x_2, x_3)$,

$$\psi(\mathbf{r}, 0) = \sum_{n_1, n_2, n_3} a_{n_1 n_2 n_3} \varphi'_{n_1 n_2 n_3}(x_1, x_2, x_3)$$

corresponding to the eigenvalues $E_n = (n + \frac{1}{2})\hbar\omega$, $n = n_1 + n_2 + n_3$, of the harmonic oscillator as described in Section 12.2. The $a_{n_1 n_2 n_3}$ are the spectral coefficients of the initial state in the harmonic oscillator base $\varphi'_{n_1 n_2 n_3}$. Show that the time-dependent wave function ($n = n_1 + n_2 + n_3$)

$$\psi(\mathbf{r}, t) = \sum_{n_1, n_2, n_3} e^{-i/\hbar E_n t} a_{n_1 n_2 n_3} \varphi'_{n_1 n_2 n_3}(x_1, x_2, x_3)$$

is a solution of the time-dependent Schrödinger equation

$$i\hbar \frac{\partial}{\partial t}\psi(\mathbf{r}, t) = \left(-\frac{\hbar^2}{2M}\nabla^2 + \frac{k}{2}r^2\right)\psi(\mathbf{r}, t)$$

and fulfills the initial condition.

12.10 Analyze the behavior of the three-dimensional wave packet under the influence of a harmonic force, as plotted in Figure 12.14, in terms of the behavior of three independent one-dimensional oscillators, as plotted in Figures 6.5 and 6.7. Describe the initial conditions of these independent oscillators in terms of classical mechanics.

12.11 Show that the general solution $\psi(\mathbf{r}, t)$ for the motion in a harmonic oscillator,

$$\psi(\mathbf{r}, t) = \sum_{n_1, n_2, n_3} \exp\left[-\frac{i}{\hbar}E_n t\right] a_{n_1 n_2 n_3} \varphi'_{n_1 n_2 n_3}(x_1, x_2, x_3)$$

with the energy of the state $\varphi'_{n_1 n_2 n_3}$,

$$E_n = (n + \tfrac{3}{2})\hbar\omega, \qquad n = n_1 + n_2 + n_3$$

possesses the following periodicity property:

$$\psi\left(\mathbf{r}, t + m\frac{2\pi}{\omega}\right) = e^{-im\pi}\psi(\mathbf{r}, t), \qquad m = 1, 2, 3, \ldots$$

The periodicity property implies that

$$\left|\psi\left(\mathbf{r}, t + m\frac{2\pi}{\omega}\right)\right|^2 = |\psi(\mathbf{r}, t)|^2$$

This result can be read off Figure 12.14.

12.12 Calculate the minimum of the effective potential, $V_{2,\min}^{\text{eff}}$, for $l = 2$ of the hydrogen atom,

$$V_2^{\text{eff}} = \frac{\hbar^2}{2M}\frac{6}{r^2} - \hbar c\frac{\alpha}{r}$$

Find the differences between the eigenvalues E_n of the electron in the hydrogen atom and this minimum, $E_n - V_{2,\min}^{\text{eff}}$. Explain why only states with $n \geq 3$ exist for angular momentum $l = 2$ in the hydrogen atom.

12.13 Show that the Bohr radius a as given in Section 12.4 is at the position of the maximum of $r^2 R_{10}(r)$, that is, show that, at $r = a$,

$$\frac{d}{dr}[r^2 R_{10}(r)] = 0$$

12.14 The energy of the ground state of a two-particle system bound by a Coulomb force is

$$E_1 = -\tfrac{1}{2}\mu c^2 \alpha^2$$

where $\mu = M_1 M_2 (M_1 + M_2)$ is the reduced mass of the system of two particles of masses M_1 and M_2. For $M_1 \ll M_2$, μ tends toward M_1. Using this formula, calculate the ground-state energy E_1 for the hydrogen atom and for a positronium, which is a system of an electron and a positron, that is, an electron of positive charge.

12.15 Muons are particles similar to the electron but possessing a mass

$$m_\mu = 105.6\ \mathrm{MeV}/c^2$$

The Bohr radius, that of the innermost orbit, of a system made up of a positive and negative charge is

$$a = \frac{\hbar^2}{\alpha\mu c}$$

where μ is the reduced mass of the system, as given in the preceding problem. Calculate the Bohr radii for a hydrogen atom; for a muonic hydrogen atom, whose electron has been replaced by a muon; and for positronium, a hydrogenlike system in which the proton has been replaced by a positron.

12.16 The Bohr radius a of the innermost orbit of a nucleus with atomic number Z is

$$a = \frac{\hbar^2}{Z\alpha\mu c}$$

where μ is the reduced mass of the system. For the uranium nucleus, $Z = 92$, the reduced mass can safely be taken as the mass of the particle in the innermost orbit. Calculate the Bohr radius for a muonic uranium atom and compare the result with the radius of the uranium nucleus, $r_0 \approx 6 \cdot 10^{-15}$ m.

13. Three-Dimensional Quantum Mechanics: Resonance Scattering

13.1 Scattering by Attractive Potentials

We now return to the discussion of scattering in three dimensions. In Chapter 11 we looked only at scattering by repulsive potentials. Now we shall study the effects of an attractive potential.

In Figure 13.1 the wave function $\varphi_{\mathbf{k}}^{(+)}(\mathbf{r})$ is shown in terms of its real part, imaginary part, and absolute square. The figure is analogous to Figure 11.1, except for the sign of the square-well potential in region I. In comparing Figures 11.1 and 13.1, we observe no striking differences except that in region I where the potential is nonzero the probability density $|\varphi_{\mathbf{k}}^{(+)}|^2$ is appreciably larger for the attractive potential. This larger probability density was to be expected since for the repulsive potential the particle can enter region I only by the tunnel effect.

Figure 13.2 presents the scattered spherical wave $\eta_{\mathbf{k}}(\mathbf{r})$, as defined in Section 11.2. Again we observe that the plot of $|\varphi_{\mathbf{k}}^{(+)}(\mathbf{r})|^2$ has a ripple structure whereas the plot of $|\eta_{\mathbf{k}}(\mathbf{r})|^2$ does not. As discussed at the end of Section 11.1, the ripples of $|\varphi_{\mathbf{k}}^{(+)}(\mathbf{r})|^2$ are caused by the interference of the incident wave $\exp(i\mathbf{k} \cdot \mathbf{r})$ and the scattered spherical wave $\eta_{\mathbf{k}}(\mathbf{r})$. The absolute square of $\eta_{\mathbf{k}}(\mathbf{r})$ shows no such ripples, and for larger r there is only a $|f(\vartheta)|^2/r^2$ falloff.

Comparing the two sets of pictures for the attractive and repulsive potentials (Figures 13.1 and 11.1), we realize that the forward scattering, the scattering into angles ϑ close to

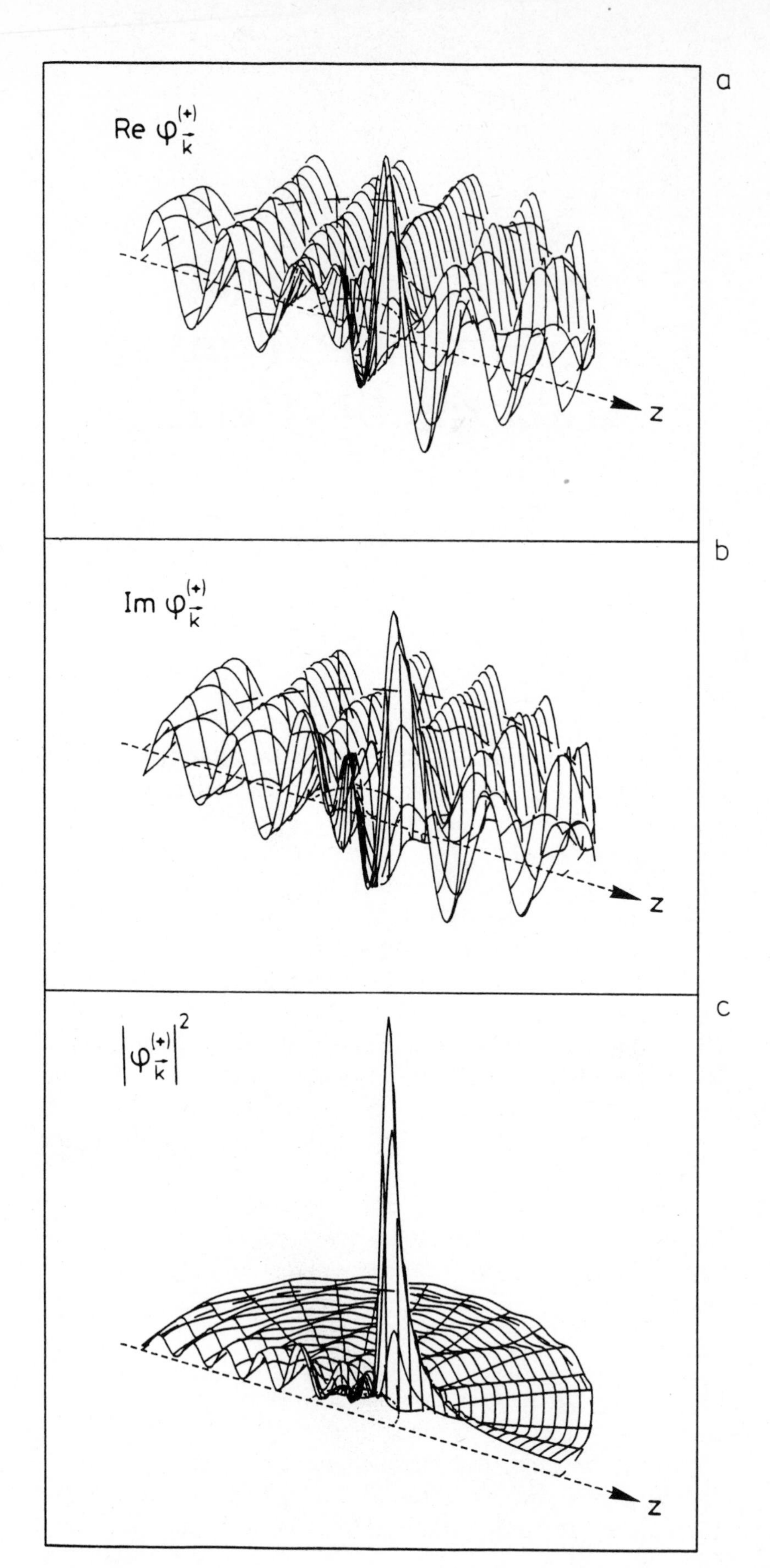

Figure 13.1 **Scattering of a plane wave incident from the left along the z-direction by an attractive potential. The potential is confined to region $r < d$ indicated by the small half circle. Shown are (a) the real part, (b) the imaginary part, and (c) the absolute square of the wave function $\varphi^{(+)}_{\mathbf{k}}$. The figure corresponds exactly to the situation of Figure 11.1, except for the change $V_0 \rightarrow -V_0$ in the scattering potential.**

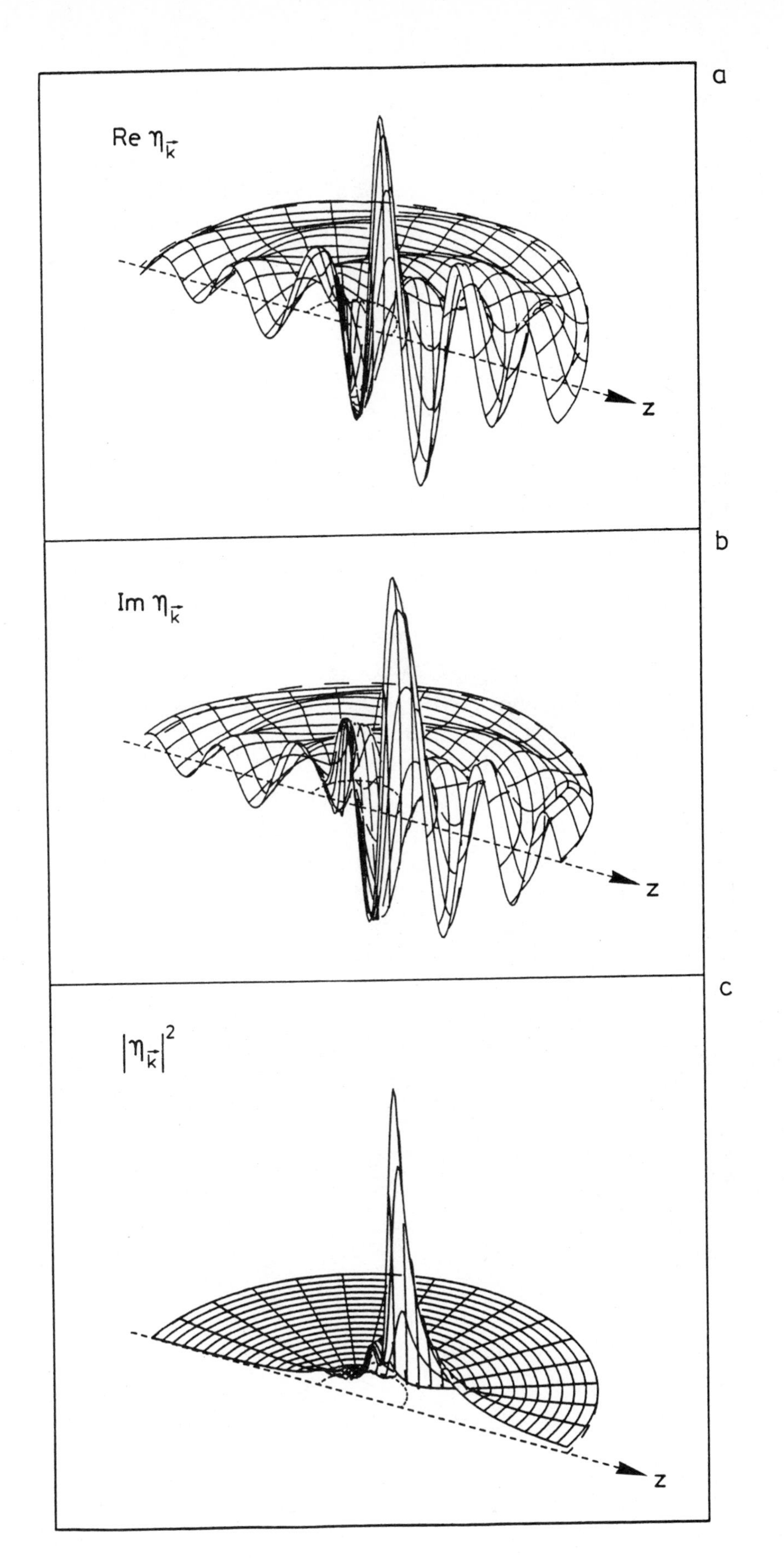

Figure 13.2 **(a) Real part, (b) imaginary part, and (c) absolute square of the scattered spherical wave η_k resulting from the scattering of a plane wave by an attractive potential, as shown in Figure 13.1.**

zero, is for the repulsive potential only shadow scattering. In other words, immediately beyond the repulsive square well there is very little probability of finding the particle, whereas there is considerable probability that the particle has transversed the attractive square-well region.

13.2 Resonance Scattering

In the preceding example the energy of the incoming wave was chosen at random. Let us now consider the scattering of a plane wave at a particular energy $E_{\rm res}$ by the attractive potential used in Figures 13.1 and 13.2. A systematic way for determining the particular energy $E_{\rm res}$ will be presented in Section 13.3.

Real and imaginary parts of the wave function $\varphi_{\mathbf{k}}^{(+)}(\mathbf{r})$ with particular energy $E_{\rm res}$ are plotted in Figure 13.3 together with $|\varphi_{\mathbf{k}}^{(+)}|^2$. Unlike the situation in Figure 13.1, there is now a rather symmetric structure in the region of the attractive potential. This symmetry is also apparent in the plots of the scattered spherical wave $\eta_{\mathbf{k}}(\mathbf{r})$ in Figure 13.4.

To clarify the origin of the dominating symmetric structure, we inspect the scattered partial waves η_l, as introduced in Section 11.2. Their real and imaginary parts are plotted in Figure 13.5, revealing the dominant contribution of the scattered partial wave for angular momentum $l = 3$. Since scattered partial waves are significantly different from zero only for low values of l—in our example for $l = 0, 1, 2, 3$,—clearly close to the potential region wave η_3 dominates the wave function $\varphi_{\mathbf{k}}^{(+)}$ as well as the scattered spherical wave $\eta_{\mathbf{k}}$.

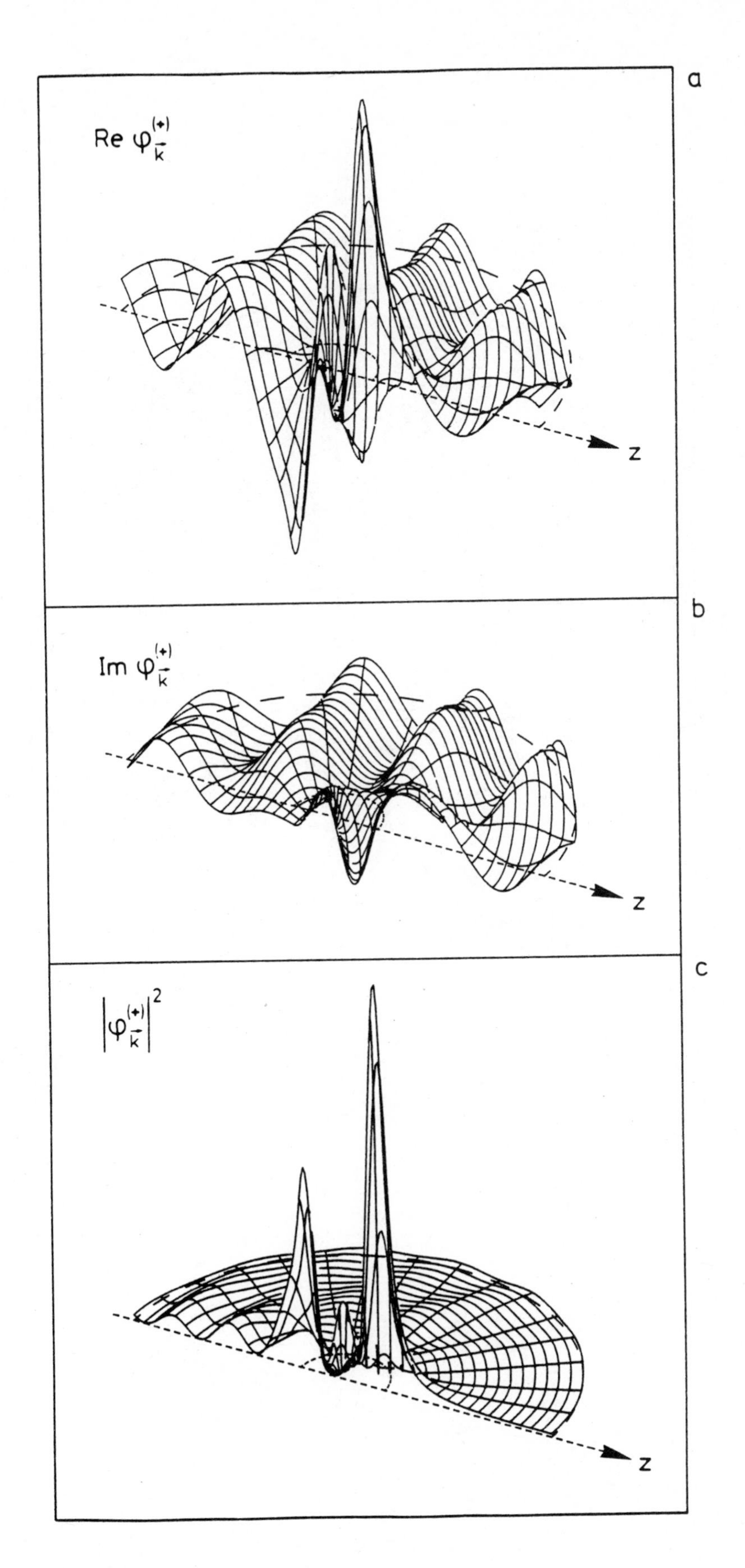

Figure 13.3 **(a) Real part, (b) imaginary part, and (c) absolute square of the wave function $\varphi_{\mathbf{k}}^{(+)}$ for the scattering of a plane wave by an attractive potential as given in Figure 13.1, but for a resonance energy $E = E_{\text{res}}$ of the wave.**

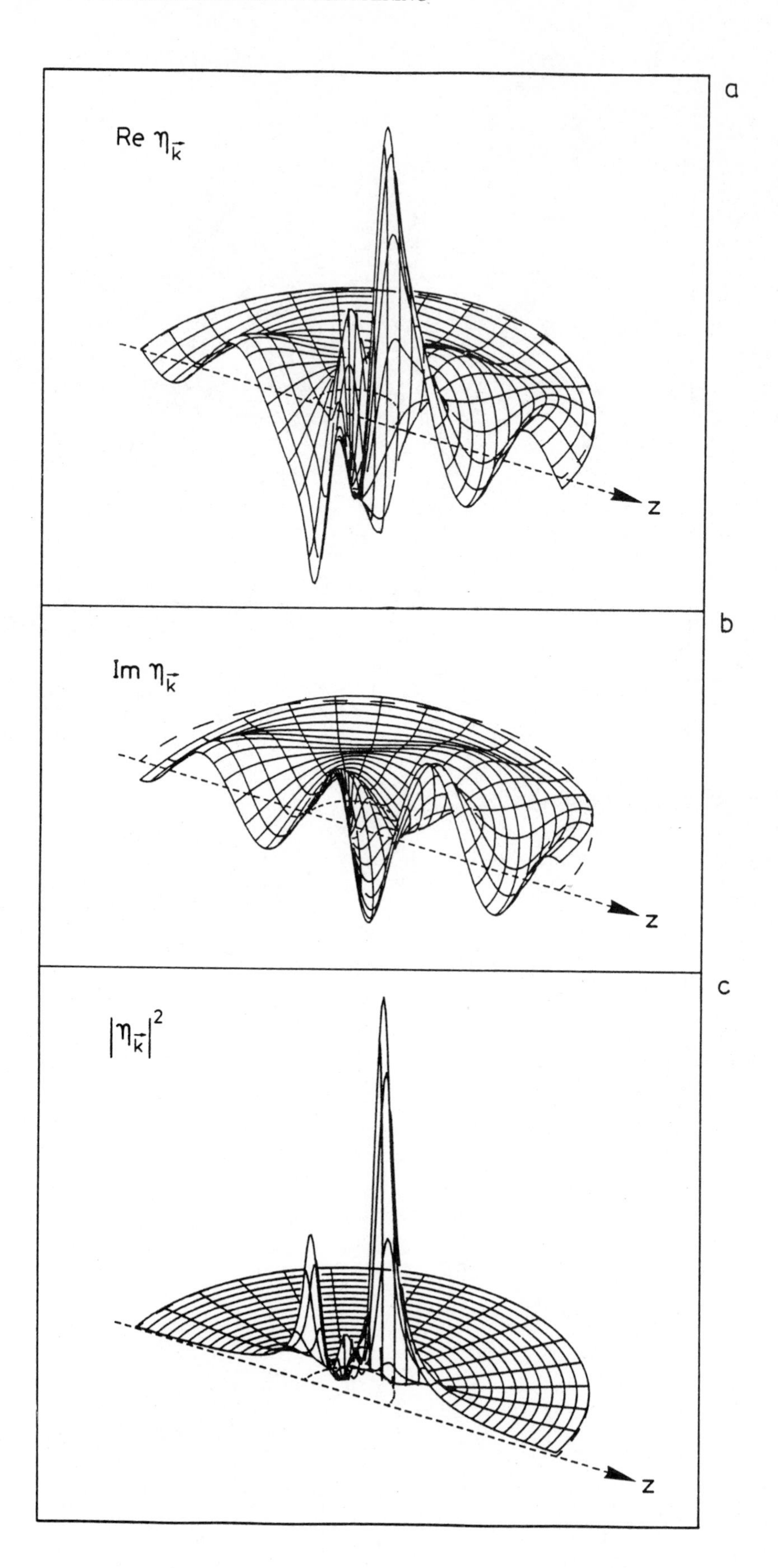

Figure 13.4 **(a) Real part, (b) imaginary part, and (c) absolute square of the scattered spherical wave η_k resulting from the scattering of a plane wave of resonance energy $E = E_{res}$ by an attractive potential.**

13.3 Phase Shift Analysis

In this section we investigate the energy dependence of the partial cross sections $\sigma_l(E)$, the phase shifts $\delta_l(E)$, and the partial scattering amplitudes $f_l(E)$ for scattering by an attractive potential. The parameters of the potential are the same as those already used in Figures 13.3 through 13.5.

In Figure 13.6a the partial cross sections are shown as a function of energy for $l = 0, 1, \ldots, 5$. The striking feature of this figure is the rather pronounced maximum in the energy dependence of σ_3. This maximum produces a peak in the total cross section σ_{tot}, shown in the background of Figure 13.6a. The energy value of this maximum is very near the energy E_{res} at which we observed the striking structure in $\eta_3(\mathbf{r})$ in Figure 13.5. It was this structure that dominated the functions $\varphi_{\mathbf{k}}^{(+)}(\mathbf{r})$ and $\eta_{\mathbf{k}}(\mathbf{r})$. To investigate this phenomenon further, we study the behavior of the phase shifts $\delta_l(E)$ in Figure 13.6b. Except for $l = 3$, the phase shifts show a rather smooth energy dependence. The phase shift δ_3, however, rises sharply in the neighborhood of E_{res}, crossing the value $3\pi/2$ at E_{res}. From the phase shifts δ_l we now construct the complex partial scattering amplitudes f_l, as described in Section 11.3.

Figure 13.7 shows the corresponding Argand diagrams for the complex functions $f_l(E)$. The Argand diagram for $f_3(E)$ shows a swift counterclockwise motion of point f_3 in the complex plane as the energy passes through the energy E_{res}. As we have learned from the examples of one-dimensional scattering (Section 5.4), this is the signature for a resonance scattering process. As the phase ascends through $\pi/2$ or $3\pi/2$, the real part passes through zero in a sharp decrease, whereas the imaginary part reaches its maximum, $\operatorname{Im} f_l = 1$. Figure 13.7 indicates that none of the scattering amplitudes f_0, f_1, and f_2 has a resonance.

Of particular interest is the differential scattering cross section

$$\frac{d\sigma}{d\Omega} = |f(\vartheta)|^2$$

with

$$f(\vartheta) = \frac{1}{k} \sum_{l=0}^{\infty} (2l + 1) f_l(k) P_l(\cos\vartheta)$$

The differential cross section is used to measure the angular momentum of resonances. If near the resonance energy the

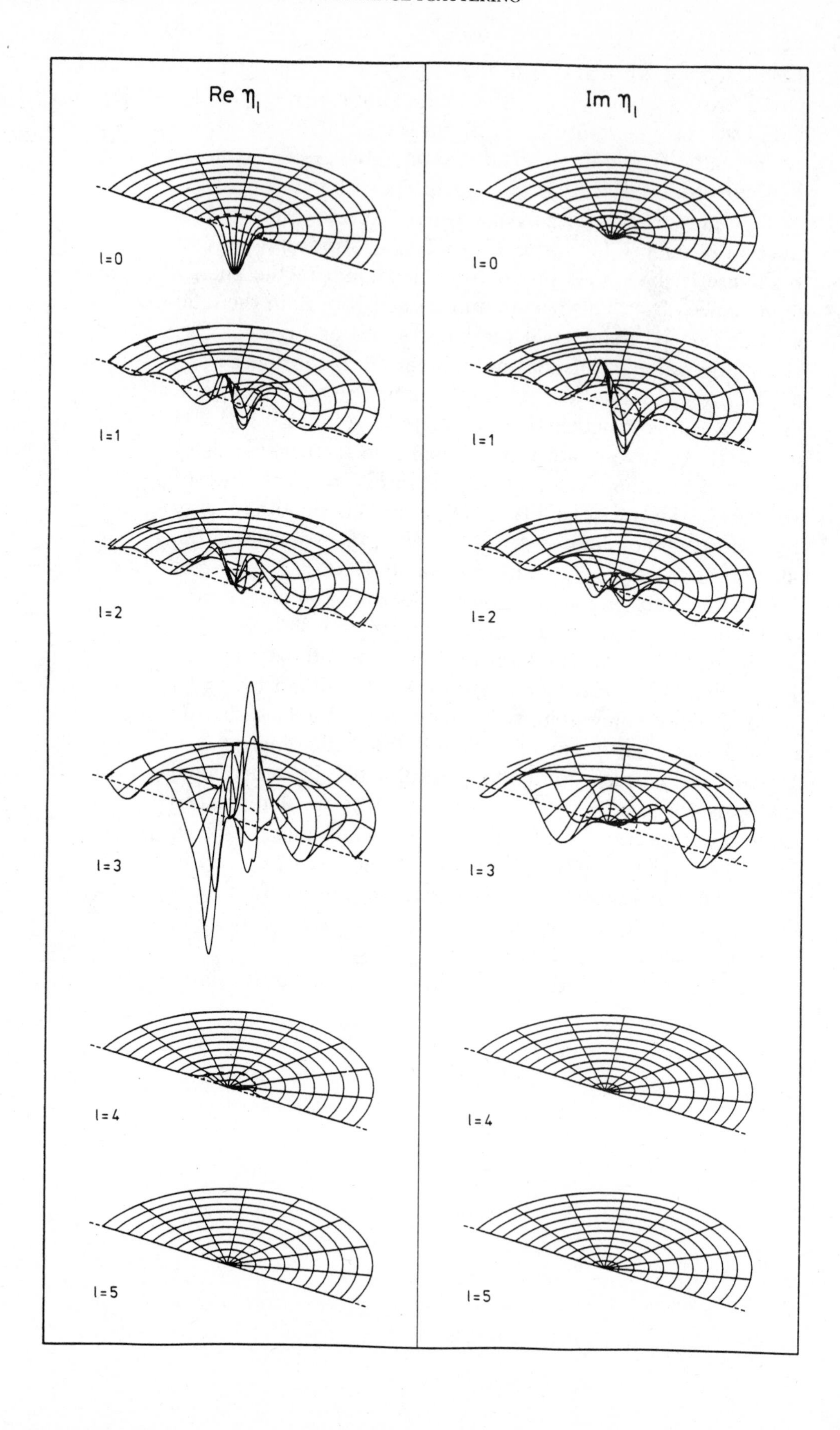
Re η_l
Im η_l
l=0
l=0
l=1
l=1
l=2
l=2
l=3
l=3
l=4
l=4
l=5
l=5

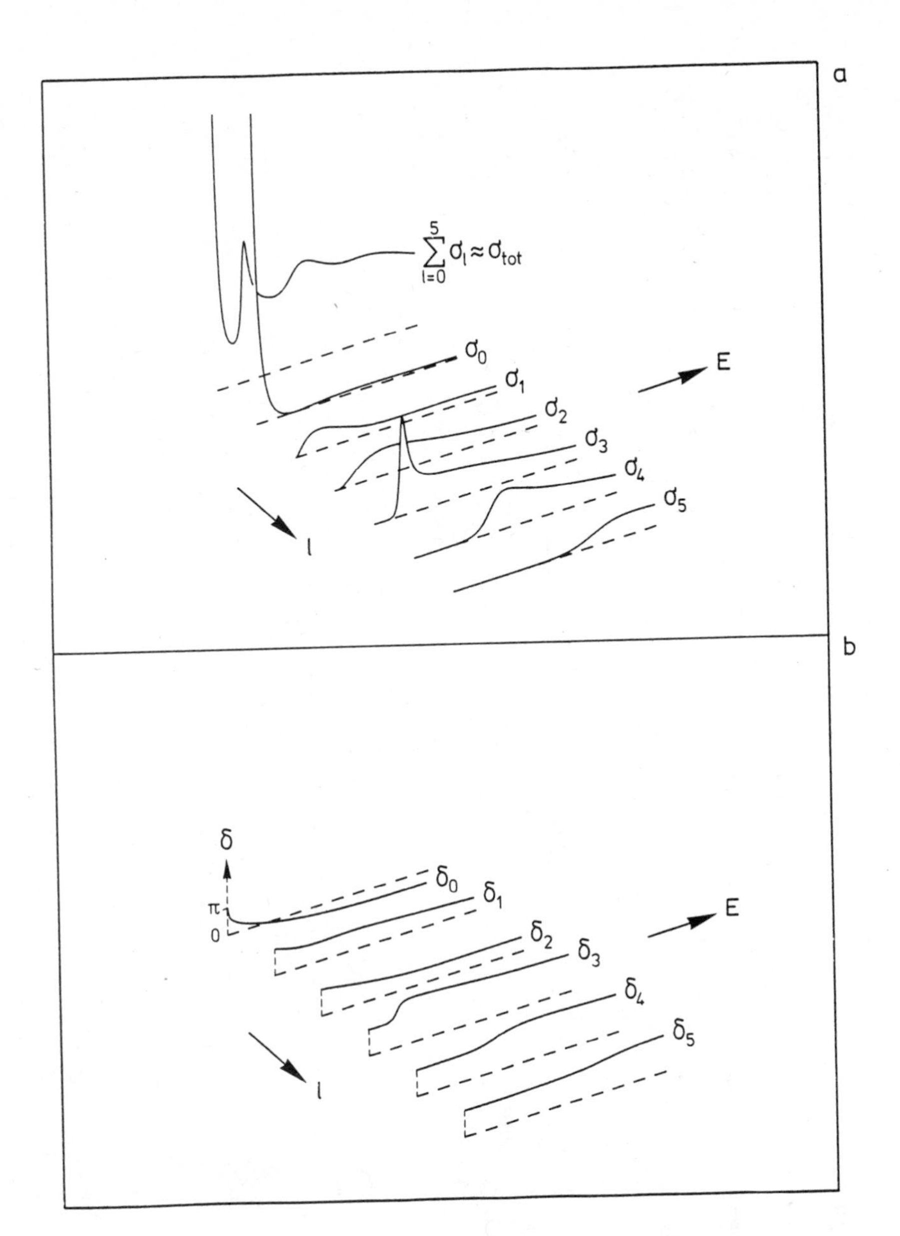

Figure 13.6 **(a) The partial cross sections $\sigma_l(E)$ and the total cross section $\sigma_{tot}(E)$ approximated by the sum over the first few partial cross sections for the scattering of a plane wave of energy E by the attractive potential used in Figures 13.1 through 13.5. For resonant energy $E = E_{res}$ there is a sharp maximum in σ_3, which is reflected in σ_{tot}. (b) The corresponding phase shifts $\delta_l(E)$. For $E = 0$ we put $\delta_l(0) = \pi$. All phase shifts except δ_3 vary only slowly with energy. Near $E = E_{res}$ the phase shift $\delta_3(E)$ rises sharply, passing through $\delta_3(E = E_{res}) = 3\pi/2$.**

Figure 13.5 **Real and imaginary parts of the scattered partial waves η_l resulting from the scattering of a plane wave of resonance energy $E = E_{res}$ by an attractive potential. The resonance is in the partial wave for $l = 3$.**

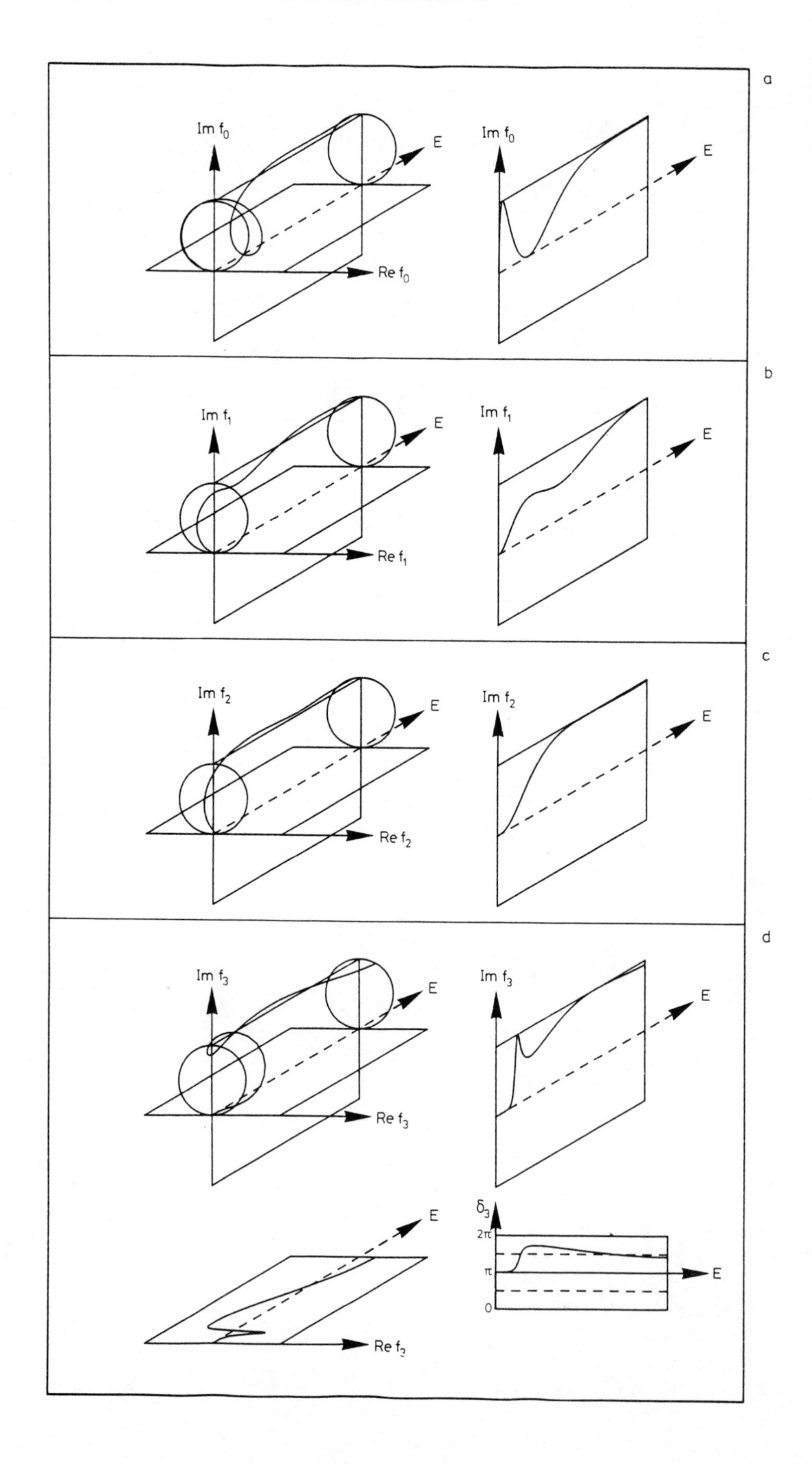
a
Im f_0
E
Re f_0
Im f_0
E
b
Im f_1
E
Re f_1
Im f_1
E
c
Im f_2
E
Re f_2
Im f_2
E
d
Im f_3
E
Re f_3
Im f_3
E
E
Re f_3
δ_3
2π
π
0
E

absolute values of all partial scattering amplitudes f_l except the resonant one are small, the differential cross section is determined by the square of the Legendre polynomial corresponding to the angular momentum of the resonance. This is the case for our example. At the resonance energy we expect the differential cross section to be approximately proportional to $[P_3(\cos\vartheta)]^2$. Figure 13.8b shows the differential cross section as a function of $\cos\vartheta$ for various energies. For the resonance energy it is indeed very similar to $(P_3)^2$, as we can see by comparing this figure with Figure 9.2. In Figure 13.8a the intensity of the scattered wave at the resonance energy is plotted over a half circle in a plane containing the z-axis. The detectors measuring the flux of scattered particles could be situated on this half circle.

Often the background from nonresonant amplitudes is not small. Then the resonant amplitude varies rapidly with energy close to E_{res}. By careful analysis of the angular distribution, it is often possible to separate resonant and nonresonant partial-wave contributions in the differential cross section and thus to measure the angular momentum of a resonance that has already been seen in the total cross section.

So far in this section we have studied the phase shifts δ_l and the quantities derived from them that describe the scattering globally. We now turn to the detailed features of the radial wave functions $R_l(k,r)$. Here $k = \sqrt{2ME}/\hbar$ is the wave number for a vanishing potential, as introduced in Section 10.1. In Figure 13.9 the energy dependence of these functions is shown for $l = 0, 1, \ldots, 3$. We observe that R_0, R_1, and R_2 do not change appreciably with energy except for the decrease in the wavelength that is clearly visible in the region outside the potential. The wave function R_3, however, changes its shape rapidly as the energy varies. At the resonance energy it has a pronounced maximum within the

Figure 13.7 Argand diagrams, that is, diagrams of the energy dependence of the complex partial scattering amplitudes $f_l(E)$, for the scattering of a plane wave of energy E by the attractive potential used in Figures 13.1 through 13.6 for $l = 0, 1, 2, 3$. For the nonresonant partial waves, $l = 0, 1, 2$, only the Argand diagram itself and its projection on the $\text{Im}\, f_l$, E-plane are shown. The function $\text{Im}\, f_l(E)$ is closely related to the partial cross section $\sigma_l(E)$. For resonant wave $l = 3$ both $\text{Im}\, f_l(E)$ and $\text{Re}\, f_l(E)$ projections and the phase shift $\delta_3(E)$ are shown. Near resonance energy $E = E_{\text{res}}$ the partial scattering amplitude $f_3(E)$ performs a swift counterclockwise motion through point 0, 1 in the complex plane, giving rise to (1) the pronounced maximum in $\text{Im}\, f_3(E_{\text{res}})$; (2) the steep drop of $\text{Re}\, f_3(E)$ through $\text{Re}\, f_3(E_{\text{res}}) = 0$, and (3) the sharp rise of $\delta_3(E)$ through $\delta_l(E_{\text{res}}) = 3\pi/2$.

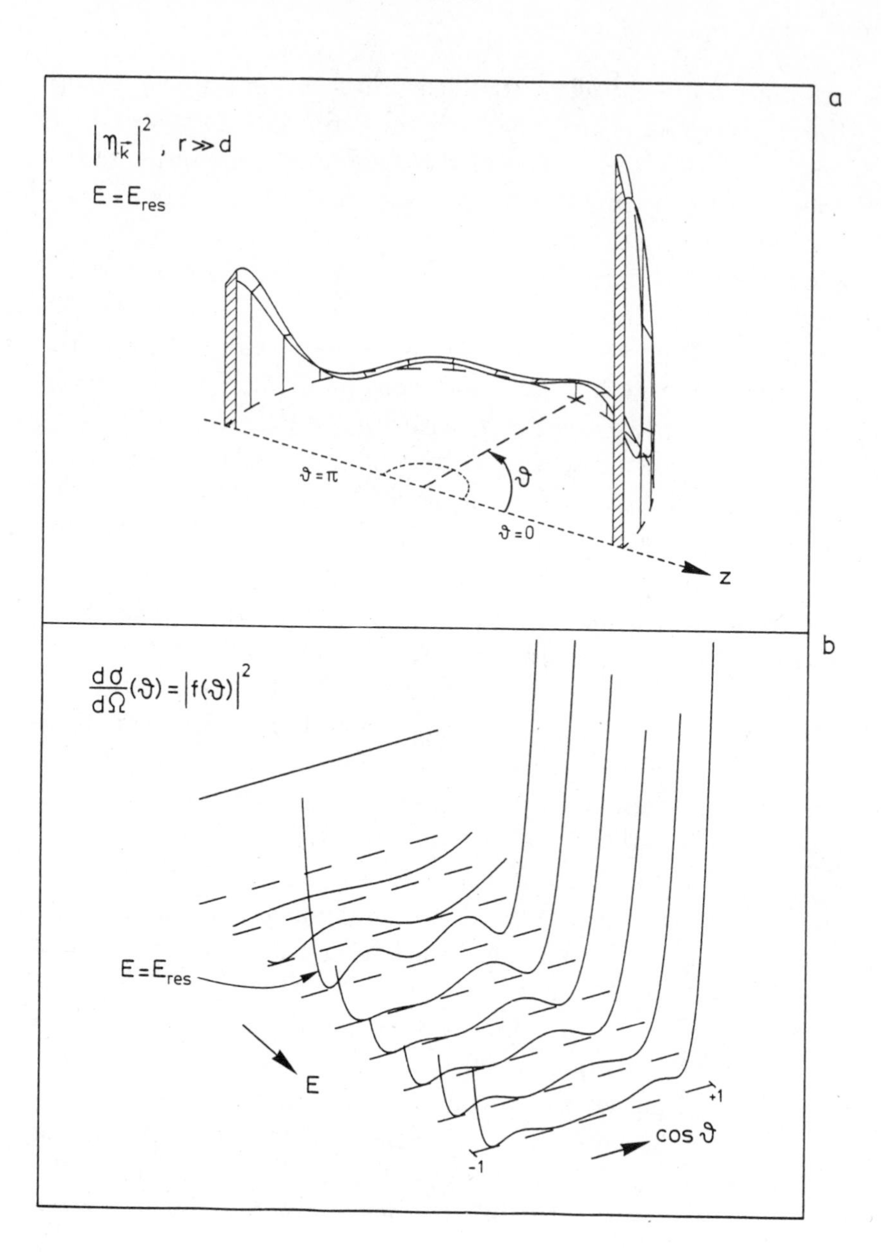

Figure 13.8 (a) Intensity of the scattered spherical wave resulting from the scattering of a plane wave incident in the z-direction onto an attractive potential restricted to a small region $r < d$, indicated by the small dashed half circle. The intensity far outside the potential region is a function of the scattering angle ϑ. The energy of the incident wave is the resonance energy $E = E_{res}$.

(b) Energy dependence of the differential scattering cross section $d\sigma / d\Omega$ shown over a linear scale in $\cos\vartheta$. The differential cross section is constant in $\cos\vartheta$, indicating isotropic scattering, for $E \approx 0$ (background). At resonance energy $E = E_{res}$ it is given approximately by the square of the Legendre polynomial $P_3(\cos\vartheta)$, since the partial scattering amplitude f_3 dominates the cross section.

Figure 13.9 Energy dependence of the radial wave function $R_l(k, r)$ for scattering by an attractive square-well potential. The form of the potential is indicated by the long-dash line, the wave energy by the short-dash line which also serves as zero line for the wave function. Whereas R_0, R_1, and R_2 change very little within the potential region, near the energy E_{res} the wave function R_3 of the resonant partial wave develops a very pronounced maximum. Outside the potential region all wave functions show trivial shortening of the wavelength with growing energy.

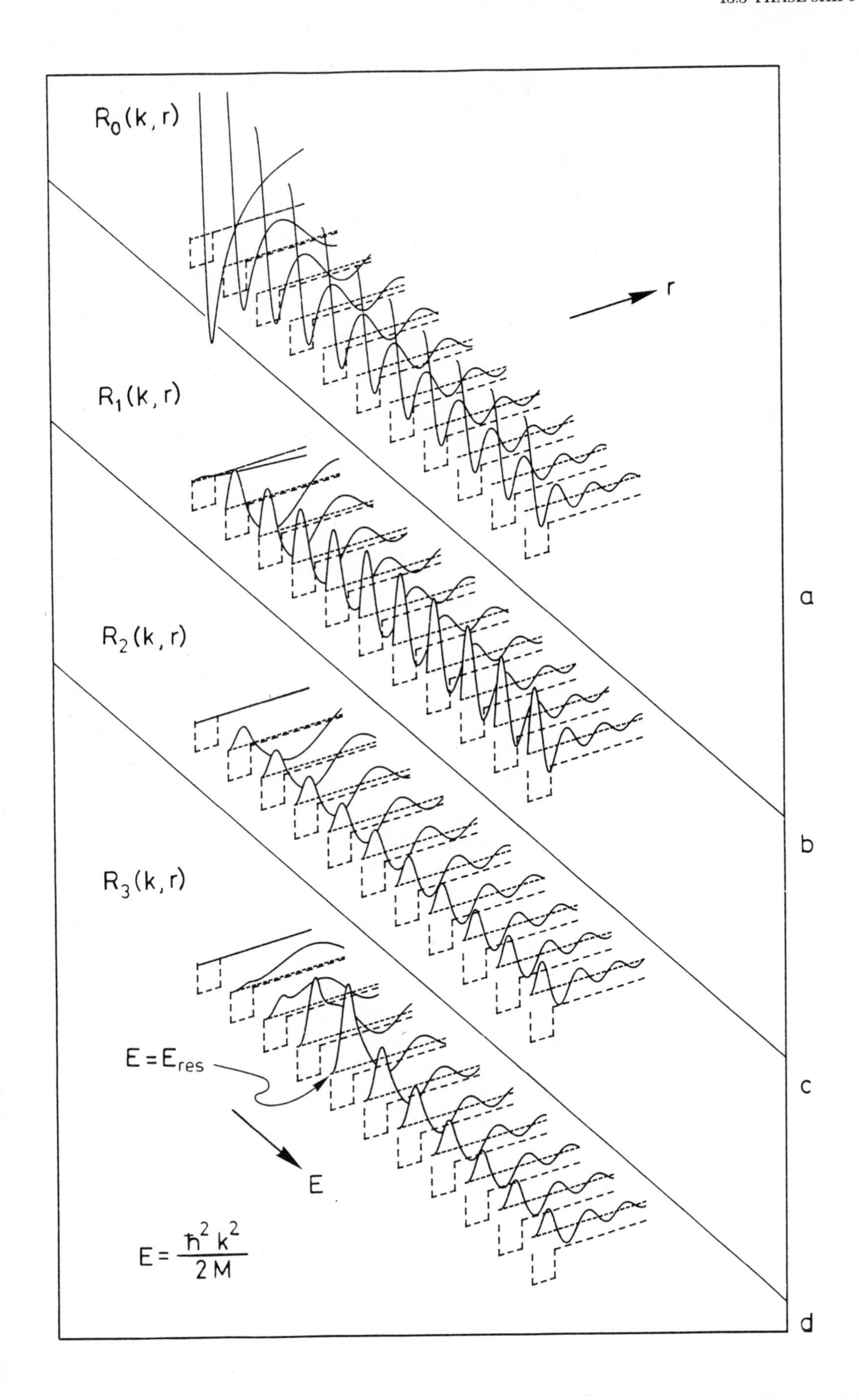

$R_0(k, r)$
r
$R_1(k, r)$
a
$R_2(k, r)$
b
$R_3(k, r)$
$E = E_{res}$
c
E
$E = \frac{\hbar^2 k^2}{2M}$
d

attractive square-well potential. It is this maximum that characterized the wave function $\varphi_{\mathbf{k}}^{(+)}(\mathbf{r})$ and the scattering wave function $\eta_{\mathbf{k}}(\mathbf{r})$, which were shown in Figures 13.3 and 13.4 to introduce the resonance phenomena.

13.4 Bound States and Resonances

The pronounced maximum of the radial wave function $R_3(k_{\text{res}}, r)$, $k_{\text{res}} = \sqrt{2ME_{\text{res}}}/\hbar$, in the range of the attractive square-well potential signifies that the particle in a resonant state has a rather large probability of being in the potential range. This situation resembles to some extent that of a particle bound within a square-well potential. The relation between bound states and resonances is indeed intimate, and we shall try to indicate their connection. We start with Figure 13.10d. It shows the attractive square-well potential used throughout this chapter, the effective potential for $l = 3$, the energy of the resonance, and the radial wave function $R_3(k_{\text{res}}^{(1)}, r)$. This plot reveals the reason why the resonance phenomenon occurs. We remember from the introduction to Chapter 10 that the effective potential is the sum of the potential $V(r)$ and centrifugal potential $\hbar^2 l(l+1)/(2Mr^2)$. Thus the effective potential has a wall just outside the square well, that is, a region where V_{eff} is larger than E_{res}. This wall keeps the particle from leaving the potential region except by the tunnel effect. The wall is of finite thickness, however, so that the particle, unlike the particle in the bound state, can also populate the outside region. For this depth of the potential there is no bound state for $l = 3$. We now increase the depth of the potential well. There is a continuous decrease of the resonance energy E_{res} as the potential increases in depth, and eventually the resonant state turns into a bound state with negative energy and with the radial wave function $R_{13}(r)$. This situation is shown in Figure 13.10c. If we increase the potential depth even further (Figure 13.10b), a new resonance with the wave function $R(k_{\text{res}}^{(2)}, r)$ appears possessing one node within the potential region, just as the second bound state would have. When the depth is increased even further (Figure 13.10a), this resonance too becomes a bound state with the wave function $R_{23}(r)$.

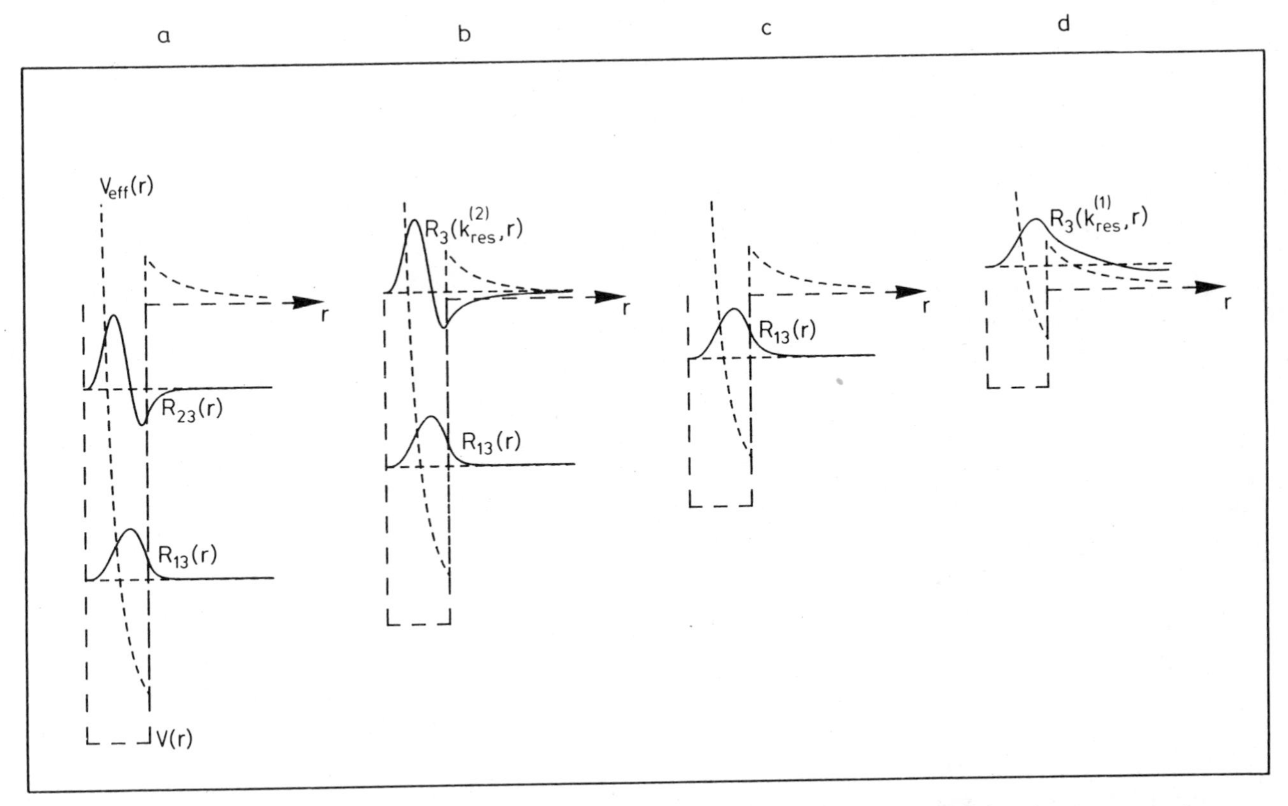

Figure 13.10 **Bound states and resonances of an attractive square-well potential for angular momentum quantum number $l = 3$. The potential wells have constant fixed widths but different depths. The potential $V(r)$ is shown as a long-dash line. The effective potential is also shown. (a) For a rather deep potential well there are two bound states with negative energies indicated by the horizontal short-dash lines. The lower bound state has no radial nodes; the second has one node. (b) A somewhat shallower well has only one bound state but it does have a resonance. The resonance energy corresponds to the horizontal line of positive energy. The radial wave function $R_3(k^{(2)}_{\text{res}}, r)$ has one node in the potential region, just as the second bound state in part a has. (c) This potential well has only one bound state. (d) The bound state in part c now reappears as a resonance. Its wave function is $R_3(k^{(1)}_{\text{res}}, r)$. The resonance is the same as that in Figures 13.3 through 13.9.**

13.5 Resonance Scattering by a Repulsive Shell

We have found that resonances occur when there is a repulsive wall in the effective potential. In our example of an attractive square-well potential, this wall originated from the centrifugal force. We can also study resonance scattering on a repulsive shell potential:

$$V(r) = \begin{cases} 0, & 0 < r < d_1 \\ V_0, & d_1 < r < d_2 \\ 0, & d_2 < r < \infty \end{cases}$$

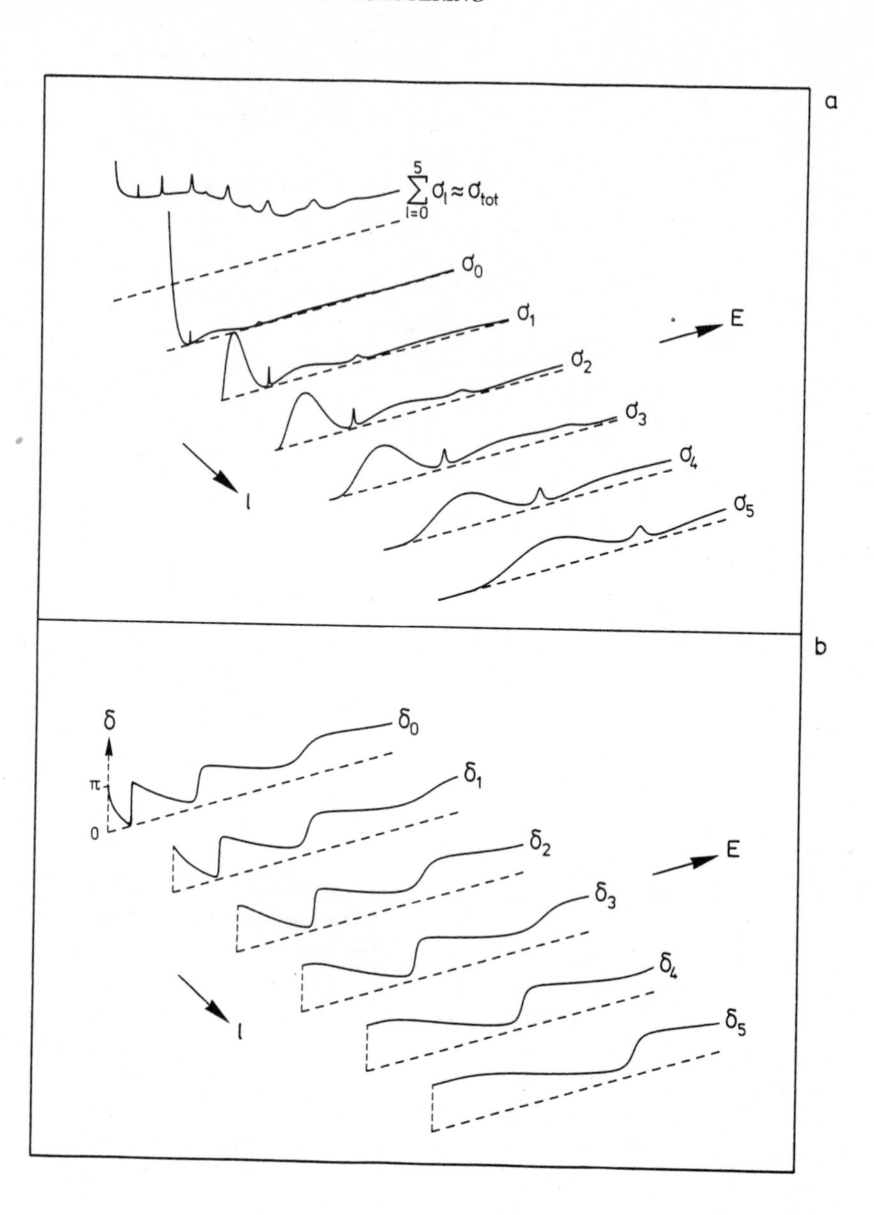

Figure 13.11 **Scattering of a plane wave by a repulsive shell. The potential is $V = V_0 > 0$ within a spherical shell $d_1 \leq r \leq d_2$ and $V = 0$ for all other r. In this figure the energy of the plane wave ranges from $E = 0$ to $E = 2V_0$. (a) Energy dependence of the partial cross sections $\sigma_l(E)$ and of the total cross section $\sigma_{tot}(E)$, which is approximated by the sum over the first few partial cross sections. Resonances for the different partial waves are visible as maxima in σ_l and σ_{tot}. The maxima are rather sharp for the first resonance and broader for the second. The resonances shift systematically to higher energies as angular momentum quantum number l increases.**

(b) Energy dependence of the phase shifts $\delta_l(E)$. At a resonance energy the corresponding phase shift rises steeply and passes through $\pi/2$.

Here V_0 is positive and denotes the height of the potential within the shell. The shell potential provides a spherical potential wall of height V_0, of inner radius d_1, and of outer radius d_2 around the origin. We can expect that this wall will produce resonances quite independent of a centrifugal force.

Figure 13.12 **Argand diagrams for the complex partial scattering amplitudes $f_1(E)$ and $f_2(E)$ for scattering by a repulsive shell. As in Figure 13.11, the energy ranges from $E = 0$ to $E = 2V_0$. The resonances have a swift counterclockwise motion of f_l through the point 0, 1 in the complex plane and the characteristic resonance patterns in Im $f(_l(E)$, Re $f_l(E)$, and $\delta_l(E)$ already familiar from Figure 13.7d. Because of the shell structure of the potential, there are now more resonances.**

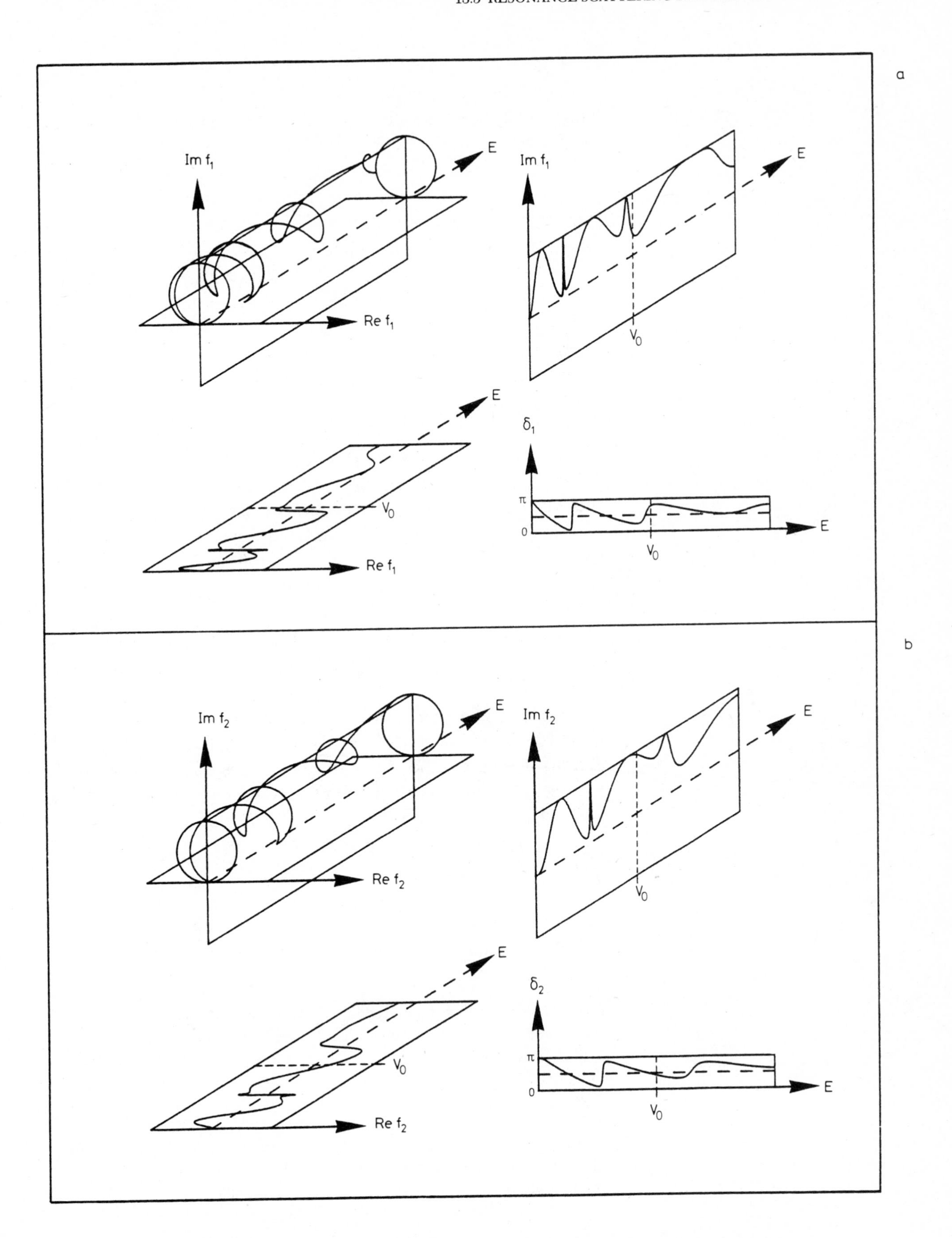
a
Im f_1
E
Re f_1
Im f_1
E
V_0
E
V_0
Re f_1
δ_1
π
0
E
V_0
b
Im f_2
E
Re f_2
Im f_2
E
V_0
E
V_0
Re f_2
δ_2
π
0
E
V_0

Figure 13.11 plots the total cross section σ_{tot}, the partial cross sections σ_l, and the phase shifts δ_l for $l = 0, 1, \ldots, 5$. The resonances are clearly visible as peaks in σ_l and as jumps in δ_l. For $l = 0, 1, 2, 3$ there are two resonances at two different energies. We shall refer to them as first and second resonances. For $l = 4, 5$ only the first resonance is visible in the energy range of the figure. The second resonance is much wider in energy than the first. The width of both resonances increases with angular momentum l. There is also a striking regularity between the angular momentum and the energy of the first resonance. In a plane spanned by energy E and angular momentum l, the first resonances fall on a curved, smooth line called a *Regge trajectory*. There is a similar trajectory for the second resonances.

In the total cross section, given in the background of Figure 13.11a, the various resonances in different partial waves appear as peaks. When they are sufficiently narrow, they can easily be separated from the smooth background.

Figure 13.12 gives the Argand diagrams for the partial-wave amplitudes f_1 and f_2. It shows the resonance structure known from Figure 13.7d. In both amplitudes there are now two resonances indicated by the swift counterclockwise motion of f_l through the top of the unitarity circle, that is, point $\operatorname{Re} f_l = 0$, $\operatorname{Im} f_l = 1$. The unitarity circle is also shown in Figure 11.8a.

In Figure 13.13 the energy dependence of the radial wave functions $R_0(k, r)$ and $R_1(k, r)$ is shown in energy intervals around the first and second resonances. The radial wave functions have the typical enhancement at resonance energies. Since the potential vanishes inside the shell, the wave function has no node in r for the first resonance and one node for the second resonance.

We conclude this section by showing for some resonances the full stationary wave function $\varphi_{\mathbf{k}}^{(+)}(\mathbf{r})$, the scattered spherical wave $\eta_{\mathbf{k}}(\mathbf{r})$, and the scattered partial waves $\eta_l(\mathbf{r})$ for

Figure 13.13 **Energy dependence of the radial wave functions $R_l(k,r)$ within restricted energy intervals surrounding the resonances (a, b) in $l = 0$ and (c, d) in $l = 1$ for scattering by a repulsive shell. The form $V(r)$ of the potential is indicated by the long-dash line, the energy E of the wave by the short-dash line. The middle diagram of each series corresponds to the resonance energy. The wave functions $R_l(k_{\text{res}}, r)$ shown in these middle diagrams display no node and one node inside the shell for the first and second resonance, respectively.**

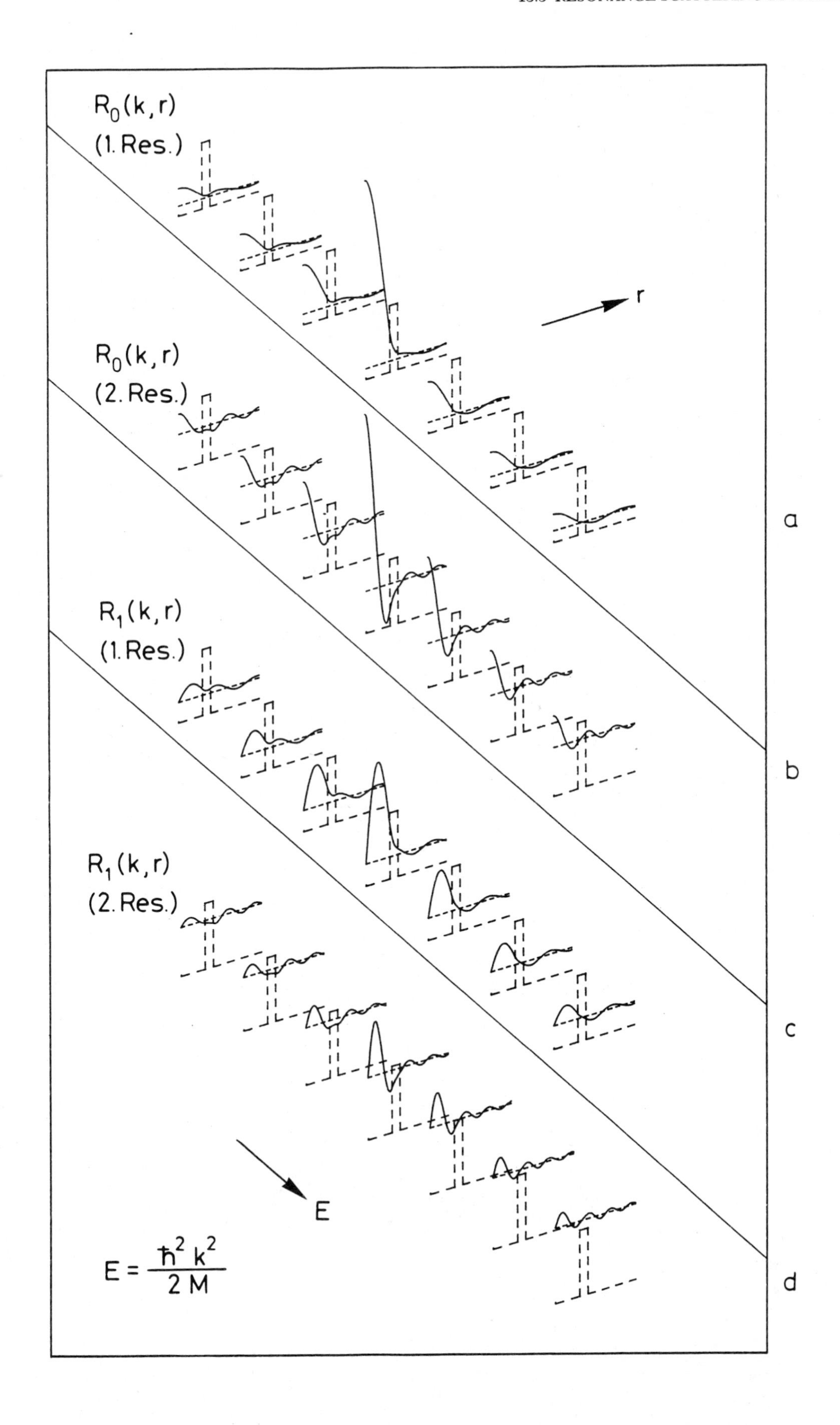

$R_0(k,r)$
(1. Res.)
r
$R_0(k,r)$
(2. Res.)
a
$R_1(k,r)$
(1. Res.)
b
$R_1(k,r)$
(2. Res.)
c
E
$E = \frac{\hbar^2 k^2}{2M}$
d

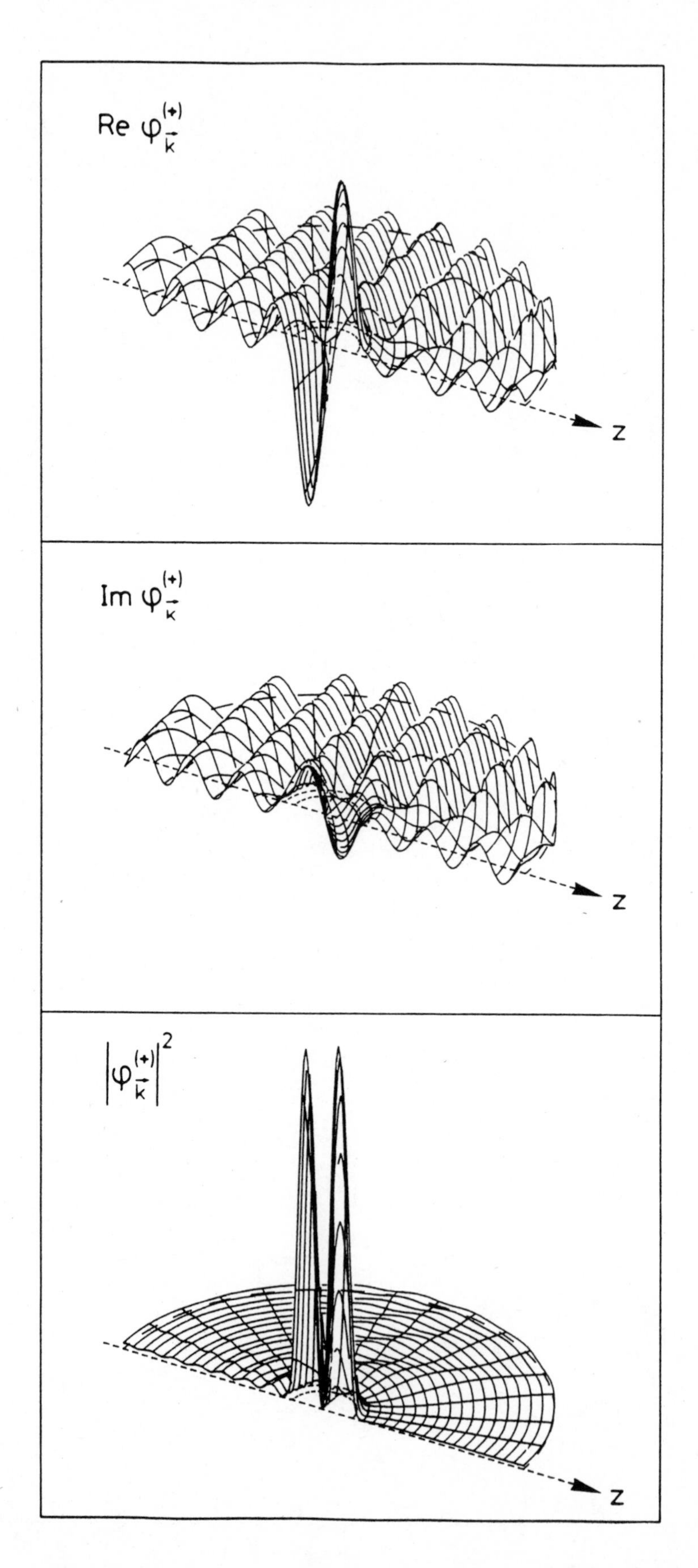

Figure 13.14 **Wave functions $\varphi_{\mathbf{k}}^{(+)}$ for the scattering of a plane wave incident along the *z*-direction by a repulsive shell. The energy is that of the first resonance in partial wave $l = 1$. The two half circles near the center indicate the inner and outer boundaries of the spherical potential shell.**

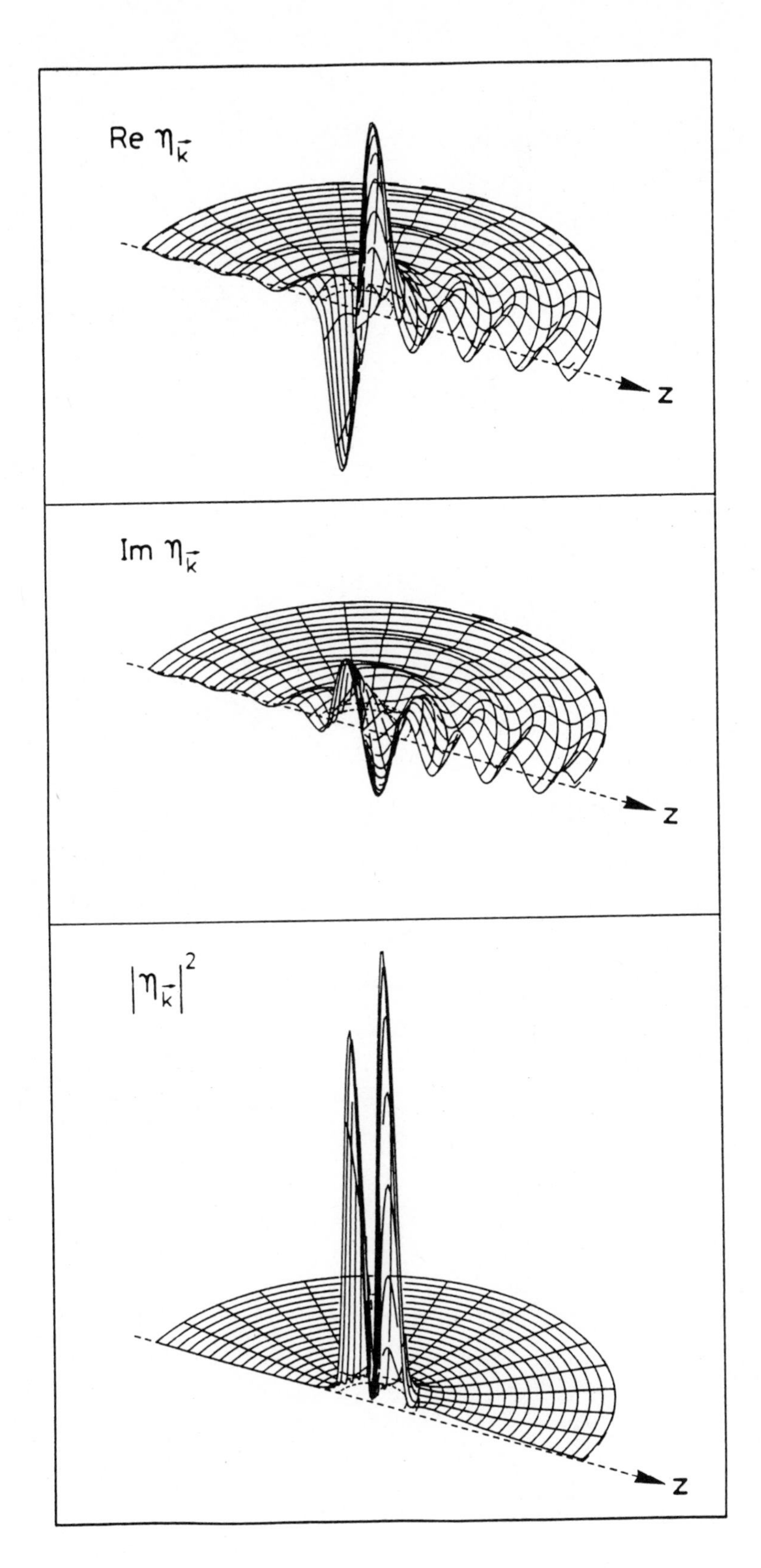

Figure 13.15 **The scattered spherical wave η_k resulting from the scattering of a plane wave incident along the z-direction by a repulsive shell. The energy is that of the first resonance in partial wave $l = 1$.**

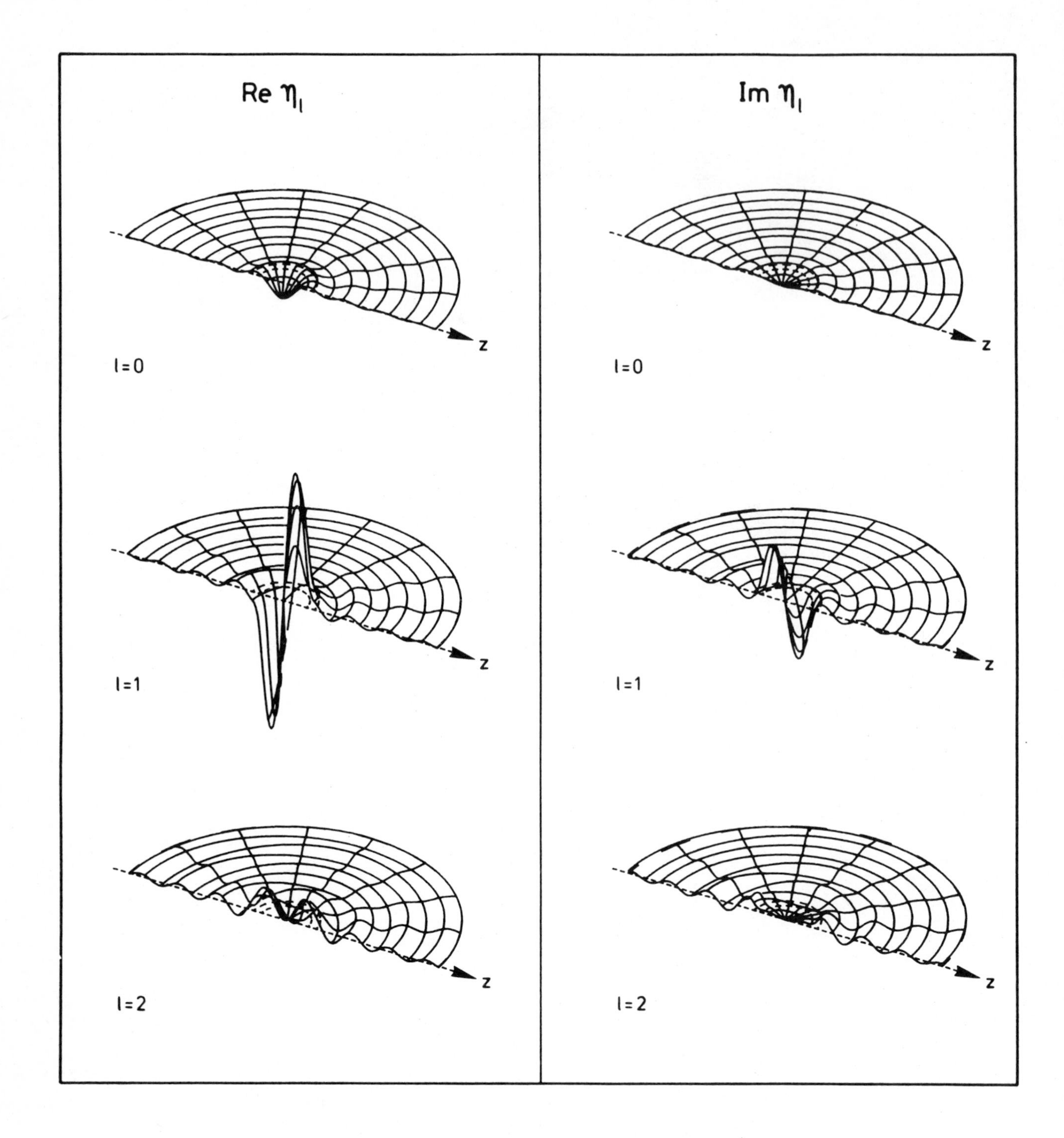

Re η_l
Im η_l
z
l = 0
l = 1
l = 2

$l = 1, 2, 3$. In each figure the size of the spherical potential shell is indicated by the two half circles near the origin. They correspond to the inner and outer boundaries of the shell. The first resonance with angular momentum $l = 1$ is illustrated in Figures 13.14 through 13.16. In Figure 13.16 we observe that only the partial scattered wave for $l = 1$, that is η_1, shows a resonance structure. It has no node in r, indicating the first resonance. There is one node in the polar angle ϑ at $\vartheta = \pi/2$. It is caused by the Legendre polynomial $P_1(\cos\vartheta) = \cos\vartheta$, which determines the ϑ-dependence of η_1 as discussed at the beginning of Section 11.2. The scattered spherical wave $\eta_{\mathbf{k}}(\mathbf{r})$ in Figure 13.15 is obtained by summing up the partial scattered waves $\eta_l(\mathbf{r})$. Since the dominating term in this sum is $\eta_1(\mathbf{r})$, it is not surprising that the structure of $\eta_{\mathbf{k}}(\mathbf{r})$ in the central region is that of $\eta_1(\mathbf{r})$, displaying no node in r but one node in ϑ. Even the full stationary wave function $\varphi_{\mathbf{k}}^{(+)}(\mathbf{r})$, which is shown in Figure 13.14 and is a superposition of the incoming harmonic plane wave and the scattered spherical wave $\eta_{\mathbf{k}}(\mathbf{r})$, retains much of this structure.

The second resonance for $l = 1$ is illustrated in Figures 13.17 through 13.19. Its higher energy is immediately apparent from the shorter wavelength of the incoming wave dominating the outer regions of Figure 13.17. The inner region in all three figures is determined by the resonating scattered partial wave η_1. It still has one node in ϑ, since the angular momentum has not changed. But it now also has a node in r indicating the second resonance.

Figure 13.16 Scattered partial waves η_l, $l = 0, 1, 2$, resulting from the scattering of a plane wave incident along the z-direction by a repulsive shell. The partial wave η_1 has its first resonance at this particular energy of the incident plane wave.

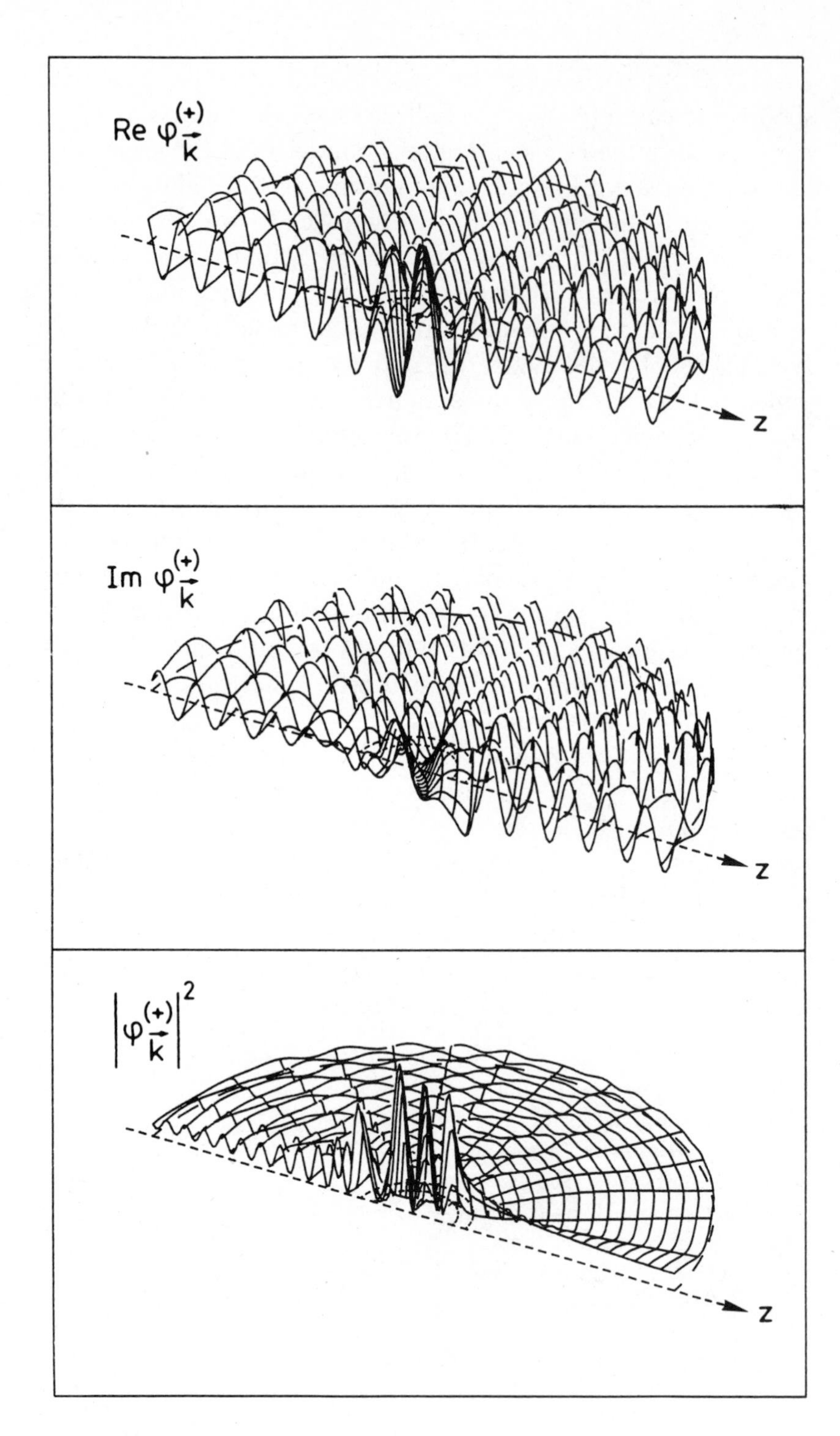

Figures 13.17–13.19 **Wave function $\varphi^{(+)}_{k}$, scattered spherical waves η_k, and scattered partial waves η_l for the second resonance in $l = 1$ produced by the scattering of a plane wave incident along the *z*-direction by a repulsive potential shell.**

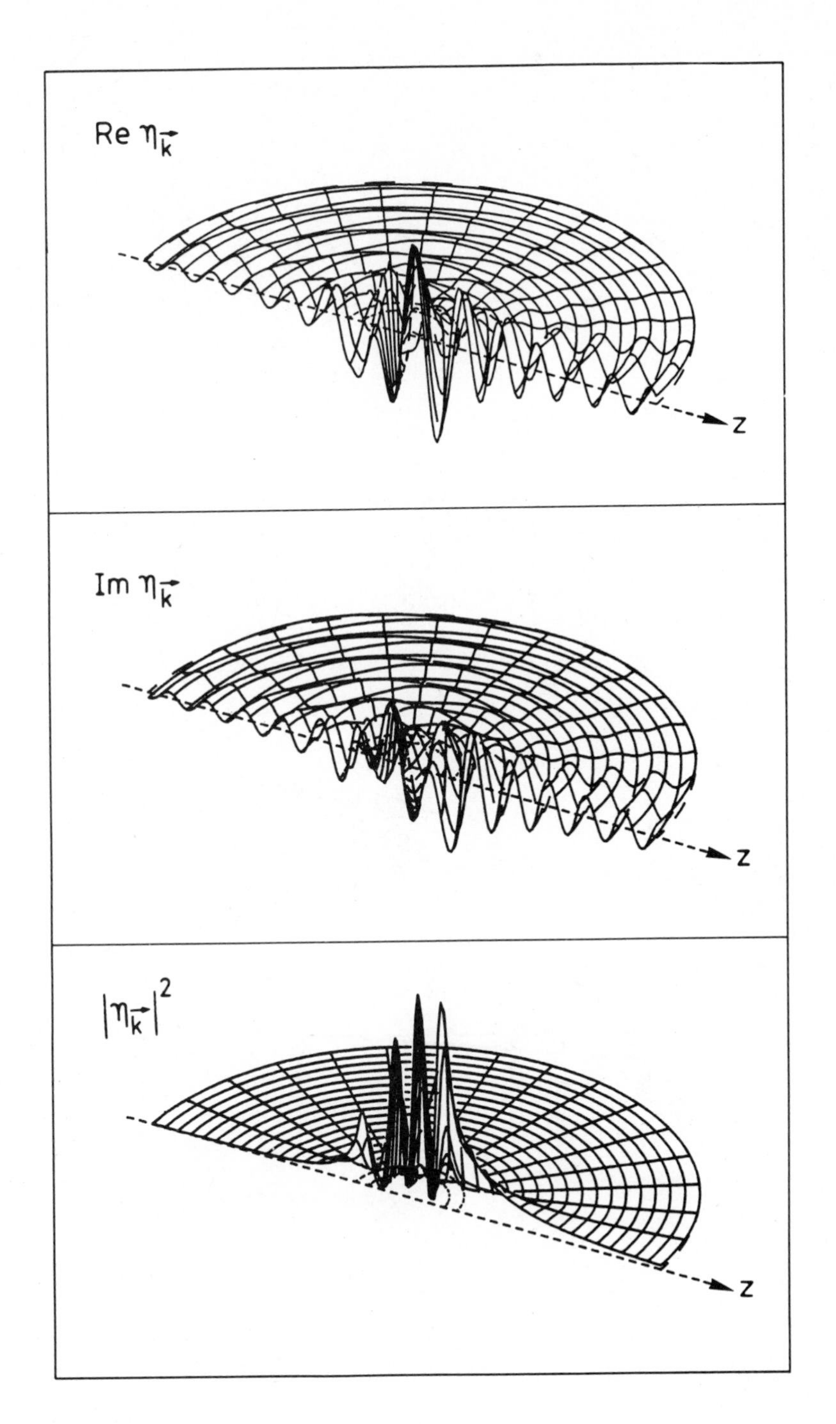

Figure 13.18

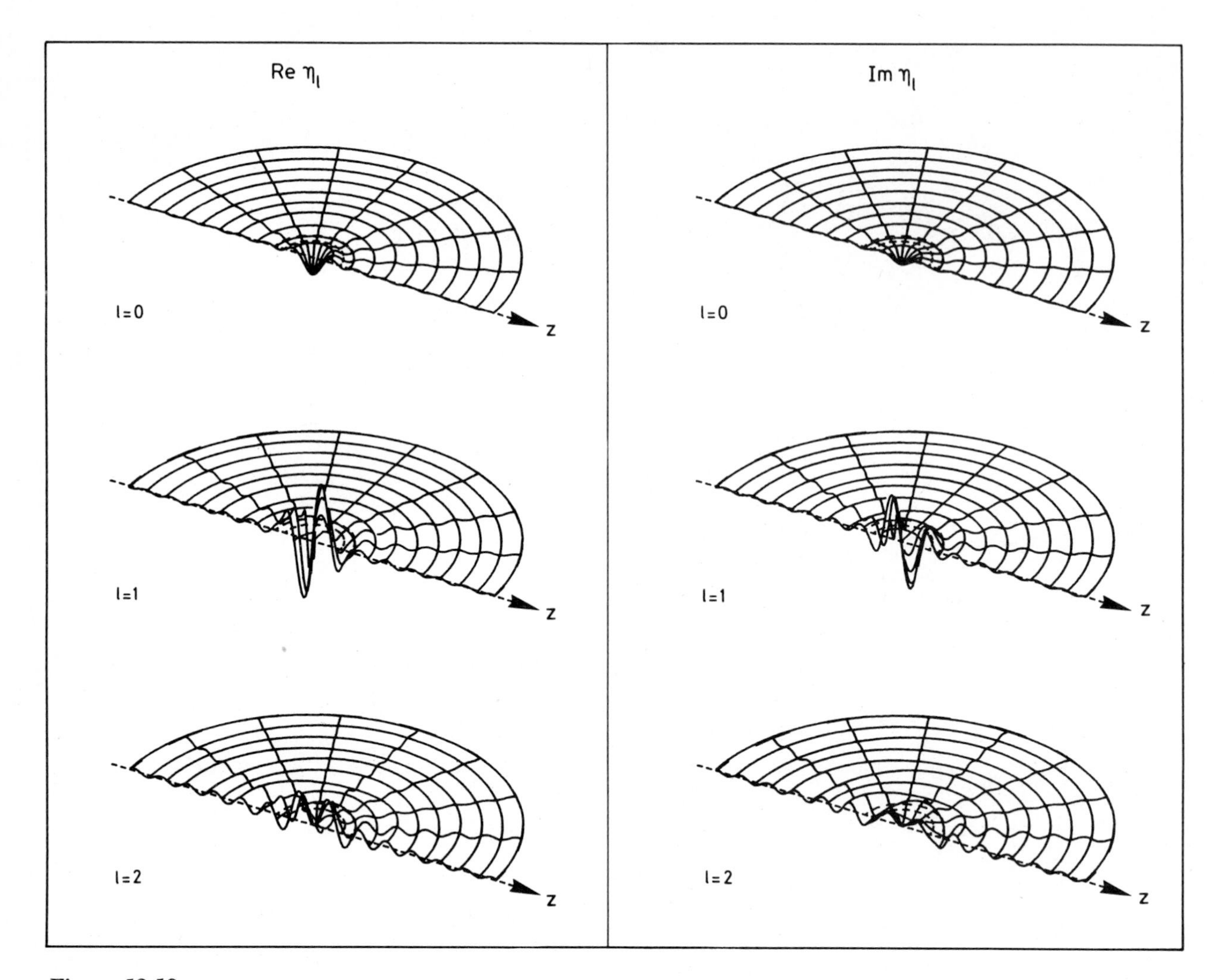

Figure 13.19

Finally, we turn to the resonances of angular momentum zero. Figures 13.20, 13.21, and 13.22 show the functions $\varphi_{\mathbf{k}}^{(+)}(\mathbf{r})$, $\eta_{\mathbf{k}}(\mathbf{r})$, and $\eta_l(\mathbf{r})$ for the first resonance. The resonating partial wave is now η_0. It has no node in ϑ since the Legendre polynomial P_0 does not depend on ϑ. As a first resonance it also has no node in r. These simple features of η_0 are very clearly retained in the scattered spherical wave $\eta_{\mathbf{k}}(\mathbf{r})$, Figure 13.21, and in the stationary wave function $\varphi_{\mathbf{k}}^{(+)}(\mathbf{r})$, Figure 13.20.

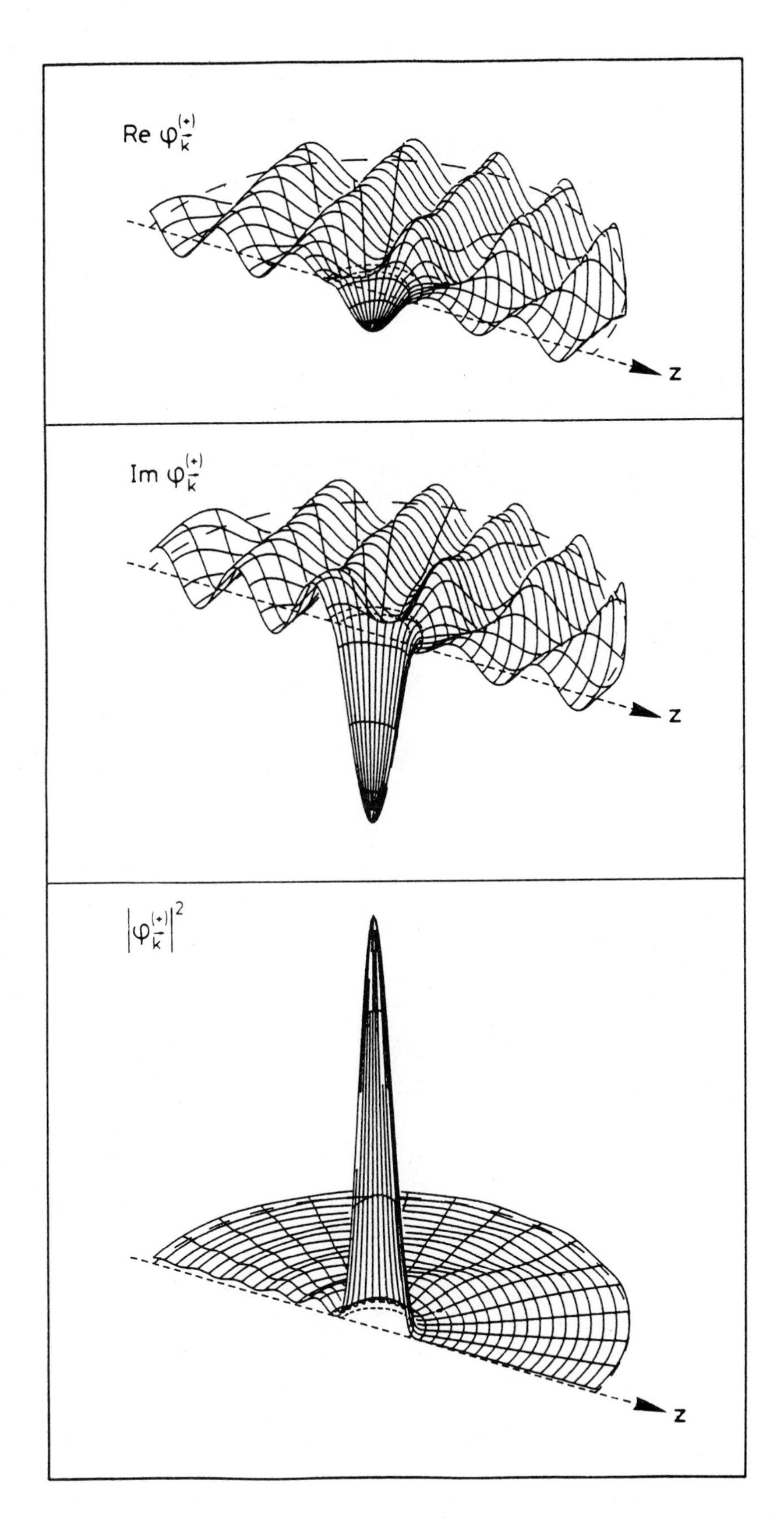

Figures 13.20–13.22 **Wave function $\varphi_{\mathbf{k}}^{(+)}$, scattered spherical waves $\eta_{\mathbf{k}}$, and scattered partial waves η_l for the first resonance in $l = 0$ produced by the scattering of a plane wave incident along the *z*-direction by a repulsive potential shell.**

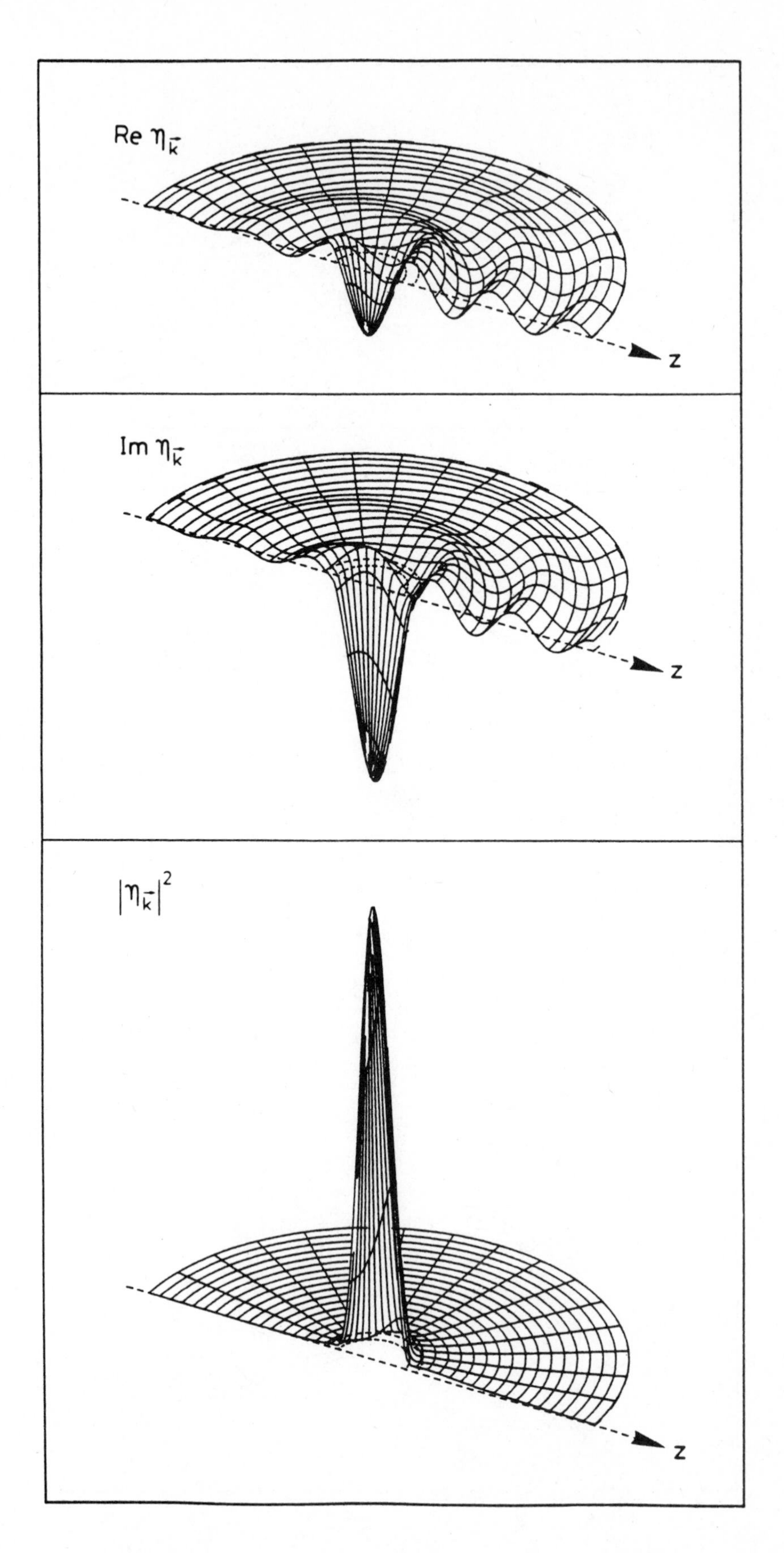

Figure 13.21

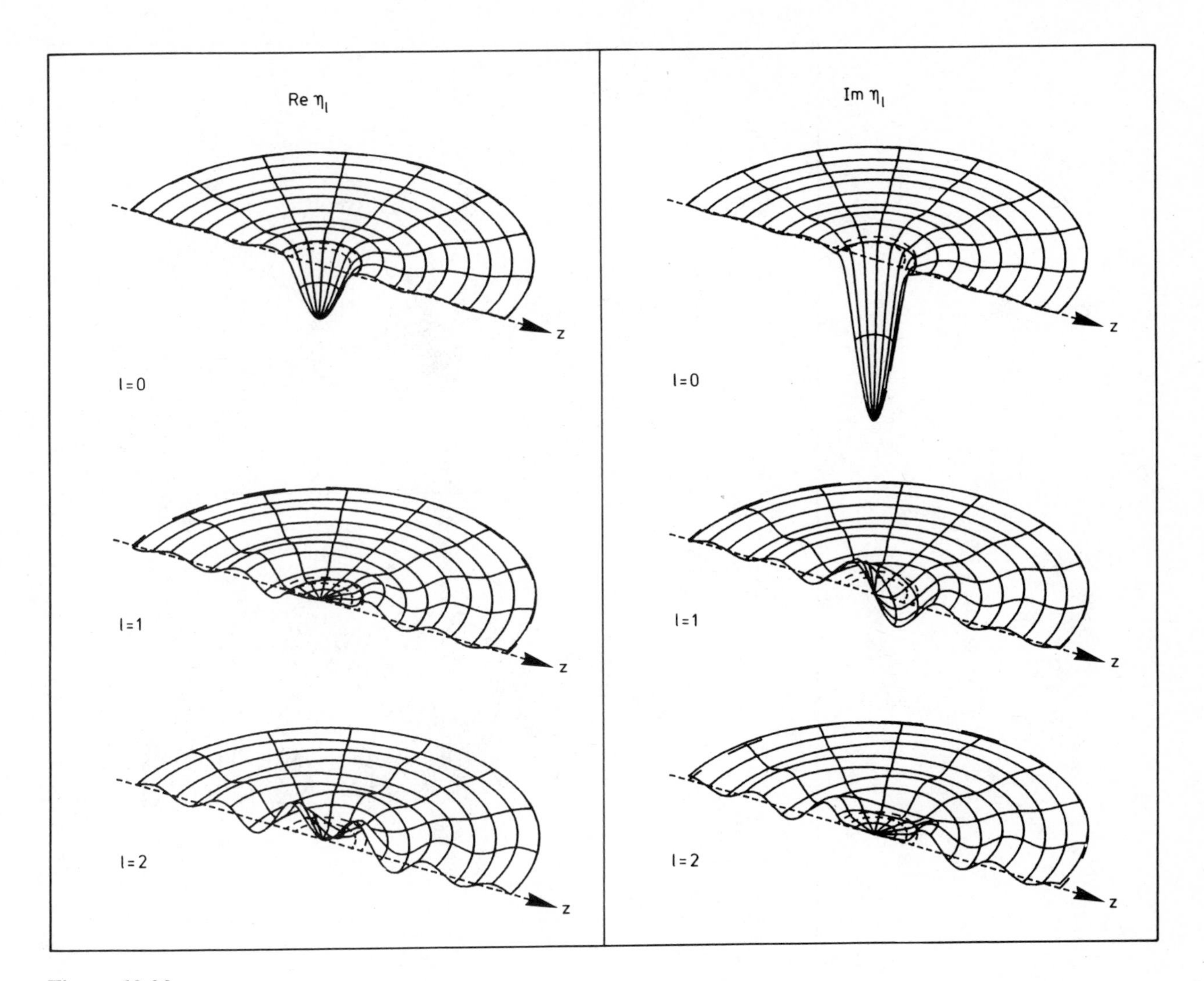

Figure 13.22

The corresponding plots for the second resonance of angular momentum zero are shown in Figures 13.23, 13.24, and 13.25. The higher energy of the resonance again results in a shorter wavelength of the incoming wave, which is particularly visible in the outer regions of Figure 13.23. The resonant wave η_0 now has one radial node in the inner region but no node in the polar angle ϑ.

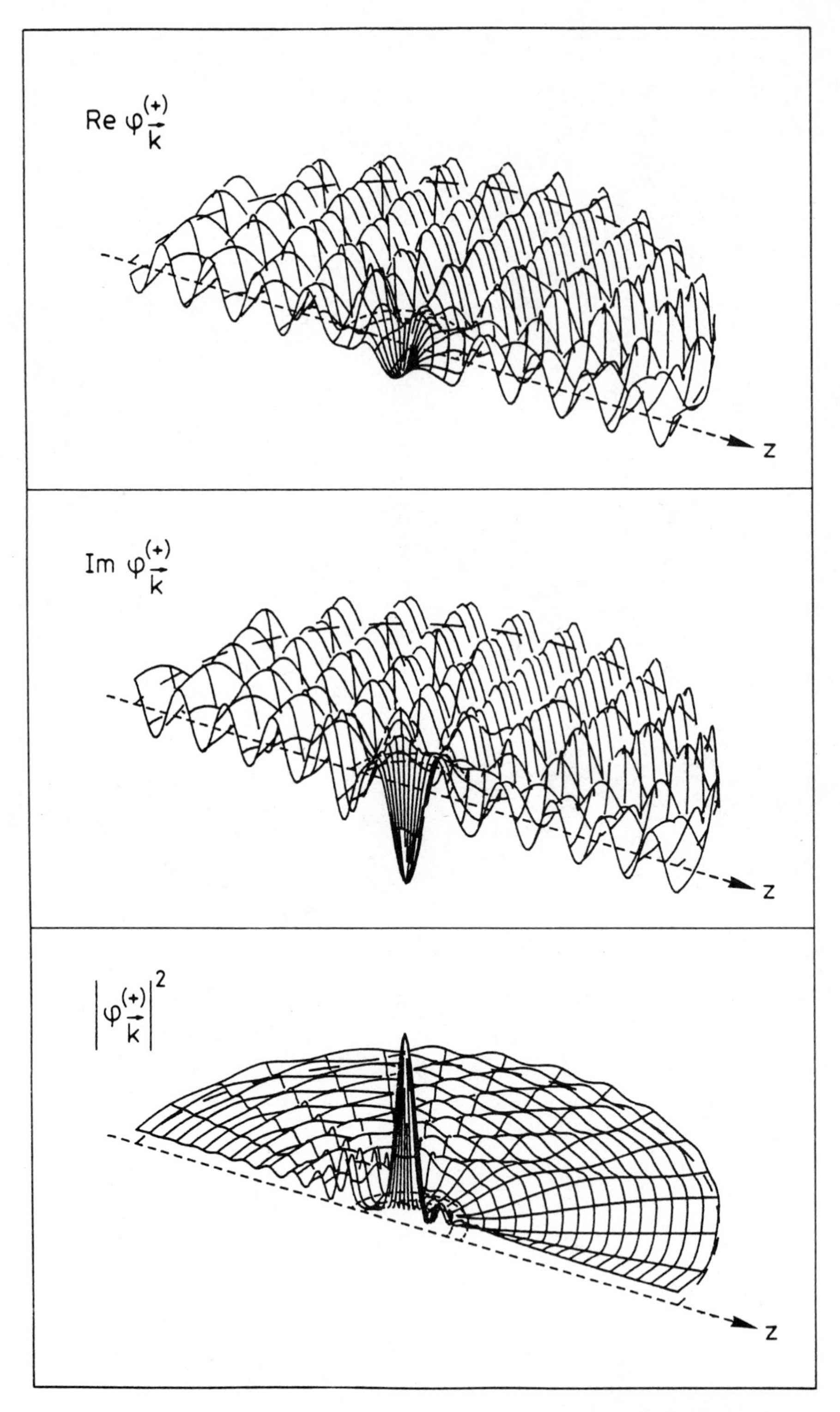

Figures 13.23–13.25 **Wave function $\varphi_{\mathbf{k}}^{(+)}$, scattered spherical waves $\eta_{\mathbf{k}}$, and scattered partial waves η_l for the second resonance in $l = 0$ produced by the scattering of a plane wave incident along the *z*-direction by a repulsive potential shell.**

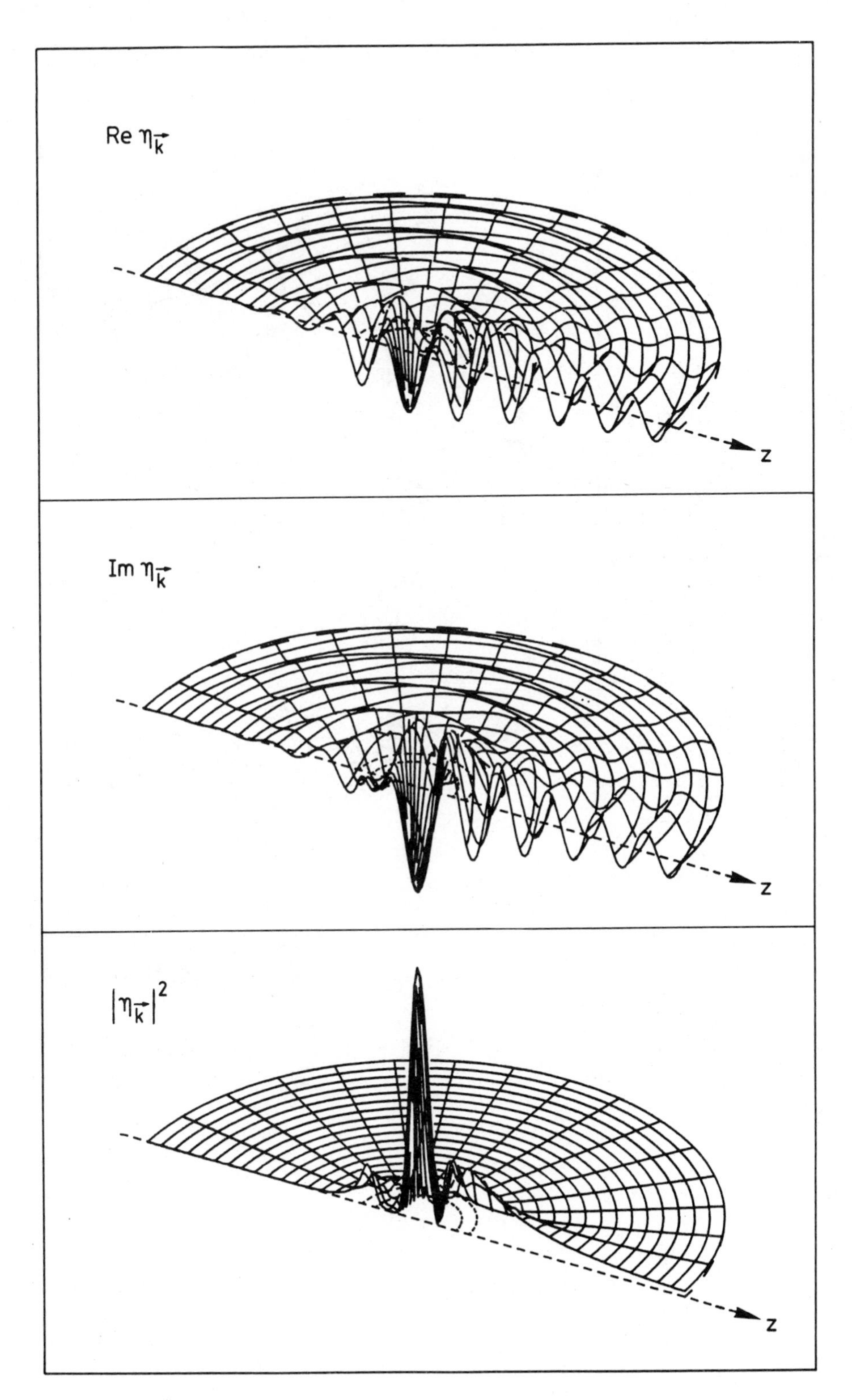

Figure 13.24

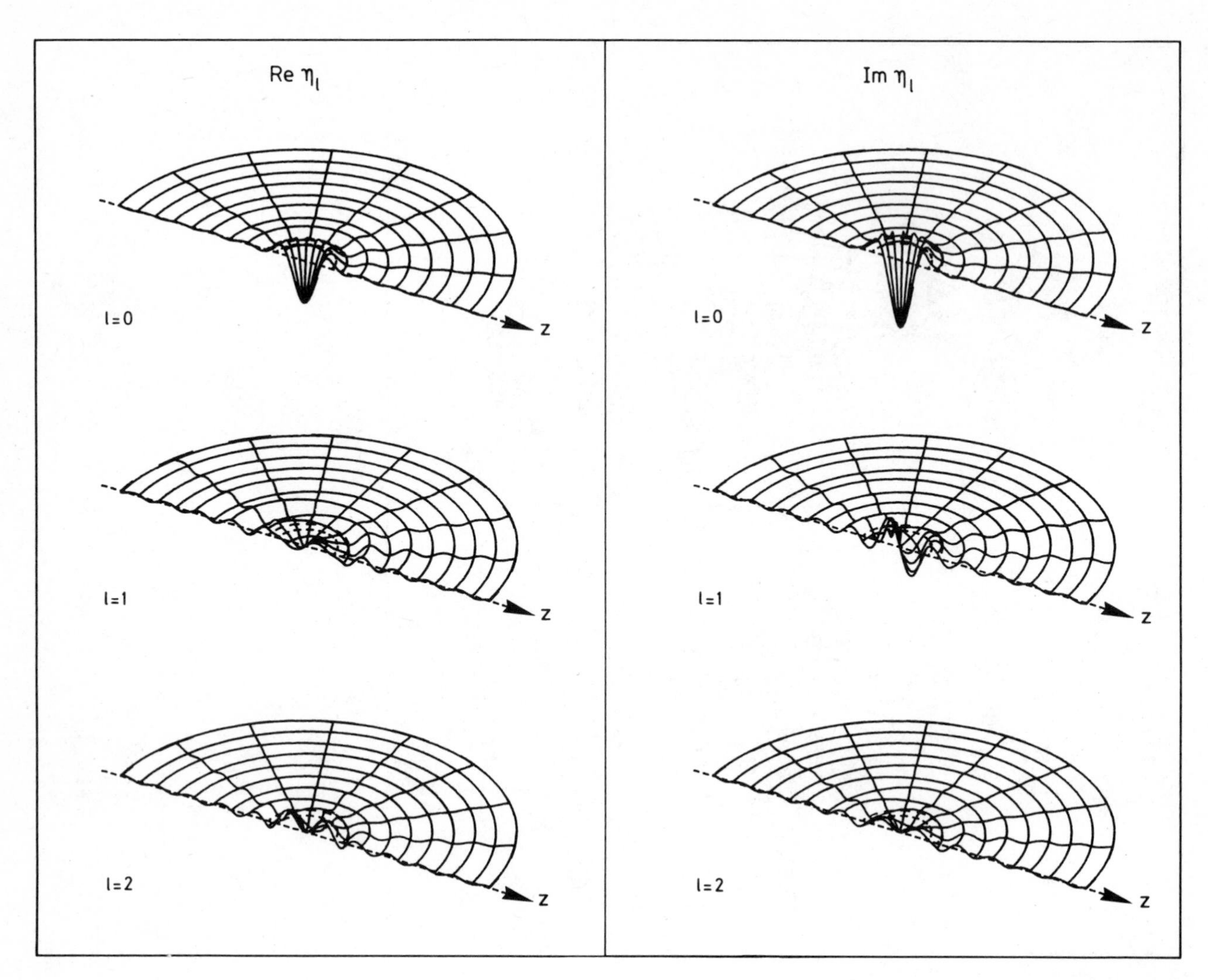

Figure 13.25

Problems

13.1 What is the relation between the wavelengths inside and outside the potential region of Figure 13.1?

13.2 Explain the features of Figures 13.1a, b, and c with the help of Figures 13.2a, b, and c in terms of the initial plane wave, the scattered spherical wave, and their interference pattern.

13.3 Find the features in the r- and ϑ-dependence of the resonant partial wave η_3 in Figure 13.5 that are characteristic for angular momentum $l = 3$.

13.4 Given the form of the resonance partial wave η_3 in

Figure 13.5, is there for $l = 3$ a bound state or resonance with lower energy than the one plotted?

13.5 Relate the appearance of the backward peak in the differential scattering cross section at resonance energy in Figure 13.8 to the partial-wave decomposition of the scattered wave in Figure 13.5 and to $|\eta_k|^2$ in Figure 13.4.

13.6 Describe the behavior, for large values of r, of the bound-state wave function of the first excited state in Figure 13.10a and of the resonance wave function in Figure 13.10b.

13.7 Compare the criteria for a resonance, as found in Figure 13.7d, with the resonances indicated in the Argand diagrams in Figures 13.12a and b. Which peaks correspond to resonance phenomena and why?

13.8 Are the energies of the resonances in Figure 13.13 higher or lower than the energies of bound states in an infinitely deep potential well?

14. Examples from Atomic, Molecular, Solid-State, Nuclear, and Particle Physics

So far we have investigated mechanical systems using the description and tools of quantum mechanics. In this final chapter we look at actual systems as they occur in nature. We shall discuss scattering phenomena, bound systems, and metastable states as they play a role in rather different fields of science.

Before discussing the results of actual experiments, we need to spend a little time on the units in which the data are given. The velocities of several of the particles studied are not much slower than the speed of light. To describe them we therefore have to use the Einstein's theory of relativity. It states that, if E is the total energy and p the magnitude of the momentum of a particle, the quantity

$$E^2 - p^2c^2 = m^2c^4$$

has the same value in any frame of reference in which E and p are measured. Here $c = 3 \cdot 10^8$ m/s is the speed of light in vacuum. In the particular frame of reference in which the particle is at rest, $p = 0$, we have

$$E = mc^2$$

Therefore the constant m is called the *rest mass* of the particle. The quantity mc^2 is the *rest energy* of the particle. In a frame of reference in which the particle is not at rest,

$p \neq 0$, the total energy is larger:

$$E = \sqrt{m^2c^4 + p^2c^2} = mc^2 + E_{\text{kin}}$$

The additional term is called the *kinetic energy* of the particle.

In the experiments discussed in this section, the particles are characterized by their momentum p, their total energy E, or their kinetic energy E_{kin}. The energies are measured in electron volts (eV). A particle that carries the *elementary charge*

$$e = 1.602 \cdot 10^{-19}\ \text{C}$$

and that has traversed an accelerating potential difference of 1 V has gained the kinetic energy

$$1\ \text{eV} = 1.602 \cdot 10^{-19}\ \text{W s} = 1.602 \cdot 10^{-19}\ \text{J}$$

A convenient notation for higher energies is 1 keV $= 10^3$ eV, 1 MeV $= 10^6$ eV, 1 GeV $= 10^9$ eV. Since mc^2 is an energy, masses can be measured in electron volts per c^2:

$$1\frac{\text{eV}}{c^2} = \frac{1.602 \cdot 10^{-19}}{(3 \cdot 10^8)^2}\ \text{kg} = 1.78 \cdot 10^{-36}\ \text{kg}$$

The rest mass of the electron is

$$m_e = 511\ \text{keV}/c^2$$

The rest masses of the proton and the neutron are nearly 2000 times larger,

$$m_p = 938.3\ \text{MeV}/c^2, \qquad m_n = 939.6\ \text{MeV}/c^2$$

It is important to remember that a proton with kinetic energy of $E_{\text{kin}} = 10$ MeV has a total energy of $E = m_pc^2 + E_{\text{kin}} = 948.3$ MeV. Often the momentum p is easiest to measure. Since the product pc is an energy, the momentum is measured in electron volts per c:

$$1\frac{\text{eV}}{c} = \frac{1.602 \cdot 10^{-19}}{3 \cdot 10^8}\ \text{kg m/s} = 5.3 \cdot 10^{-28}\ \text{kg m/s}.$$

Once the momentum p and the rest mass m of a particle are known, its total energy E and its kinetic energy E_{kin} are easily computed.

14.1 Scattering of Atoms, Electrons, Neutrons, and Pions on Different Targets

In Chapters 11 and 13 we have discussed the scattering of a particle incident on a spherically symmetric potential, which was assumed to be fixed in space. In actual experiments projectile particles scatter on target particles. In "colliding beam" experiments the projectile and target particles both move in opposite directions within a storage ring and scatter on each other in a head-on collision. An example for such an arrangement is given at the end of Section 14.4, where the production of elementary particles J/ψ and Υ in colliding beams of electrons and positrons is discussed. In "fixed target" experiments the target particles are at rest before the scattering process; however, they move after the collision. As in classical mechanics, this two-body process can be reduced to a one-body problem if the coordinate $\mathbf{r}$ in the wave function is taken to be the distance between the two particles and the mass appearing in the one-body Schrödinger equation is taken to be the reduced mass $M = m_1 m_2/(m_1 + m_2)$ of the two bodies. It is customary to present results of scattering experiments as differential cross sections $d\sigma/d\vartheta^*$ with respect to the scattering angle ϑ^* in the center-of-mass system (CMS). In this reference frame target and projectile have initially equal and opposite momenta (Figure 14.1a).

Figures 14.1b through e show results obtained in scattering experiments in entirely different fields of physics using completely different experimental techniques. Figure 14.1b shows the differential cross section for the scattering of sodium atoms by mercury atoms. The kinetic energy in the laboratory frame is only a fraction of an electron volt. The momentum is of the order of 100 keV$/c$ corresponding to a de Broglie wavelength of about 10^{-11} m, which is one order of magnitude below the atomic radius. Scattering experiments such as this one provide information about the electric potential acting between atoms. Such investigations are helpful in studying problems of chemical bonds.

Nuclear forces can be investigated by using neutrons, which carry no electric charge, as projectiles incident on nuclei. The differential cross section for the scattering of neutrons on lead nuclei is given in Figure 14.1c for two energies $E_{\rm kin}^{\rm lab} = 7$ MeV and 14.5 MeV. The corresponding momenta are $p^{\rm lab} = 110$ MeV$/c$ and 160 MeV$/c$. They in turn corresponds to de Broglie wavelengths of roughly $11 \cdot 10^{-15}$ m and $7.6 \cdot 10^{-15}$ m. These wavelengths are of the same order of magnitude as the radius of the lead nucleus, which is roughly $7 \cdot 10^{-15}$ m. As expected, there are more

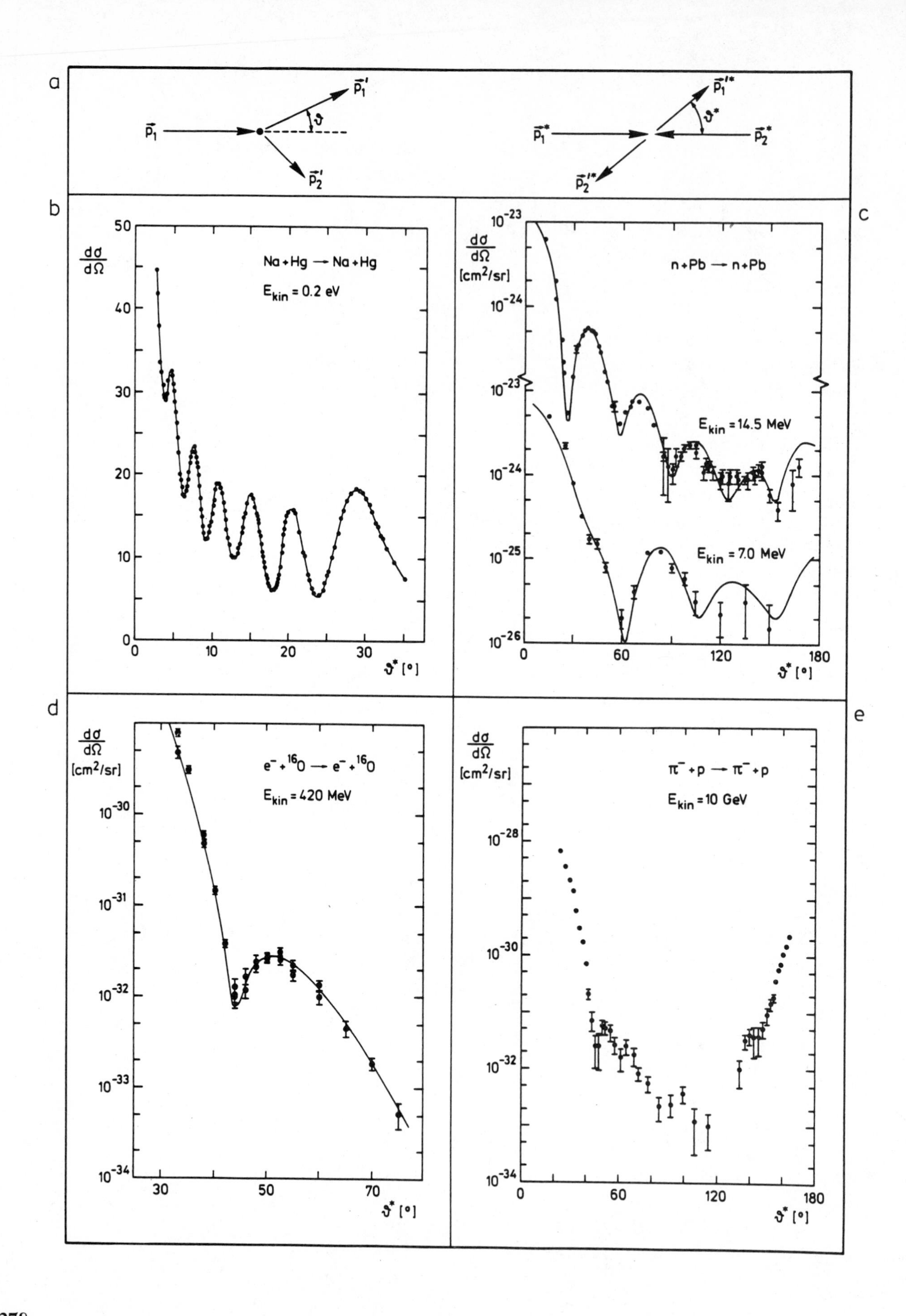

a
b
c
d
e
Na + Hg → Na + Hg
E_{kin} = 0.2 eV
n + Pb → n + Pb
E_{kin} = 14.5 MeV
E_{kin} = 7.0 MeV
$e^- + {}^{16}O \to e^- + {}^{16}O$
E_{kin} = 420 MeV
$\pi^- + p \to \pi^- + p$
E_{kin} = 10 GeV
$\frac{d\sigma}{d\Omega}$
[cm²/sr]
ϑ^* [°]

minima in the differential cross sections for the higher energy, that is, for the shorter wavelength, of the incoming particles (see Figure 13.8b).

To investigate the electric potential of nuclei, we choose electrons as projectiles because they are not affected by the nuclear forces of the nucleus. Figure 14.1d shows the differential cross section of electrons with a laboratory energy of 420 MeV scattered by oxygen nuclei. Because the electron mass is small, the electron momentum is 420 MeV/c, its corresponding wavelength about $3 \cdot 10^{-15}$ m. From such experiments the electric-charge distribution of the oxygen nucleus was found to have a characteristic radius of about $3 \cdot 10^{-15}$ m.

The nuclei are composed of protons and neutrons, often referred to by the collective term *nucleons*. If we want to study the internal structure of protons or neutrons, we can perform scattering experiments using either electrons or particles with nuclear interaction as projectiles. Results of an experiment using particles with nuclear interaction as projectiles are shown in Figure 14.1e. Here the projectiles are π-mesons. These particles exert nuclear forces and have a mass of about 140 MeV/c^2. The experiment was performed with laboratory energies of 10 GeV, that is, a momentum of 10 GeV/c corresponding to a wavelength of $0.1 \cdot 10^{-15}$ m, which is one order of magnitude below the proton radius.

Figure 14.1 **(a) Scattering of a projectile particle 1 on a target particle 2. In the laboratory the target particle is initially at rest, $\mathbf{p}_2 = 0$. In the center-of-mass system (CMS) the particles have initially equal and opposite momenta, $\mathbf{p}_1^* = -\mathbf{p}_2^*$. For elastic scattering, considered here, the momenta are also equal and opposite after the scattering process, $\mathbf{p}_1'^* = -\mathbf{p}_2'^*$.**

(b) Sodium atoms scattered on mercury atoms, (c) neutrons on lead nuclei, (d) electrons on oxygen nuclei, and (e) π-mesons on protons. The differential cross section $d\sigma / d\Omega$ for the elastic scattering of two particles is given as a function of the CMS scattering angle ϑ^*. The laboratory kinetic energy E_{kin} of the projectile is given on each figure. For part b the ordinate is a linear scale given in arbitrary units. For parts c, d and e it is a logarithmic scale given in square centimeters per steradian.

Sources: (b) From U. Buck and H. Pauly, *Zeitschrift für Naturforschung* **23a** (1968) 475, copyright © 1968 by Verlag der Zeitschrift für Naturforschung, Tübingen, reprinted by permission. (c) From F. Perey and B. Buck, *Nuclear Physics* **32** (1962) 352, copyright © 1962 by North-Holland Publishing Company, Amsterdam, reprinted by permission. (d) From R. Hofstadter, Nuclear and Nucleon Scattering of Electrons at High Energies, reproduced with permission from the *Annual Review of Nuclear and Particle Science*, Volume 7, copyright © 1957 by Annual Reviews Inc. (e) From a conference contribution by J. Orear et al. as reported by G. Belletini, Intermediate and High Energy Collisions, in *Proceedings of the 14th International Conference on High Energy Physics at Vienna* (J. Prentki and J. Steinberger, editors), copyright © 1968 by CERN, Geneva, reprinted by permission.

The results given in Figure 14.1 bear a qualitative resemblance to the differential scattering cross sections shown in Figures 11.4b and 13.8b. No quantitative comparison is justified, for the forces acting in the collisions in Figure 14.1 cannot be described by simple square-well potentials. Moreover, effects attributable to the spin of the target and the projectiles were not taken into account in the calculations of Chapters 11 and 13.

14.2 Spectra of Bound States in Atoms, Nuclei, and Crystals

The first striking success of quantum mechanics was the explanation of the *hydrogen spectrum*. Sufficiently heated atomic hydrogen emits light with a characteristic wavelength spectrum consisting of discrete wavelengths. In Section 12.4 we found that the energy levels of the electron bound in the hydrogen atom are

$$E_n = -\tfrac{1}{2}Mc^2\frac{\alpha^2}{n^2}, \qquad n = 1, 2, \ldots,$$

Here M is the electron mass, c is the speed of light, and $\alpha = 1/137$ is the fine-structure constant. A transition from one level to another is effected by the emission or absorption of the energy difference

$$\Delta E = E_{n_1} - E_{n_2} = -\tfrac{1}{2}Mc^2\alpha^2\left(\frac{1}{n_1^2} - \frac{1}{n_2^2}\right)$$

in the form of a light quantum of frequency ν corresponding to

$$\Delta E = h\nu$$

or to the wavelength

$$\lambda = \frac{c}{\nu} = \frac{hc}{\Delta E}$$

A set of transitions for a fixed value of n_1 but variable n_2 is called a *spectral series* (see Figure 14.2a). In particular, the one with $n_1 = 2$ and $n_2 > 2$ is called the *Balmer series*. Its wavelengths are in the region of visible light and can be easily measured with a prism spectrograph. The spectral lines of the Balmer series are observed in the light emitted by electric discharges in hydrogen gas but also in the light emitted by some stars, proving that there is hot hydrogen in the atmospheres of such stars. Outside the region of some stars that

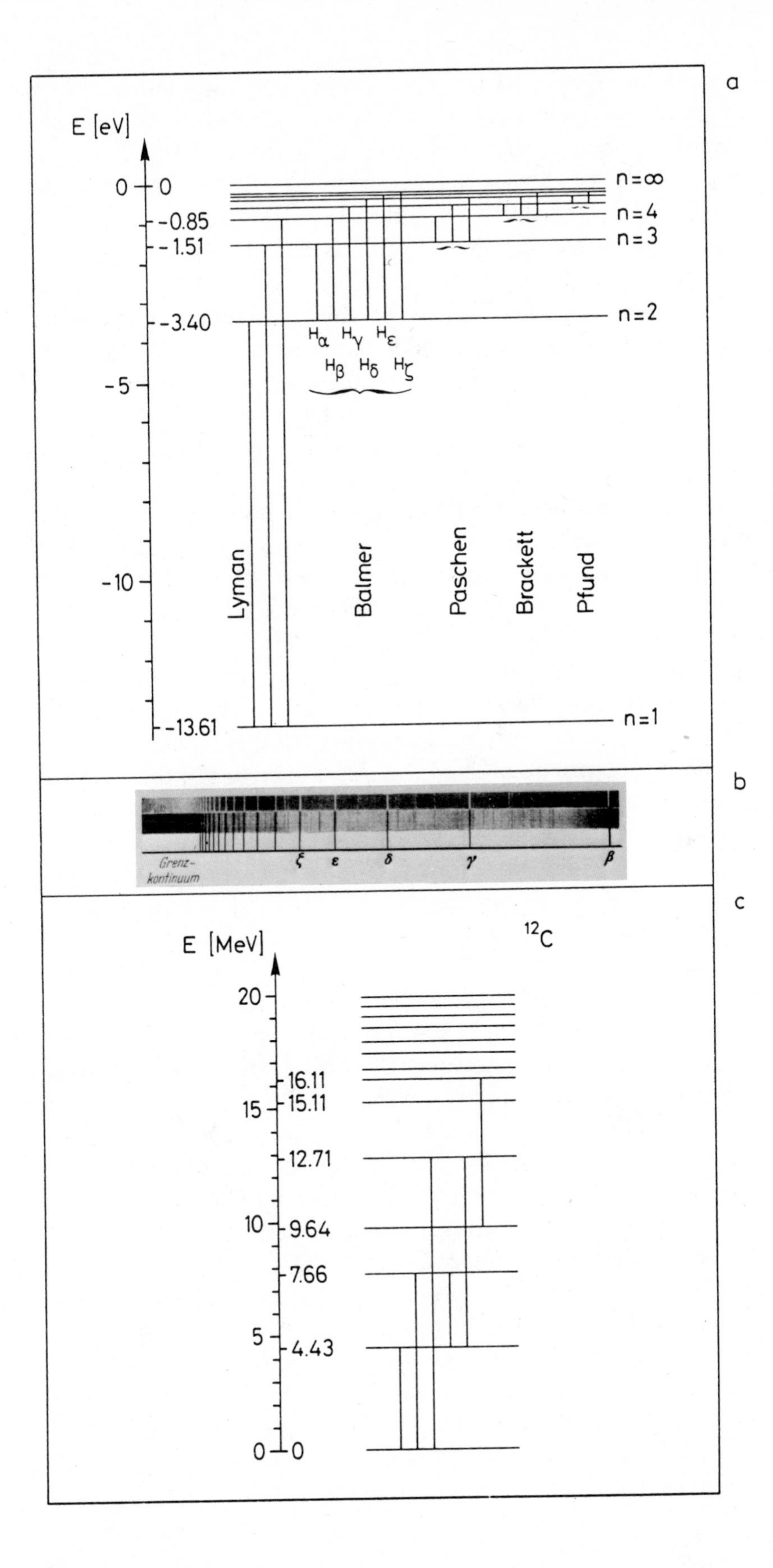

Figure 14.2 (a) The energy levels that the electron of the hydrogen atom can take are indicated by horizontal lines and enumerated by the principal quantum number n. Vertical lines indicate the energies at which transitions between different energy levels take place. Transitions to or from the same lower energy level form a series. For example, transitions to or from energy level $n = 1$ make up the Lyman series. Those to or from energy level $n = 2$ make up the Balmer series. Transitions to a lower level consist of the emission of a light quantum corresponding to the transition energy. Those to a higher level consist of the absorption of a light quantum.

(b) Wavelength spectra of light from different stars show the Balmer series in emission (top) and absorption (bottom). The stars are α Cassiopeiae and β Cygni. From R. W. Pohl, *Optik und Atomphysik*, ninth edition, copyright © 1954 by Springer-Verlag Berlin, Göttingen, Heidelberg, reprinted by permission.

(c) The different energy levels of the carbon nucleus ^{12}C. The ground state of the nucleus has been chosen to be the zero point of the energy scale. Some of the observed transitions between energy levels are indicated. These transitions, like those for the hydrogen atom in part a, consist of the emission or absorption of a photon.

emit light, the hydrogen gas is cold. Then we observe dark lines in the spectrograph for the wavelengths of the Balmer series, indicating that hydrogen atoms of the cold gas have absorbed light. Stellar spectra showing the Balmer series in emission and absorption are given in Figure 14.2b. The energy spectrum shown in Figure 14.2a has already been obtained in Section 12.4. It is characteristic of the Coulomb potential acting between the nucleus of the hydrogen atom and its electron. It possesses an infinite number of levels accumulating at the upper end of the spectrum, $E = 0$. The spectra of more complicated atoms which contain more electrons in the atomic shell become more involved but retain these general features.

Transitions between different energy levels effected by the absorption or emission of photons are also observed in *atomic nuclei*. A typical energy scale for these photons is 1 MeV, compared to 1 eV in atoms. Nuclear spectra are complex because the nucleus usually consists of a large number of protons and neutrons bound together by nuclear forces. Some of the low-lying levels of nuclei can be explained by the following model. Every nucleon moves in the nuclear potential owing to the presence of all the other nucleons in the nucleus. Since nucleons are fermions and obey the Pauli exclusion principle, they fill up the lowest states in a common nuclear potential, forming the ground state of the nucleus. The simplest states of higher energy are those in which a single nucleon occupies a higher state. Figure 14.2c shows the energy spectrum of the low-lying states of the carbon nucleus ^{12}C. The nucleus contains six protons and six neutrons, that is, twelve nucleons. Since the carbon nucleus is a twelve-particle system, its spectrum, as might be expected, is rather different from the energy spectrum of the hydrogen atom.

In Section 6.6 we saw that the energy levels of periodic potentials form bands. Because a crystal is a regular lattice of atoms and therefore has spatial periodicity, the energy levels of the electrons in a crystal form such bands. Figure 6.11 indicates that the number of levels inside each band is equal to the number of single potentials, that is, to the number of atoms in the crystal. Since this is a very large number indeed, we do not expect to resolve the single energy levels within a band. Experimentally, the band hypothesis can be verified using the photoelectric effect. Monoenergetic photons of high energies, that is, monochromomatic X-rays, are directed onto a crystal surface. The energy of the electrons liberated from the

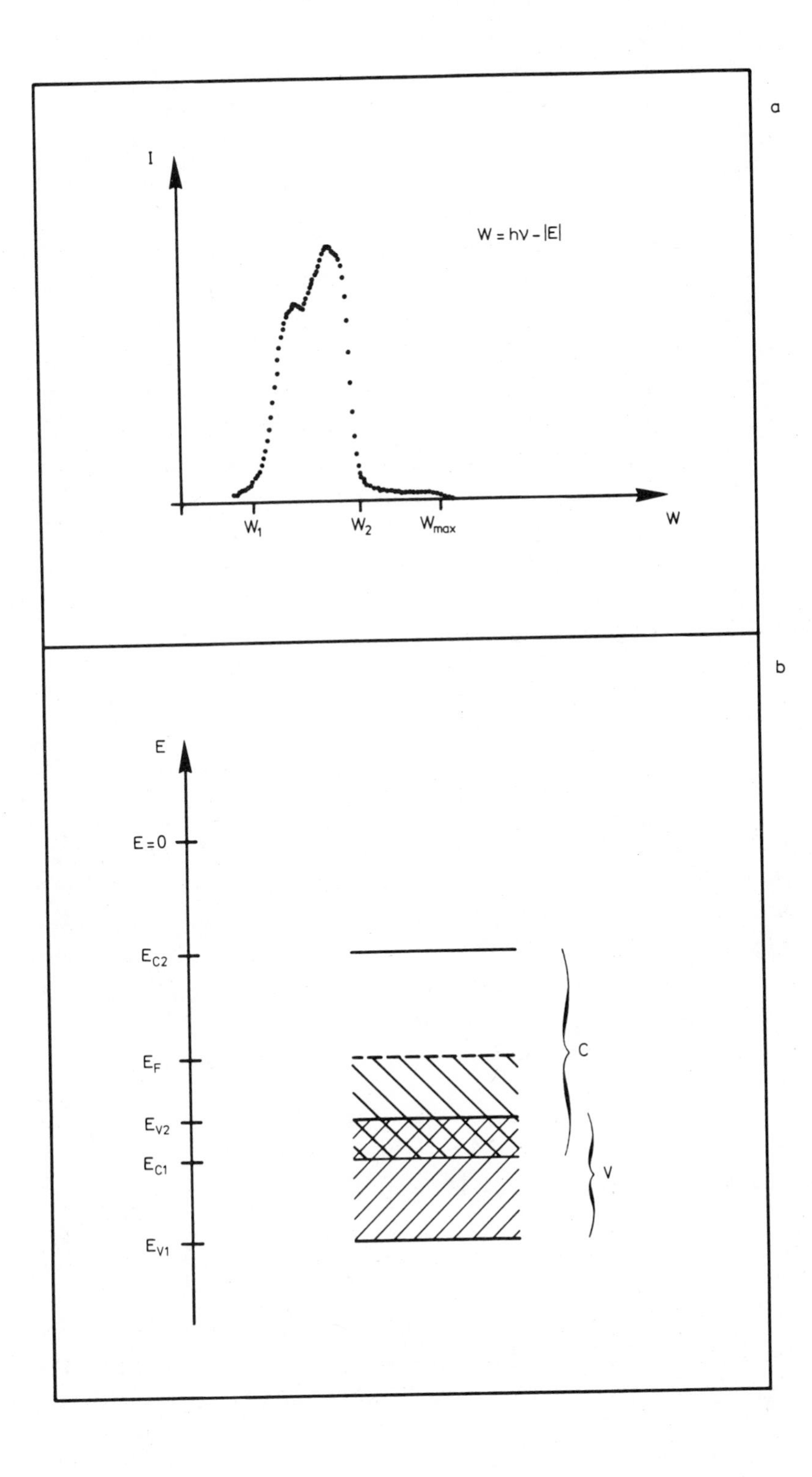

Figure 14.3 (a) **Current I of photoelectrons emitted by a silver crystal, which has been irradiated by monochromatic X-rays, as a function of the kinetic energy W of the photoelectrons. By energy conservation we have $W = h\nu - |E|$, where $h\nu$ is the energy of the X-ray photon and E the energy with which the electron was originally bound in the crystal.** Adapted from K. H. Hellwege, *Einführung in die Festkörperphysik*, copyright © 1976 by Springer-Verlag, Berlin, Heidelberg, New York, reprinted by permission.

(b) **Energy bands of electrons in the silver crystal shown schematically. The conduction band C is only partly filled with electrons, indicated by the hatched area. The valence band V, which is completely filled, partly overlaps with the conduction band. Photoelectrons with the highest energy originate from the region of highest energy in the conduction band, that is, $W_{max} = h\nu - |E_F|$.**

crystal by the photoelectric effect can be measured using the principle illustrated in Figure 1.1 or more refined techniques.

In Figure 14.3a the energy spectrum of electrons obtained by directing monochromatic X-rays on silver is shown. The bulk of photoelectrons appears in the low-energy range between W_1 and W_2, which has a width of about 5 eV. A small fraction is emitted with an energy range between W_2 and W_{max}, which has a width of about 4 eV. This result is taken as evidence that there are two different energy bands in the silver crystal. They are shown schematically in Figure 14.3b. These bands are the *conduction band* with edges E_{C1}, E_{C2} and the *valence band* with edges E_{V1}, E_{V2}. The valence band is completely filled with electrons. The conduction band is only partly filled; the electrons with maximum energy in this band have Fermi energy E_F. It is therefore clear that the minimum energy needed to free an electron is equal to Fermi energy; a photoelectron with energy W_{max} originates from the Fermi edge in the conduction band. We now identify photoelectrons with energies W_2 and W_1 as originating from the upper, E_{V2}, and lower, E_{V1}, edges of the valence band. The number of electrons freed from the valence band is much larger than the number freed from the conduction band because the valence band contains many more electrons.

14.3 Shell Model Classification of Atoms and Nuclei

The only atom we have studied in some detail is the hydrogen atom, which consists of a proton of charge $+e$ as nucleus and an electron of charge $-e$. The heavier atoms have Z protons, and additional uncharged neutrons, in their nucleus and Z electrons in their hull. This Z number representing the positive charge of the nucleus of the atom of an element is its *atomic number*. The potential energy of a single electron in the electric field of the nucleus of a heavier atom is

$$V(r) = -Z\alpha\hbar c\frac{1}{r}$$

Consequently, the energy levels are

$$E_n = -\tfrac{1}{2}Mc^2\frac{\alpha^2}{n^2}Z^2$$

Here the forces acting between the electrons have been neglected. In Section 8.1 we learned that fermions obey the Pauli exclusion principle, which says that two identical fermions cannot populate the same state. Let us now count the number of different states for a given value n of the principal quantum number. The angular momentum quantum number l can take the values $l = 0, 1, \ldots, n-1$. For a given l there are $2l + 1$ states of different quantum number m, which measures the z-component of angular momentum, $m = -l, -l+1, \ldots, l$. Thus the total number of states for a given n is

$$\sum_{l=0}^{n-1} (2l + 1) = n^2$$

This number has still to be multiplied by 2 since the electron possesses an intrinsic angular momentum $s = \frac{1}{2}$, called spin, with the z-component $S_z = \pm \frac{1}{2}$. An electron with given "orbital" quantum numbers n, l, and m can therefore still exist in the two different spin states, $S_z = \frac{1}{2}, -\frac{1}{2}$, so that the total number of states for a given n is equal to $2n^2$.

In our simplified description, all electrons in an atom that have the same principal quantum number n have the same energy. They are said to be in the same *shell*. There can be two electrons in the innermost shell which has $n = 1$, eight electrons in the next shell with $n = 2$, and so on. In this way the *periodic table of elements* is easily explained.

For hydrogen ($Z = 1$) and helium ($Z = 2$) the electrons have principal quantum numbers $n = 1$. For lithium ($Z = 3$) both states with $n = 1$ are filled; therefore the third electron has to be in state $n = 2$ and in a second shell. When all the states with $n = 2$ are filled, the element is the noble gas neon ($Z = 10 = 2 \cdot 1^2 + 2 \cdot 2^2$). The element sodium ($Z = 11$) has an additional electron with $n = 3$, which goes in the third shell, and so on. The electrons in shells that are filled up are chemically inactive, which is seen from the chemical inertia of the noble gases helium and neon. Elements with the same number of electrons in an unfilled shell possess similar chemical properties, for example, lithium, sodium, and so on. The consecutive filling of the $n = 3$ shell continues only until the $l = 0$ and $l = 1$ states are all occupied. The element is argon ($Z = 18$), which again has the chemical properties of a noble gas. After argon the shell with $n = 4$ and $l = 0$ begins filling, forming potassium ($Z = 19$) and calcium ($Z = 20$). Only then are the so-far vacant states with $n = 3$ and $l = 2$ filled. The reason for this irregularity is that the states with $n = 4$, $l = 0$ are situated at lower energy than the states $n = 3$, $l = 2$. This situation is in contrast to our simple scheme, in which we have totally neglected the forces between the electrons in an atom.

Because of the forces acting between electrons, the energy levels of atoms with more than one electron are not simply the energy levels of a hydrogenlike atom with Z protons in its nucleus and with its lowest states filled with Z electrons. In fact, the actual calculation of the levels of many-electron atoms is complicated and can be carried out only with simplifying approximations. The levels least influenced by interactions between the electrons are the innermost levels for $n = 1$ and $n = 2$. Their Z-dependence is given by the formula

$$E_n = -\tfrac{1}{2}Mc^2\alpha^2\frac{Z^2}{n^2}$$

The difference between the energy of the state with the principal quantum number n_2 and that of the ground state with $n_1 = 1$ for an atom with atomic number Z is then

$$\Delta E = -\tfrac{1}{2}Mc^2\alpha^2\left(\frac{1}{n_2^2} - \frac{1}{n_1^2}\right)Z^2 = \tfrac{1}{2}Mc^2\alpha^2\left(1 - \frac{1}{n_2^2}\right)Z^2$$

This difference can be measured in an experiment in which electrons accelerated to some 10 keV knock an electron out of the ground state of an atom with atomic number Z. The unoccupied state ($n_1 = 1$) can be filled by an electron jumping from state $n_2 = 2$, $n_2 = 3$, and so on in the atom to the ground state.

The energy difference between the two states is radiated off as an X-ray quantum of frequency

$$\nu = \frac{1}{h}\Delta E$$

With this formula for ΔE, we find a linear relation between the atomic number Z and the square root of the frequency, $\sqrt{\nu}$, of the emitted X-ray;

$$\sqrt{\nu} = \sqrt{Mc^2/(2h)}\,\alpha\left(1 - \frac{1}{n_2^2}\right)^{1/2} Z$$

Henry G. J. Moseley first measured these transitions in 1913. His results are reproduced in Figure 14.4a. They allow the simple interpretation that the atomic number Z is the number of positive charges on the nucleus of the atom, since the data take the expected line in a $Z, \sqrt{\nu}$-plot. Actually, the line of the data does not follow our formula exactly. The deviation is caused by the screening—even though small for the inner atomic shells—of the nuclear Coulomb field by other inner electrons.

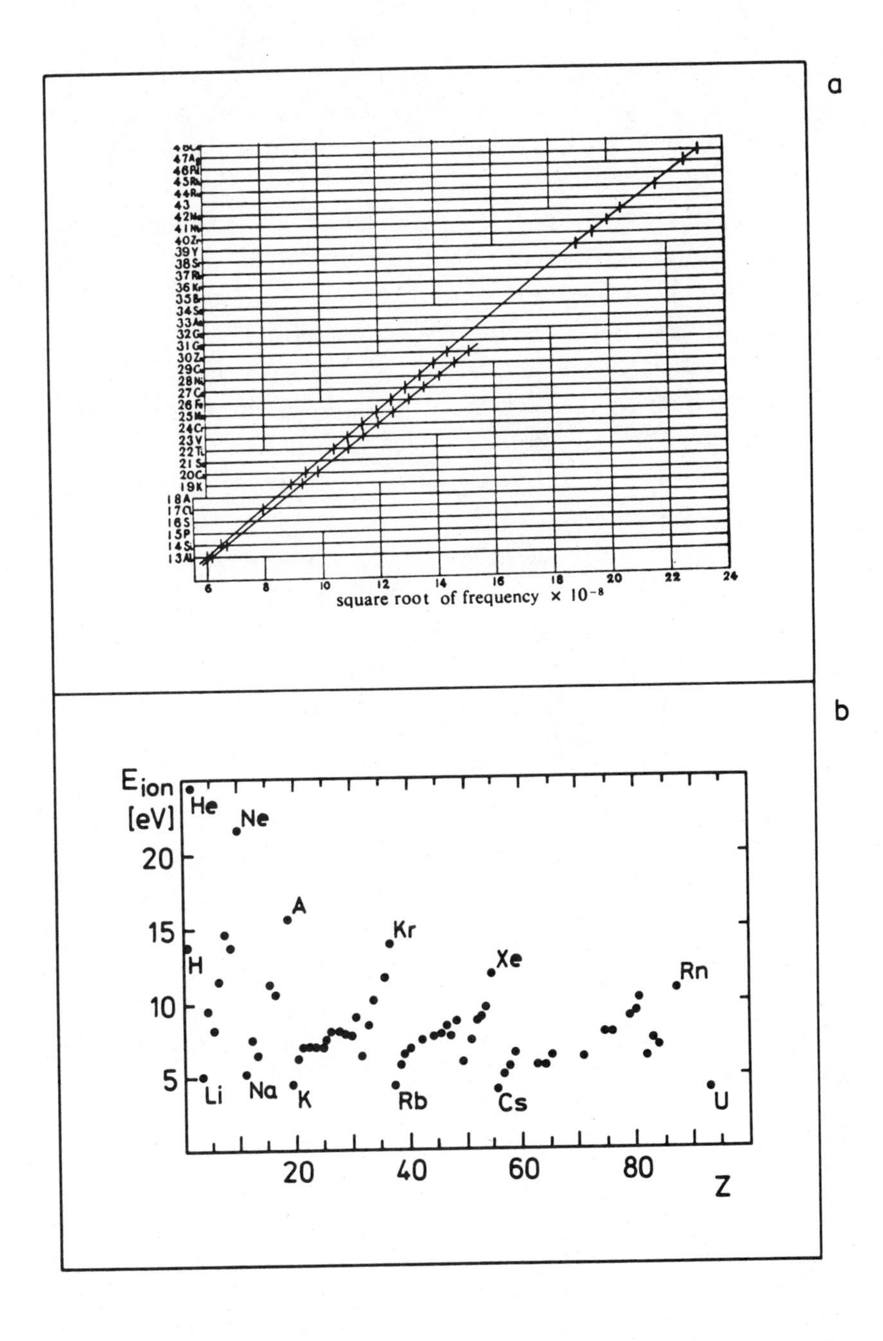

Figure 14.4 **(a) Moseley's plot showing the square root of the X-ray frequency versus the atomic number Z for K_α-radiation, $n_2 = 2$ (upper line) and for K_β-radiation, $n_2 = 3$ (lower line).** From H. G. J. Moseley, *The Philosophical Magazine* 27 (1914) 703, copyright © 1914 by Taylor and Francis, Ltd., London, reprinted by permission.

(b) Ionization energies for atoms as a function of the atomic number Z. The maxima for noble gases, which have closed shells with $Z = Z_c$ electrons—$Z_c = 2$ for helium, $Z_c = 10$ for neon, and so on—are pronounced, and the drop from $Z = Z_c$ to $Z = Z_c + 1$ —that is, from helium to lithium, from neon to sodium, and so on—is sharp.

Another test for the viability of the shell model of the atomic hull is indicated by the formula

$$E(Z, n) = -\tfrac{1}{2}Mc^2\alpha^2\frac{Z^2}{n^2}$$

for the energy of the outermost electron with principal quantum number n in the hull of an atom with nuclear charge Z. This energy is called ionization energy. The expression for $E(Z, n)$ is actually only a rough estimate of the ionization energy, since it does not take into account the mutual interaction of the electrons in the atomic hull. Nevertheless, for atoms with low atomic number it suffices to demonstrate how the values of the ionization energies indicate the closure of atomic shells.

In the process of "constructing" the chemical elements by filling the levels with electrons, the ionization energy $E(Z, n)$ rises with Z as long as levels are filled with the same principal quantum number n. The highest value $E(Z_c, n)$ within each shell is reached in the element that has a closed shell with atomic number Z_c, that is, a noble gas. For the element with the next atomic number, a new shell with the principal quantum number $n + 1$ begins to be occupied. Even though Z increases in this step from Z_c to $Z_c + 1$, the increase from n to $n + 1$ means a definite decrease in ionization energy $E(Z_c + 1, n + 1)$ for the first element in the new shell compared to the value $E(Z_c, n)$ for the noble gas. Because there are many states belonging to each principal quantum number n, for each electron the principal quantum number is lower than Z_c. The ratio of the two ionization energies is

$$r(Z_c) = \frac{E(Z_c + 1, n + 1)}{E(Z_c, n)} = \frac{(Z_c + 1)^2}{(n + 1)^2} \frac{n^2}{Z_c^2}$$

$$= \frac{(1 + 1/Z_c)^2}{(1 + 1/n)^2} < 1$$

because Z_c is higher than n. For the jump from helium to lithium, neon to sodium, and argon to potassium, we find these values

lithium/helium	$r(2) = 0.56$
sodium/neon	$r(10) = 0.54$
potassium/argon	$r(18) = 0.63$

In contrast, the ratio of the ionization energy of an element closing a shell to the energy of the preceding element in the periodic table is

$$r'(Z_c) = \frac{E(Z_c, n)}{E(Z_c - 1, n)} = \frac{Z_c^2}{(Z_c - 1)^2} = \frac{1}{(1 - 1/Z_c)^2} > 1$$

For the corresponding closures of the atomic shells, we find

helium/hydrogen	$r'(2) = 4$
neon/fluorine	$r'(10) = 1.23$
argon/chlorine	$r'(18) = 1.12$

that is, values larger than one. The peak behavior expected by these arguments can be immediately verified by looking at the measured ionization energies plotted in Figure 14.4b, even though the experimental values for the ratios r and r' are different from the ones we have given.

In the classification of nuclei, the *nuclear shell model* has been successful in explaining observed regularities. For the electrons in a light element, it was reasonable to describe their motion in the Coulomb potential of the nucleus, neglecting the repulsion between electrons. For the protons and neutrons forming the nucleus, no analogous center of force exists. Nevertheless, it has proved useful in describing the motion of a single nucleon in the nuclear potential created by all remaining nucleons. Such a potential has, as does the nuclear force of a single nucleon, short range. For our simple discussion we assume that the potential is that of a harmonic oscillator. The lowest states in this potential are filled by the nucleons. Since protons and neutrons have spin $\frac{1}{2}$ according to Pauli's exclusion principle, every state characterized by n, l, and m can be occupied by two protons and two neutrons. The lowest state in the harmonic oscillator (see Section 12.2) has quantum numbers $n = 0$, $l = 0$; therefore it can accommodate at most two protons and two neutrons. This is the case for the nucleus of the element helium. This nucleus, also called the α-particle, is the most stable nucleus known; for its disintegration the largest amount of energy is needed. The helium nucleus has a closed proton shell and a closed neutron shell.

For the nuclei of the next heavier elements, the $n = 1$, $l = 1$ shell of the oscillator potential is successively filled. It offers $2 \cdot (2l + 1) = 6$ states for protons as well as six states for neutrons, so that the next closure of the proton shell, as well as of the neutron shell, is reached for $Z = 8$ and $N = 8$. Here Z, as before, gives the number of protons in the nucleus and N gives the number of neutrons. The *nucleon number* $A = Z + N$ together with the chemical symbol which itself contains the information about Z is commonly used to characterize the nucleus. The shells $Z = 8$ and $N = 8$ are those of the oxygen nucleus, ^{16}O. As we know from Section 12.2, in the harmonic oscillator potential the states with principal quantum number $n = 2$ are degenerate for $l = 0$ and $l = 2$. The nuclear shell with $n = 2$ contains $2 \cdot 1 + 2 \cdot 5 = 12$ states for protons and for neutrons. Thus the next closed shell is reached for $Z = N = 20$, the nucleus of the element cadmium, ^{40}Ca. As was true of the atomic hull, this simple constructive scheme for finding the closed nuclear shells works for the lighter nuclei only.

It was the achievement of Maria Goeppert-Mayer and of Otto Haxel, Hans Jensen, and Hans Suess to discover the physical reason for the structure of the higher closed shells. They are reached at the higher *magic numbers* 28, 50, 82, 126, which cannot be obtained from the oscillator potential. In fact, these numbers are "magic" because they denote a large spin-orbit interaction, that is, a large interaction between spin s and angular momentum l of the nucleons. This coupling gives rise to an additional potential energy term in the Schrödinger equation. Evidence for nuclear shells comes from experiments in nuclear spectroscopy. We do not present them here, for their interpretation would require discussing additional details of nuclear physics.

14.4 Resonance Scattering off Molecules, Atoms, Nuclei, and Particles

In Chapter 13 we studied resonance phenomena in some detail. We have seen, in particular, that the total cross section for elastic scattering of a particle by a spherically symmetric potential may have pronounced maxima, as a function of the energy of the particle (see Figures 13.6a and 13.11a). Such resonance phenomena are not restricted to simple potential scattering. They are observed in a variety of physical situations. In a more general situation, the collision of two particles, the total cross section is a measure of the probability that they will react. One or both particles may even be compound systems. The total cross section then is a measure of the probability for a reaction between these systems. In fact, we have seen evidence for such reactions earlier in this chapter when looking at the absorption spectrum of hydrogen (see Section 14.2, Figure 14.2b). The process is actually a collision between a photon and the hydrogen atom, which excites the electron in the atom into a higher energy level. The photograph of the spectrum shows that the absorption probability, that is, the total cross section, has pronounced maxima at particular photon energies. These energies correspond to the differences between the bound-state energies of the hydrogen atom. It turns out that in this process the higher bound states of the hydrogen atom are not absolutely stable. After excitation by absorbing a photon, a higher bound state, through photon emission, decays with a certain average lifetime into a state of lower energy and finally into the ground state. In our original calculations of the hydrogen atom (Section 12.4), only the Coulomb interaction between electron and proton was taken into account. Now we are also considering the interaction of photons and electrons. The total process of absorption and emission of a photon is nothing but the resonance scattering of a photon by the atom. We expect the process to show the qualitative features of resonance scattering discussed in Section 5.3 and Chapter 13.

Of course, similar resonance structures in total cross sections can be observed in more complicated atoms and even in molecules. Figure 14.5a shows the absorption spectra of infrared light by different paraffin molecules,

n-pentane $CH_3—CH_2—CH_2—CH_2—CH_3$,

n-hexane $CH_3—CH_2—CH_2—CH_2—CH_2—CH_3$,

⋮

There is a strong similarity in the absorption spectra, indicating the excitation of very similar resonances in the different molecules. They correspond to vibrations between neighboring CH_2 groups, which are common to all the paraffin molecules.

Figure 14.5b presents an example from nuclear physics, the total cross section of neutrons scattered off lead nuclei. The many resonances indicate that the nuclei can exist in a variety of metastable states, covering a rather wide range of energies.

We have seen that resonance scattering reveals the presence of excited states in molecules, atoms, and nuclei. Single nucleons, for example, protons, can also be investigated by scattering different projectiles on them. We choose here positive pions, also called π-mesons. These particles are lighter than protons but heavier than electrons. They play an important role in explaining nuclear forces. In Figure 14.5c the total cross section for the scattering of positive pions on protons is shown as a function of the pion energy. The pronounced resonance at the left side of the picture we interpret as a metastable state. Actually, it corresponds to a short-lived particle called the Δ^{++}-baryon. The sequence of its production in a pion-proton collision and its subsequent decay into a pion and a proton is written as

$$\pi^+ p \to \Delta^{++} \to \pi^+ p$$

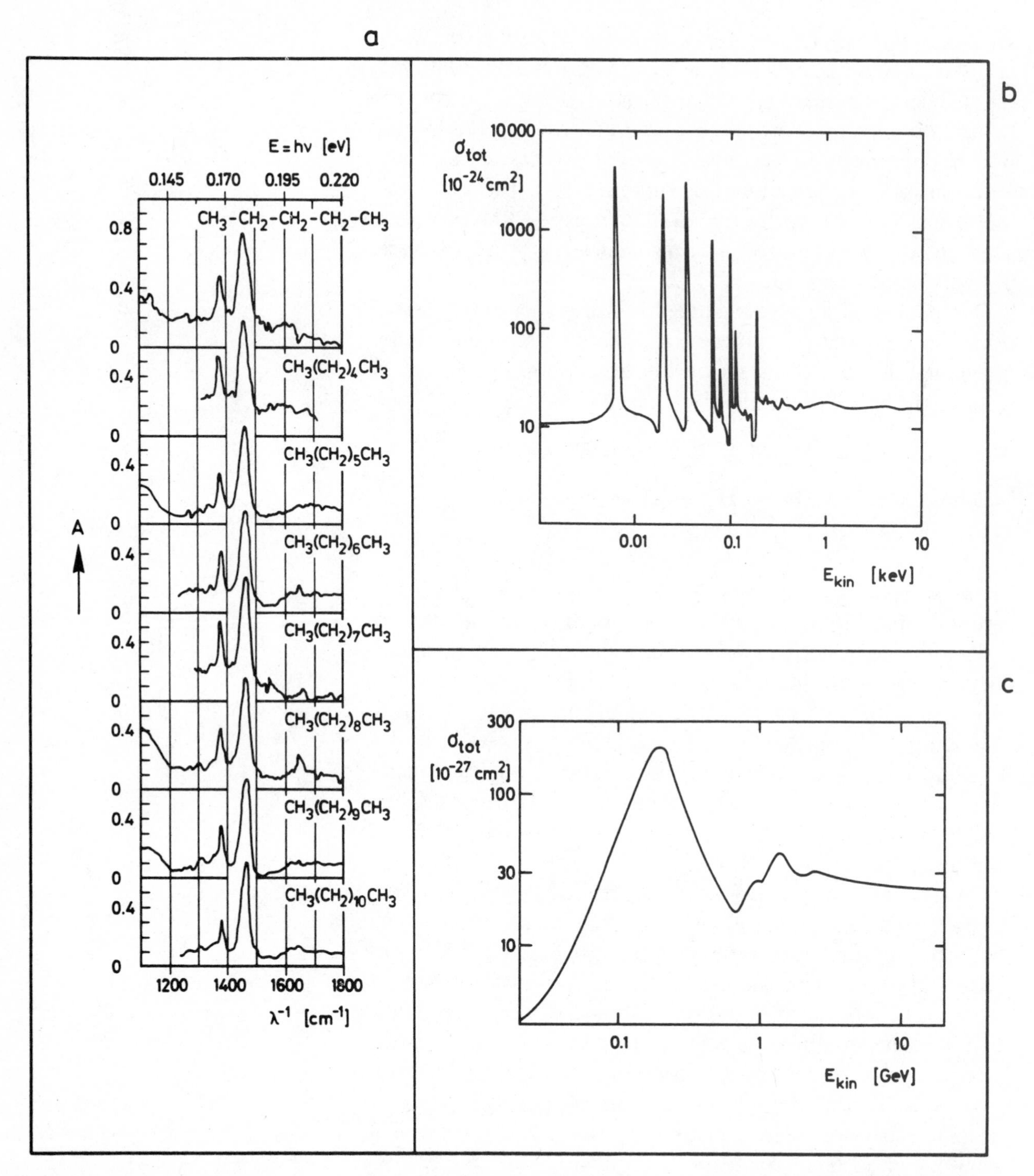

Figure 14.5 **Total cross sections for various reactions as a function of the kinetic energy of the incident particle in the laboratory. (a) The absorption coefficient A for infrared light passing through a layer of paraffin 0.02 mm thick. The abscissa is the wave number $\lambda^{-1} = \nu / c$ (bottom), which is proportional to the energy $E = h\nu$ of the light quanta (top). A high rate of absorption corresponds to a large total cross section. Thus the graphs can be interpreted as measurements of the total cross sections as a function of energy. Two characteristic resonances near $E = 0.17$ eV, which are associated with the vibrations of neighboring CH_2 groups, are present in all paraffins considered.**

Electrons and positrons can be accelerated to very high energy, more than 20 GeV; they can be accumulated in a storage ring and be brought to head-on collisions. The total cross section as a function of the center-of-mass energy of the e^+e^- system has characteristic resonances. Figure 14.6 shows two series of resonances which are located near 3 GeV and near 10 GeV. They are evidence for short-lived particles called the J/Ψ-family and the Υ-family. The first one found is the J/Ψ-particle with a mass of 3.1 GeV. Its production and subsequent decay into electron and positron is a resonance scattering of the form

$$e^+e^- \to J/\Psi \to e^+e^-$$

Besides this elastic process, the inelastic one,

$$e^+e^- \to J/\Psi \to \text{hadrons}$$

is also observed. Hadrons are particles that interact strongly in the nucleus, in particular pions, protons, and neutrons. All hadrons are assumed to be composed of only a few constituent particles called quarks q and antiquarks $\bar{q}$. The J/Ψ-particle is composed of the very heavy charm-quark c and its antiparticle $\bar{c}$, so that the reaction above reads

$$e^+e^- \to (c\bar{c}) \to \text{hadrons}$$

where $(c\bar{c})$ symbolizes the metastable state J/Ψ of c and $\bar{c}$. The next resonance, Ψ' at 3.7 GeV, is another resonance of the $c\bar{c}$-system that can be regarded as an exited state of the J/Ψ-particle. In fact, these and the other observed $(c\bar{c})$-states can be explained as bound states in a potential describing the interaction of c and $\bar{c}$. The discovery of these states has led to a much better understanding of quark bound states and to deeper insight into the structure of matter.

A similar series of resonances in electron-positron scattering is observed at 9.46 GeV and beyond. The family of Υ-particles are understood to be bound states of the even heavier beauty-quark b and its antiquark $\bar{b}$.

(b) Total cross section for neutrons scattered off lead nuclei. There are many resonances at low energies corresponding to the formation of various metastable states of lead isotopes.

(c) Total cross section for positive pions scattered on protons. The wide resonance near $E_{kin} \approx 0.2$ GeV corresponds to the excitation of the metastable state $\Delta^{++}(1232)$.

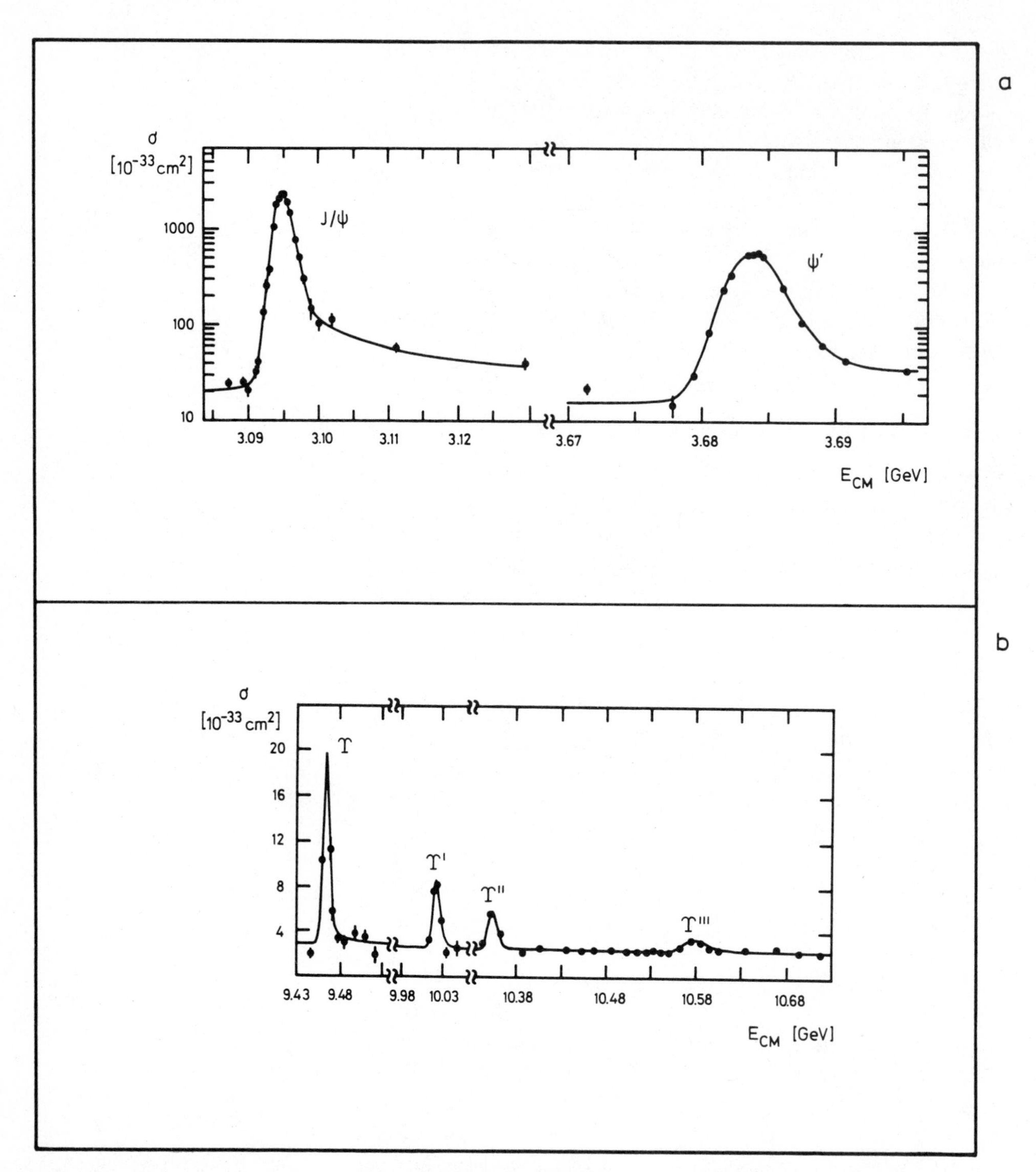

Figure 14.6 **Total cross section observed for the reaction in which an electron (e^-) and a positron (e^+) annihilate each other to form a number of strongly interacting particles, such as π-mesons. The cross section shows very sharp resonances near (a) $E_{CM} \approx 3$ GeV and (b) $E_{CM} \approx 10$ GeV. Here E_{CM} is the total energy in the center-of-mass system, the system in which e^+ and e^- have equal and opposite momenta. The unexpectedly sharp resonances are interpreted as evidence that metastable states consisting of a quark-antiquark pair have formed. The J/ψ-family of states is composed of a "charm"-quark and its antiparticle. The ϒ-family of states**

It is interesting to note that the quantum-mechanical phenomena studied in this section span an energy range of eleven orders of magnitude, from infrared radiation at 0.2 eV to high-energy electron storage rings at 10 GeV.

14.5 Phase Shift Analysis in Nuclear and Particle Physics

In the preceding section we identified a resonance in the total cross section as evidence for the existence of a metastable state. Figure 13.6a showed that a maximum in the total cross section is usually an indication for a resonance in a single partial wave. The quantum numbers of the resonant partial wave are therefore those of the metastable state which, in elementary particle physics, we have also called a particle. To determine the quantum numbers of such a particle, we use the method of phase shift analysis outlined in Section 13.3. We decompose the measured differential cross section into partial waves, obtaining the complex partial-wave amplitudes f_l as a function of the energy or equivalently the momentum of the incident particle. From the different partial-wave amplitudes Argand diagrams analogous to those in Figure 13.12 can be constructed.

Figure 14.7a shows the differential cross section for the elastic scattering of positive pions on protons for various pion energies E. Near $E = 200$ MeV the cross section has a simple parabolic form, indicating the dominance of the Legendre polynomial $P_1(\cos\vartheta) = \cos\vartheta$ in the expression

$$\frac{d\sigma}{d\Omega} = |f(\vartheta)|^2 \sim |P_1(\cos\vartheta)|^2$$

for the differential scattering cross section. This parabola indicates a resonance with angular momentum $l = 1$. The proton has an intrinsic angular momentum, that is, a spin s of

is a bound system of a "beauty"-quark and the corresponding antiquark.

Sources: (a) From A. M. Boyarski et al., *Physical Review Letters* **34** (1975) 1357 and from V. Lüth et al., *Physical Review Letters* **35** (1975) 1124, copyright © 1975 by Americal Physical Society, reprinted by permission. (b) From D. Andrews et al., *Physical Review Letters* **44** (1980) 1108 and **45** (1980) 219, copyright © 1980 by American Physical Society, reprinted by permission.

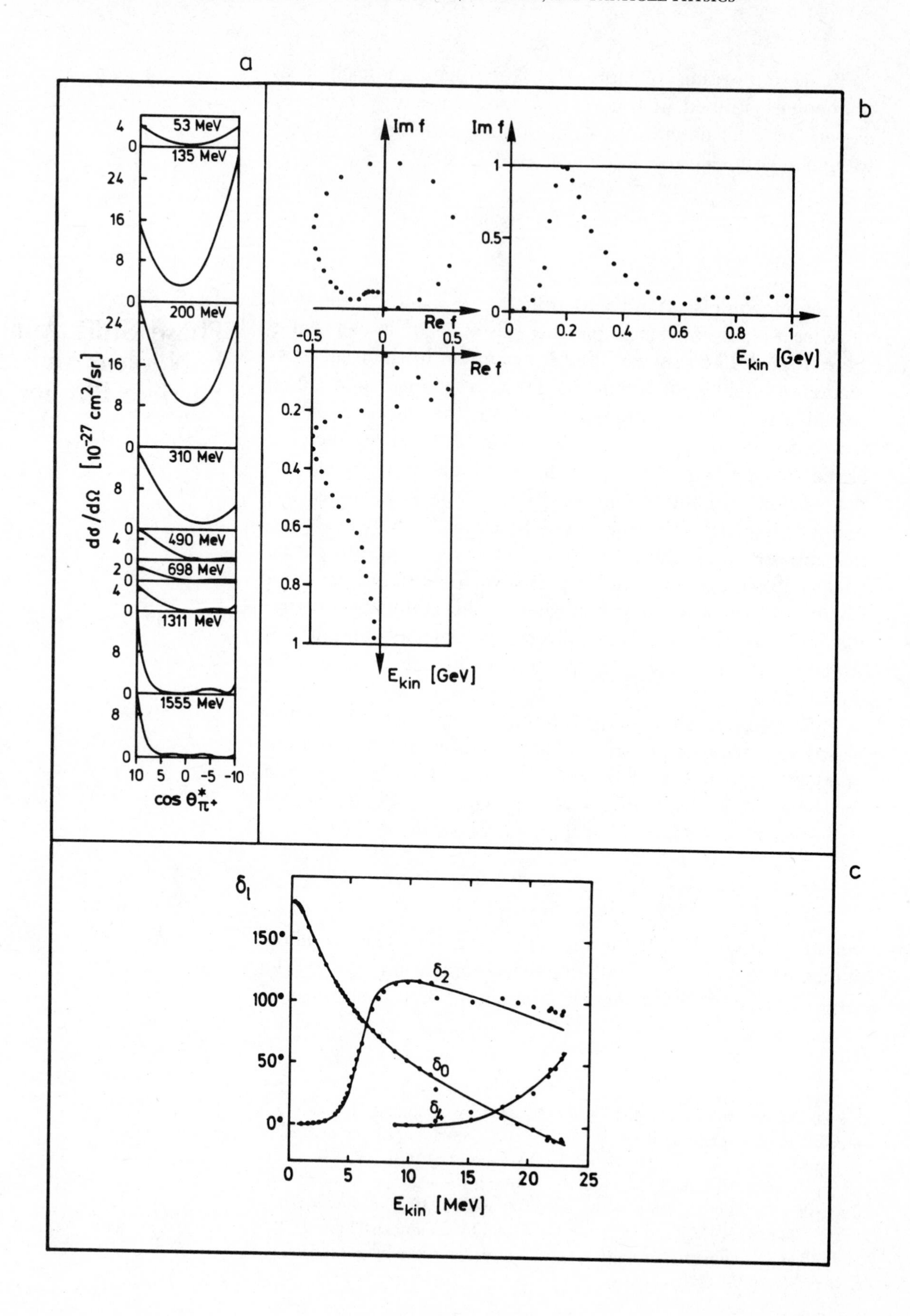

a
b
c
53 MeV
135 MeV
200 MeV
310 MeV
490 MeV
698 MeV
1311 MeV
1555 MeV
dσ/dΩ [10^-27 cm^2/sr]
cos θ*π+
Im f
Re f
Ekin [GeV]
δl
δ2
δ0
δ4
Ekin [MeV]

$\frac{1}{2}$. It turns out that the metastable state has total angular momentum $\frac{3}{2}$. Figure 14.7b gives the Argand diagram for the corresponding partial-wave amplitude. As in Figure 13.12, the partial scattering amplitude is recognized to move, as a function of energy, on the unitarity circle in the complex plane. The deviation of the experimental points from the unitarity circle designate an inelastic process. Not only elastic scattering but also the production of one or more additional pions is possible at these energies. The real and imaginary parts of the partial-wave amplitude have the characteristic features of a resonance at a center-of-mass energy of 1.232 GeV in the pion-proton system. By this phase shift analysis the intrinsic angular momentum of the Δ^{++}-hadron, which we first observed in the total cross section (Figure 14.5c), is found to be $\frac{3}{2}$.

The method of phase shift analysis, which has proved very successful in particle physics, had already been used earlier in nuclear physics. The elastic scattering of α-particles on helium nuclei is an interesting example. In Figure 14.7c the phase shifts δ_0, δ_2, and δ_4 are given directly as functions of the energy of the incident particle. The phase shift δ_2 shows a typical resonance, a quick rise through the value $\pi/2$. The resonance corresponds to a metastable state of angular momentum 2 of the beryllium nucleus, ^{8}Be, which is formed by two ^{4}He nuclei colliding.

Figure 14.7 **Phase shift analysis. (a) The differential cross section for the elastic scattering of positive π-mesons on protons, shown for various kinetic energies of the meson, has a simple parabolic form at $E_{kin} = 200$ MeV, indicating a resonance at this energy with angular momentum $l = 1$.**

(b) The Argand diagram of the corresponding partial scattering amplitude, reconstructed from measured data. All the features of a resonance at $E_{kin} = 200$ MeV are evident. The phase shift passes swiftly through 90 degrees, while the imaginary part goes through a maximum and the real part vanishes.

(c) A resonance at much lower energies. Various phase shifts for the elastic scattering of an α-particle on a helium nucleus, that is, another α-particle, are plotted as a function of the kinetic energy of the incoming particle. The resonance in δ_2 indicates that both particles form a resonance with angular momentum $l = 2$.

Source: (a) From Robert C. Cence, *Pion-Nucleon Scattering*, copyright © 1969 by Princeton University Press, Figure 5.2, p. 62, reprinted by permission of Princeton University Press. (b) Adapted from G. Höhler in Landolt-Börnstein, *Numerical Data*, New Series, volume 9b2 (H. Schopper, editor), Figure 2.2.6, p. 58, copyright © 1983 by Springer-Verlag, Berlin, Heidelberg, New York, reprinted by permission. (c) From T. A. Tombrello and L. S. Senhouse, *The Physical Review* **129** (1963) 2252, copyright © 1963 by American Physical Society, reprinted by permission.

14.6 Classification of Resonances on Regge Trajectories

Figure 13.11a indicated that there is a striking regularity between the energies of the lowest-lying resonances of a system and their angular momenta. In a plane spanned by energy E and angular momentum l, the resonances lie on a curve in such a way that energy E of the resonance increases monotonically with angular momentum l. The correlation between the energies of a family of resonances and their angular momenta in potential scattering has been derived by Tullio Regge. In elementary particle physics families of particles lying on the same Regge trajectory are observed. As an example, Figure 14.8 shows the Regge trajectory containing the Δ^{++}-hadron, already discussed in Sections 14.4 and 14.5. We now call it more specifically $\Delta(1232)$ by indicating in brackets its mass in MeV. On the same trajectory four more resonances are shown. In this diagram, in which the square of the resonance mass is plotted on the abscissa and its spin on the ordinate, the trajectory is a straight line. From resonance

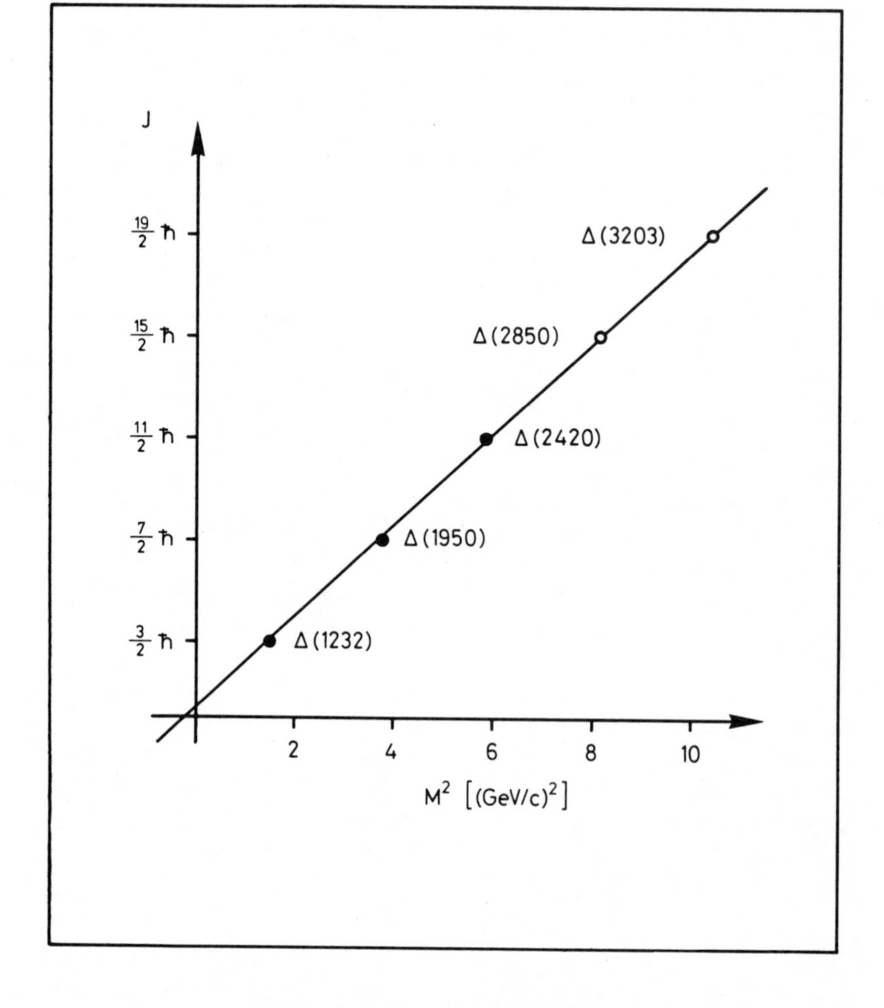

Figure 14.8 **Regge trajectory of the Δ-particles, which can be understood as resonances formed by a proton and a π-meson. The square of the resonance mass, M^2, is plotted against the angular momentum J of the resonance. For the three lowest-lying resonances (black points) both M and J have been experimentally determined. For the last two (open circles) only the mass has been measured so far.**

to resonance, the spin is increased by two units, that is, it takes the values $\frac{3}{2}\hbar$, $\frac{7}{2}\hbar$, and so on. This complication is attributable to the half-integer spin of these resonances.

14.7 Radioactive Nuclei as Metastable States

The disintegration of a radioactive nucleus by the emission of an α-particle can be considered as the decay of a metastable state. George Gamow has given a quantum-mechanical analysis of how the α-particle behaves in the potential of the other protons and neutrons in the nucleus. The effect of the short-range nuclear forces can be approximated by a square-well potential. In addition, the α-particle, which carries the electric charge $+2e$, experiences the repulsive long-range Coulomb force of the other protons. The total potential of both nuclear and Coulomb forces is attractive for small radii but repulsive for greater distances, as indicated in Figure 14.9a. Such a potential can contain bound states of negative energies that are stable as well as metastable states of positive energies that have a finite lifetime. In particular, metastable states with energies lower than the height of the repulsive wall are expected to have long lifetimes. An α-particle in such a metastable state can leave the nucleus only by tunneling through the potential barrier. For an α-particle in a metastable state, only the repulsive barrier is important. The repulsive shell studied in detail in Chapter 13 can therefore serve as a model for the potential. Figure 13.13 shows the radial wave functions of several metastable states in this potential. Figures 13.11 and 13.12 contain the total cross sections and Argand diagrams. They indicate that resonance widths increase with resonance energy. The repulsive shell of Chapter 13 has its one-dimensional analog in the two-potential barriers of Section 5.3. The widths of the metastable states confined between two potential barriers also grow with energy, as indicated in Figure 5.11. Figures 5.8 and 5.9 examined the time dependence of the decay of metastable states and revealed that the lifetime decreases with increasing energy, that is with increasing width. Indeed, the probability for the penetration of the barrier grows with the energy of the particle, which is tantamount to saying that the lifetime decreases with the energy of the α-particle.

This phenomenon is observed experimentally. The energy of α-particles is easily measured by their range in air. Figure

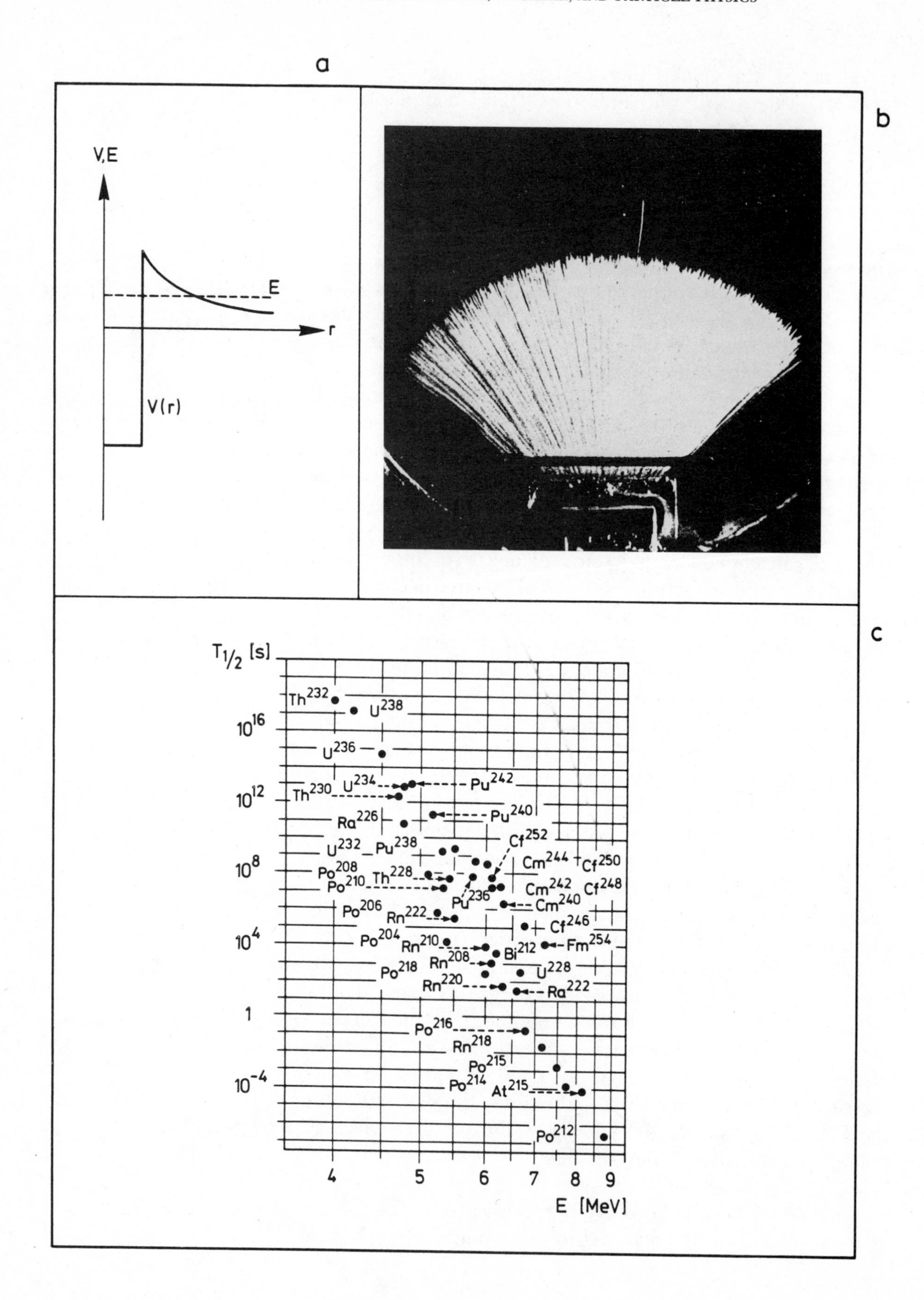
a
b
c
V,E
E
r
V(r)
T1/2 [s]
Th232
U238
U236
Th230
U234
Pu242
Ra226
Pu240
U232
Pu238
Cf252
Cm244
Cf250
Po208
Po210
Th228
Cm242
Cf248
Pu236
Cm240
Po206
Rn222
Cf246
Po204
Rn210
Bi212
Fm254
Po218
Rn208
U228
Rn220
Ra222
Po216
Rn218
Po215
Po214
At215
Po212
10^16
10^12
10^8
10^4
1
10^-4
4
5
6
7
8
9
E [MeV]

14.9b shows a cloud chamber photograph displaying the tracks of α-particles emitted by radioactive polonium, ^{214}Po. All tracks except one have very similar ranges, indicating the energy of the lowest-lying metastable state. A single track in the photograph has a considerably greater length. Its energy is that of a higher metastable state, which is already much depopulated because of its shorter lifetime. A systematic study of the relation between energy and the lifetime of α-decays of nuclei was first carried out by Hans Geiger and J. M. Nuttall. Figure 14.9c shows this correlation for many radioactive elements.

Figure 14.9 **α-Decay. (a) Potential energy $V(r)$ of an α-particle in a nucleus. Although the total energy E (dashed line) of an α-particle may be positive, the particle can leave the nucleus only by tunneling through the potential barrier created by the Coulomb attraction between nucleus and α-particle. Therefore metastable states of positive energy can exist.**

(b) Cloud chamber photograph of tracks of α-particles from the decay of the polonium nucleus, ^{214}Po. All particles except one have approximately the same range in the chamber gas, indicating that they possess equal energies. The single, long-range track was caused by the decay of an exited state of ^{214}Po possessing a higher energy. From K. Phillip, *Naturwissenschaften* 14 (1926) 1203, copyright © 1926 Verlag von Julius Springer, Berlin, reprinted by permission.

(c) Geiger-Nuttall diagram showing the relation between the half-life $T_{1/2}$ and the energy of the emitted α-particles for the lowest-lying states of radioactive nuclei. The diagram indicates that the lifetime decreases very rapidly with energy.

Index